Solid Fuel Blending

Solid Fuel Blending
Principles, Practices, and Problems

David A. Tillman
Dao N. B. Duong
N. Stanley Harding

AMSTERDAM • BOSTON • HEIDELBERG • LONDON
NEW YORK • OXFORD • PARIS • SAN DIEGO
SAN FRANISCO • SINGAPORE • SYDNEY • TOKYO

Butterworth-Heinemann is an imprint of Elsevier

Butterworth-Heinemann is an imprint of Elsevier
The Boulevard, Langford Lane, Kidlington, Oxford, OX5 1GB UK
225 Wyman Street, Waltham, MA 02451 USA

Notices
Knowledge and best practice in this field are constantly changing. As new research and experience broaden our understanding, changes in research methods, professional practices, or medical treatment may become necessary.

Practitioners and researchers must always rely on their own experience and knowledge in evaluating and using any information, methods, compounds, or experiments described herein. In using such information or methods, they should be mindful of their own safety and the safety of others, including parties for whom they have a professional responsibility.

To the fullest extent of the law, neither the Publisher nor the authors, contributors, or editors assume any liability for any injury and/or damage to persons or property as a matter of products liability, negligence, or otherwise, or from any use or operation of any methods, products, instructions, or ideas contained in the material herein.

Library of Congress Cataloging-in-Publication Data
Tillman, David A.
    Solid fuel blending : principles, practices, and problems / David A. Tillman, Dao Duong, N. Stanley Harding. – 1st ed.
        p. cm.
    Includes index.
    ISBN 978-0-12-380932-2
    1. Coal-fired power plants—Fuel. 2. Coal-fired power plants—Environmental aspects. 3. Coal—Combustion. 4. Biomass energy. 5. Waste products as fuel. I. Duong, Dao (Dao N.B.) II. Harding, N. S. III. Title.
    TJ254.5.T55 2012
    662.6'23–dc23   2011046946

British Library Cataloguing-in-Publication Data
A catalogue record for this book is available from the British Library.

For information on all Butterworth-Heinemann publications, visit our website at *www.books.elsevier.com*

Printed in the United States

12 13 14 15 16   10 9 8 7 6 5 4 3 2 1

To the friends and colleagues at DTE Energy,
particularly Monroe Power Plant; at Foster Wheeler;
and at EPRI who supported the development of many
of the fuel blending concepts put forward in this book

Contents

Preface   xiii
Acknowledgments   xv

1. Introduction to Fuel Blending   1

1.1. Overview   1
1.2. Fuel Blending for Solid Fuels   1
    1.2.1. Blending System Considerations   1
    1.2.2. Where Blending Can Occur   2
1.3. Objectives for Blending   5
    1.3.1. Economic Considerations with Fuel Blending   7
    1.3.2. Environmental Considerations with Fuel Blending   8
    1.3.3. Historical and Technical Considerations for Fuel Blending   11
1.4. Blending for the Steel Industry—The Development of Petrology   13
    1.4.1. Basics of Macerals   14
    1.4.2. Petrography Applied to the Steel Industry   15
    1.4.3. Conclusions on Blending for the Steel Industry   16
1.5. Typical Fuel Blends   16
    1.5.1. Coal–Coal Blends   16
    1.5.2. Coal–Biomass Blends   18
    1.5.3. Coal–Opportunity Fuel Blends   21
1.6. Blends and Firing Systems   22
    1.6.1. Types of Firing Systems   23
    1.6.2. Types of Boilers   26
1.7. Conclusions   27
References   27

2. Principles of Solid Fuel Blending   31

2.1. Introduction: Blending for Dollars   31
2.2. Designing the Most Favorable Fuel   32
2.3. Influences on the Most Favorable Fuel Blend   34
    2.3.1. Firing Method Considerations   34
    2.3.2. Market Considerations   36
2.4. Developing a Fuel Blending Strategy   36
    2.4.1. Blend Fuel Considerations   37
    2.4.2. Combustion Characteristics of Binary and Ternary Blends   37
    2.4.3. Reactivity, Ignition, and Flame Characteristics of Fuel Blends   42

2.5.   Formation of Pollutants   48
2.6.   Fuel Blending Characteristics Influencing Deposition   53
2.7.   Fuel Blending and Corrosion   55
2.8.   Blending's Impact on the Physical Characteristics
       of Solid Fuels   59
2.9.   Management and Control of Fuel Blending   62
2.10. Conclusions   66
References   66

3. Blending Coal on Coal   71

3.1.   Introduction and Basic Principles   71
3.2.   Blending of Coal for Combustion and/or Gasification
       Purposes   75
3.3.   Combustion and Gasification Processes   77
       3.3.1.   Combustion Processes and Fuel Blending   78
       3.3.2.   Coal Blending and the Combustion Process   80
       3.3.3.   Gasification Processes   84
3.4.   Coals Used in Commercial Applications and Their
       Blending Potential   84
       3.4.1.   Characteristics of Various Commercially
                Significant Coals   84
       3.4.2.   Relationship of Chemical Composition
                to Petrography   93
       3.4.3.   Chemical Composition and Calorific Value   96
3.5.   Kinetics and the Analysis of Coal Blend Reactivity   97
       3.5.1.   Devolatilization Kinetics   97
       3.5.2.   Reactivity and Ignition Temperature of Coal Blends   102
       3.5.3.   Char Oxidation Kinetics   102
3.6.   The Behavior of Inorganic Constituents   103
       3.6.1.   Slagging and Blended Coals   104
       3.6.2.   Fouling and Blended Coals   109
       3.6.3.   Quantifying the Inorganic Interaction   111
3.7.   Managing the Coal-on-Coal Blending Process   113
       3.7.1.   Where Blending Can Occur   113
       3.7.2.   Influence of Blending on Materials Handling Issues   115
       3.7.3.   How Coal Blends Can Be Managed   116
       3.7.4.   Other Considerations   118
3.8.   Conclusions   119
References   120

4. Blending Coal with Biomass: Cofiring Biomass with Coal   125

4.1.   Introduction   125
4.2.   Biomass and Coal Blending   125
       4.2.1.   Properties of Biomass and Coal   126
4.3.   Cofiring: Reducing a Plant's Carbon Footprint   134
       4.3.1.   The Carbon Cycle   135
       4.3.2.   The Role of Biomass for Coal-Fired Plants   135

4.4. Other Reasons for Cofiring 135
    4.4.1. SO$_2$ Management 136
    4.4.2. NO$_x$ Management 137
4.5. Cofiring in the United States and Europe 137
4.6. Characteristics of Biomass 138
    4.6.1. Types of Biomass 138
    4.6.2. Standard Characteristics of Biofuels 142
    4.6.3. Fuel Porosity and Its Implications 144
    4.6.4. Proximate and Ultimate Analysis and Higher Heating Value 145
    4.6.5. Ash Elemental Analysis 145
    4.6.6. Trace Elements 146
4.7. Reactivity Measures for Biomass 146
    4.7.1. Reactivity of Combustibles 146
    4.7.2. Structure and Reactivity 148
    4.7.3. Drop Tube Kinetics 149
4.8. Ratios from Other Measures 152
4.9. Comparisons of Biomass to Coal 156
    4.9.1. Central Appalachian Bituminous Coal 156
    4.9.2. Illinois Basin Coal 156
    4.9.3. Powder River Basin Coal 156
    4.9.4. Lignite 157
4.10. The Chemistry of Cofiring 157
    4.10.1. Reactivity and Cofiring 157
    4.10.2. Evolution of Specific Elements and Compounds 160
4.11. Burning Profiles of Biomass–Coal Blends 161
4.12. Implications for Biomass–Coal Cofiring Systems 169
    4.12.1. Biomass–Coal Blend Issues 171
    4.12.2. Biomass–Coal Blend Systems 173
    4.12.3. Cofiring Methods and Equipment—Mechanical Systems 176
4.13. Case Studies in Cofiring 178
    4.13.1. Cofiring Experiences 178
4.14. Conclusions 195
References 195

5. Waste Fuel–Coal Blending 201

5.1. Introduction 201
5.2. Tire-Derived Fuel 201
    5.2.1. Overview 201
    5.2.2. Typical Composition 202
    5.2.3. Physical Characteristics 204
    5.2.4. Types of Tire-Derived Fuel 205
    5.2.5. Preparation and Handling Issues 205
    5.2.6. Combustion Considerations 210
    5.2.7. Case Studies 210
    5.2.8. Conclusions Regarding Tire-Derived Fuel as a Blend Fuel 215

5.3. Petroleum Coke 216
    5.3.1. Fuel Characteristics of Petroleum Coke 218
    5.3.2. Petroleum Coke Issues 219
    5.3.3. Petroleum Coke Utilization in Boilers 221
    5.3.4. Petroleum Coke Utilization in Other Systems 230
5.4. Waste Plastics and Paper 232
    5.4.1. Waste Plastic Composition 233
    5.4.2. Waste Plastic and Paper Preparation 235
    5.4.3. Waste Plastic Utilization 237
5.5. Hazardous Wastes 238
    5.5.1. Fuel Characteristics of Hazardous Wastes 238
    5.5.2. Combustion of Hazardous Wastes in Rotary Kilns 239
    5.5.3. Waste Oil Utilization 241
5.6. Conclusions 243
References 244

6. Environmental Aspects of Fuel Blending 249

6.1. Introduction 249
6.2. Regulatory Climate as It Influences Blending and Cofiring 249
6.3. Blending for Environmental and Economic Reasons 250
6.4. Areas of Concern 250
    6.4.1. Particulates 250
    6.4.2. Sulfur Dioxide 251
    6.4.3. Nitrogen Oxides 251
    6.4.4. Mercury 251
    6.4.5. Fossil CO_2 252
6.5. Ash Management for Power Plants 252
    6.5.1. Bottom Ash 252
    6.5.2. Flyash 252
6.6. Blending for Emission Benefits 253
    6.6.1. Blending PRB Coal with Other Solid Fuels 253
    6.6.2. Emission Aspects 254
    6.6.3. Selected Case Studies 258
6.7. Cofiring Biomass with Coal 260
    6.7.1. Emission Aspects 260
    6.7.2. Cofiring in Europe 262
    6.7.3. Selected Case Studies 262
    6.7.4. Cofiring with Waste 264
    6.7.5. Emission Aspects 264
    6.7.6. Selected Case Studies 266
6.8. Conclusions 267
References 268

7. Modeling and Fuel Blending 271

7.1. Introduction 271
7.2. The Purposes of Modeling 272

7.3. Specific Applications of Modeling 272
　　7.3.1. Modeling to Reduce the Use of Physical Tests
　　　　　and Costs 273
　　7.3.2. Methods of Modeling 274
7.4. Principles of Physical Modeling 280
　　7.4.1. Some Applications of Physical Modeling 282
　　7.4.2. Computational Fluid Dynamics Modeling 283
7.5. The Basic Approach of Computational Fluid Dynamics
　　Modeling 284
　　7.5.1. Computational Fluid Dynamics Modeling
　　　　　of Combustion Processes 285
　　7.5.2. Products of Combustion Modeling 288
　　7.5.3. Other Applications of Computational Fluid
　　　　　Dynamics Modeling 290
7.6. Modeling for Blending Purposes 290
　　7.6.1. The Traditional Approach to Blending Analysis 290
　　7.6.2. The Detailed Analytical Approach to Blending 291
7.7. Limitations of Modeling 291
7.8. Conclusions 291
References 292

8. Institutional Issues Associated with Coal Blending 295

8.1. Introduction 295
8.2. Institutional Issues Associated with Fuel Blending 297
8.3. Economic Considerations Associated with Blending 300
　　8.3.1. Fuels Availability 300
　　8.3.2. Fuel Procurement 304
　　8.3.3. Fuel Transportation 305
8.4. Process Modifications 309
　　8.4.1. Coal Handling and Storage 309
　　8.4.2. Coal Blending 310
　　8.4.3. Pulverizer Performance 310
　　8.4.4. Furnace Effects 311
　　8.4.5. Convective Pass 312
　　8.4.6. Emissions 312
8.5. Future U.S. and World Coal Production 313
8.6. Conclusions 320
References 320

Index 323

Fuel blending with solid fuels is becoming an increasingly important process for electricity generation companies and installations, as well as for process industries that fire boilers and kilns to drive production of goods such as pulp and paper, expanded aggregates, and cement (and that periodically generate electricity in cogeneration applications). Many economic and environmental forces have combined to make fuel blending increasingly attractive. Blending low-sulfur/low-nitrogen subbituminous coals with traditional eastern and midwestern bituminous coals provides an inexpensive means for reducing airborne emissions and generating electricity. At the same time, maintaining a greater than 30% proportion of eastern or midwestern coal in the blend helps oxidize any mercury found in the coal. Oxidation of the mercury facilitates its capture in a fabric filter facility or an electrostatic precipitator.

Fuel blending can achieve a multiplicity of purposes. For example, it can be used to increase fuel reactivity, reduce fuel costs, address deposition or corrosion issues, reduce certain types of airborne emissions, address the capture of certain pollutants, and reduce the concentration of certain pollutants (e.g., chlorine, mercury, arsenic). Typically, fuel blending is thought of in terms of coal, but it also includes other materials from petroleum coke to biomass with coal or with another fuel type.

Blending without sufficient investigation and analysis can easily exacerbate problems such as slagging and fouling and corrosion. The behavior of inorganic constituents, and some organic constituents, is not necessarily linear with blends. Surprises—some favorable and many unfavorable—can occur with blending when insufficient data and analyses are used.

Blending techniques can also influence outcomes. Basic blending is done by the bucketful—a bucket of this and a bucket of that (or a slug of this and a slug of that). We have seen this approach at coal tipples, power plants, and pulp mills. In very carefully designed programs, bucket blending can work well. Transfer points can function as mixing stations and can facilitate relatively homogeneous blends. Frequently, however, this approach does not work, and the boiler "sees" a slug of this and a slug of that. Silos that rathole—that is, operate on a "first-in, first-out" basis—only make the problem worse. When metering conveyors are used, blending is more precise, and swings in fuel quality are basically eliminated. Metering conveyor systems can be elegant or basic, depending on the installation. With them, two-way blends are easily accomplished, and many can achieve three-way blends.

More and more blending involves weigh-belt feeders, metering conveyors, and positive controls leading to reproducible results. Such systems can again be simple or highly sophisticated, depending on the fuel yard and its design. These systems may involve sophisticated computer tracking and controls or may only consist of programmable logic controllers and dial-in controls. Variations appear to revolve around the extent to which automation replaces human activity in the fuel yard. The degree of sophistication also depends on how aggressive a plant is in controlling the blend and the outcome. Some plants are content to fire a single, constant blend of two or more fuels. Others vary the blend to respond to market conditions and the natural variability in such solid fuels as ranks of coal, biomass fuels, and numerous waste fuels.

The authors of this book have extensive experience in fuel blending. We have managed the fuels and combustion process for power plants and have consulted on fuels and combustion issues for utilities, process industries, and incinerators. Further, we have led research efforts in evaluating blends and blending processes for utilities and industries, EPRI, the U.S. Department of Energy, and numerous other organizations. These professional activities have led to the development of many observations and ideas that are presented here. This book reflects our collective experience with numerous organizations involved in the production and use of the vast array of solid fuels as well as our participation in professional societies and conferences dedicated to advancing the understanding of these materials and their utilization.

This book provides information on the issues of solid fuel blending and the principles, practices, and problems associated with it. Chapter 2 deals with the fundamentals of fuel blending, examining the blending of coal on coal, biomass on coal, petroleum coke on coal, and others. Chapter 3 looks at the blending of coals—the chemistry of blending, blending systems, and critical issues.

Chapter 4 focuses on biomass cofiring with coal—the fundamentals of biomass cofiring, its chemistry, and associated systems. It also looks at specific case studies. Chapter 5 studies the aspects of waste fuel blending with coal—for example, tired-derived fuel, petroleum coke, and waste plastics and papers. Chapter 6 looks at the environmental aspects of fuel blending, and Chapter 7 considers various aspects of modeling associated with it. Chapter 8 focuses on the institutional aspects of fuel blending.

Acknowledgments

We have not performed our work alone or in a vacuum, and we have not written this book without significant help. Among those who have helped are Anthony Widenman, David Nordstrand, Michael F. Dunlap II, Joe Robinson, and many other present and former colleagues at DTE Energy. They have provided information, ideas, and comments concerning our efforts. Similarly, Mike Santucci and Jim Scavuzzo of ECG have provided ideas and information particularly on computer controls and various forms of modeling.

Rob Mudry of Air Flow Sciences provided significant information on both cold flow and computer modeling. Donald Kawecki of Research Cottrell—formerly of Foster Wheeler—helped with many discussions. Bruce and Sharon Falcone Miller of The Energy Institute of Pennsylvania State University contributed, as did Gareth Mitchell of PSU. Richard Monts provided photographs of coal trains, and Pentrex provided significant photography of both coal mining and coal trains in the Powder River Basin.

A special thanks goes to Melody Huang, a graduate student at Lehigh University, who helped substantially with literature searches. Many other individuals contributed support, ideas, and more. With their help, we offer this effort in evaluating the principles, processes, and consequences of blending solid fuels.

David A. Tillman
Retired Chief Engineer for Fuels,
and Combustion, Foster Wheeler
and Consultant to DTE Energy

Dao Duong
Combustion Engineer, Foster Wheeler

N. Stanley Harding
Consultant

Chapter 1

Introduction to Fuel Blending

1.1. OVERVIEW

The blending of two or more solid fuels involves combining the desired materials together in a careful, reproducible manner. This book deals with solid fuel blending that is controllable and reproducible; it focuses on systems where controlled conveyers, weigh belt feeders, and other means are used to provide a consistent feed to the combustion or gasification system. Blending, as discussed in this text, requires knowledge of what is being blended, why it is being blended, and the expected outcome of the blending process.

1.2. FUEL BLENDING FOR SOLID FUELS

Blending for solid fuels involves producing a reasonably homogeneous mixture of the two or more solids to be fired in a boiler. These solids may be coals of the same or similar ranks; coals of dissimilar ranks; coals with biomass fuels such as wood, wood waste, herbaceous crops, and crop wastes; fecal matter from animals; and industrial residues from the processing of biomass. Blends may also include coals and a range of industrial materials and residues, including petroleum cokes of one or another type, by-product aromatic carboxylic acid (BACA), coal wastes such as culm or gob, municipal solid waste-derived fuels (e.g., refuse-derived fuel, waste paper, waste plastics), tire-derived fuel, selected hazardous wastes, and many more. Blending is limited only by the ingenuity of the engineers and by the regulatory environment [1].

1.2.1. Blending System Considerations

From the perspective of fuel blending mechanical system fundamentals, we have these things to consider:

1. Where in the overall process scheme should blending occur?
2. What types of mechanical systems are available to accomplish blending?
3. What type of blending controls should exist?
4. What modifications must be made to plant equipment?

Following this discussion, the overall impacts or consequences of blending can be considered. It is important to note that the discussion here is an overview.

To the extent that specific fuel blending influences these questions, more detail will be presented in subsequent chapters.

1.2.2. Where Blending Can Occur

Typically, blending occurs in the fuel yard of a utility power plant or industrial boiler; however, it can occur in an off-site fuel management facility with the blend being shipped to the power plant or industry. It can occur as part of the fuel handling process: in the feed system conveying fuel to the burners or other energy recovery and production systems (e.g., combustion or gasification systems). It can occur in the energy production equipment (e.g., the pulverized coal boiler) depending on the system design. In this case certain pulverizers are set up to handle one fuel, and others are set up to handle another fuel—the blend fuel(s).

Blending of different fuels depends on the fuels to be blended. At one extreme a utility or industry can purchase preblended fuel from a transloading facility or other similar operation. Many eastern tipples, such as Tanoma Coal Company, provide blends of fuel to their customers in order to meet specifications. When biomass cofiring was tested at the Shawville, Pennsylvania, generating station of (then) GPU Genco (now Reliant Energy), Tanoma Coal blended the woody biomass forms with the coal to meet the objectives of the blend process [2, 3]. When the Tennessee Valley Authority (TVA) tested firing up to 20% petroleum coke with coal at its Widows Creek Fossil Plant, the blend was prepared by the BRT facility in Kentucky and shipped on the river to the power plant. Other utilities and manufacturing industries have investigated this option as well [4–6].

Purchasing preblended fuel has several advantages. No capital investment is required at the plant site. In reality the use of a blend is transparent to the power plant. The blend is handled like a single coal. This also requires little if any change in operations and maintenance practices. Purchasing preblended fuel, however, also has several disadvantages: The system is rigid, and the blend cannot be changed at the plant to respond to power plant needs or the consequences of in-seam variability of coal. If the blend is not desirable (e.g., if the blend causes slagging, fouling, or corrosion), the electricity generating plant or manufacturing facility must burn it in any event and probably must suffer a derate in the process. It may also experience elevated operations and maintenance costs in the process.

Probably the most common form of blending involves mixing two or more fuels in predetermined blends in the fuel yard. This can be accomplished in any number of ways, as will be discussed subsequently. This is the approach taken at the Monroe Power Plant of DTE Energy (Figure 1.1), the Limestone Generating Station of Texas Genco, and numerous other utilities and industries blending various types of coal of dissimilar rank. TVA took this approach testing blends of petroleum coke and coal at its Paradise Fossil Plant and blends

FIGURE 1.1 Aerial view of the Monroe Power Plant of DTE Energy. Note the coal yard at the back of the site. This coal yard contains the $400 million blending facility constructed such that three coals can be fed to the boiler in varying proportions to meet operational requirements. *Source: [10].*

of tire-derived fuel with sawdust and coal at its Allen Fossil Plant [3–5]. This approach was taken during the testing of petroleum coke and wood waste cofiring at the Bailly Generating Station of NiSources [7, 8]. This approach is limited to coal–coal blends, coal/petroleum coke blends, and cofiring with woody biomass, such as what was done at Plant Hammond of Southern Company [9], as well as at the Allen Fossil Plant and the Bailly Generating Station. This approach cannot be used with blends of coal and agricultural products such as switchgrass or corn stover.

The advantages of this approach include the ability to adjust the fuel blend to utilize varying properties of the fuel—both good and bad. Also, depending on the plant information and control system, the on-site blending can be used to minimize the risks associated with slagging, fouling, and deposition. Fuel characteristics leading to those conditions and fuel characteristics leading to unacceptable levels of pollution can be addressed by blending as well. The blending process can move the fuel characteristics away from the most severe conditions, depending on the fuels available.

This blending approach, however, has limitations. It can be very capital intensive—for example, the DTE Energy blending facility had a capital cost of $400 million [10]. Other facilities such as the Bailly Generating Station blending

FIGURE 1.2 The on-site blending facility constructed at Bailly Generating Station for testing purposes. Note that this is a labor-intensive system. *Source: [7].*

system, shown in Figure 1.2, cost only $1.2 million. However, the Bailly blending system is more labor intensive, requiring two additional operating persons just to run the blending system [3, 7]. Maintenance must be vigorously pursued in order to preserve the accuracy and consistency of the blending; otherwise, it is not effective. This is discussed more extensively in Chapters 2 and 3. At the extreme is bucket blending—using two front-end loaders to build piles of the blends. This can result in blends that are "a slug of this and a slug of that," and these blends do not maximize the desired benefits of good blending.

On-site blending introduces another potential problem: preferential grinding or pulverization in the mills. When introducing two coals to a mill, the pulverizer will grind the softer (higher-HGI) coal more completely than the harder coal. For example, Monroe Power Plant was using a blend of 70% Powder River Basin (PRB)/30% Central Appalachian (CA) bituminous coal. Tests showed that the >50 mesh cut of the pulverized material was 70% CA bituminous coal [10]. The >200 mesh, >400 mesh, and residual products were where the PRB coal was concentrated. This preferential grinding—preparing the softer, more easily pulverized coal more thoroughly than the harder coal—is a common experience in blending operations. This is discussed more extensively in Chapter 3.

A third approach to blending is blending in the furnace or boiler itself. The various fuels are prepared separately and introduced into the boiler separately.

This has been used with dissimilar coals such as PRB subbituminous coal and lignite at the Limestone Generating Station. It is the preferred method for cofiring biomass with coal in pulverized coal boilers and is required when cofiring agricultural products such as switchgrass with coal, as has been demonstrated at Plant Gadsden of Southern Company, Ottumwa Generating Station of Alliant Energy, and Blount St. Station of Madison Gas & Electric [3].

Agricultural materials, such as switchgrass, do not lend themselves to blending with coal in the coal yard. This approach has been in existence for a long time and has been used in stoker firing as well as pulverized coal (PC) firing [11]. Detroit Stoker developed a fuel feeding system with a paddle wheel for coal stoker firing and a windswept spout for wood waste firing—simultaneously—in the large stoker-fired boilers of the pulp and paper industry [11].

This approach is also being used by Foster Wheeler in the design and construction of two 300-MWe circulating fluidized bed (CFB) boilers being supplied to Dominion Energy. These boilers will be fired with up to 20% wood waste and 80% coal. The mixing of wood waste with coal will occur in the CFB itself, not in the fuel yard. This approach was used in one cyclone installation: the Allen generating station of Northern States Power. Dry, finely divided sawdust from the adjacent Andersen Windows plant was fired in the secondary air plenum of 3 of the 12 cyclone barrels in a manner similar to the means for firing natural gas in cyclone boilers.

There are distinct advantages to this form of blending. If two coals are blended using this approach in a PC boiler, then the pulverizers can be set to the individual coals being fired. If biomass is being fired with coal, then the coal delivery system is not impacted. If wet coal is received and a derate is to be taken, the addition of the biomass can minimize that derate, depending on the specific design. It should be noted, however, that this blending approach is more suited to tangentially fired PC boilers than wall-fired boilers. Tangentially fired PC boilers have a single fireball, whereas wall-fired boilers have distinct flames from each burner; there is less mixing of the fuel and flame in such installations. A disadvantage of this approach is that some of the chemistry benefits of blending are not achieved with in-furnace blending.

Therefore, numerous locations can be used for blending different fuels being fired in a single boiler. Choice of the optimal location depends on the fuels being burned, the firing method, and the approach of the electric utility or process industry.

1.3. OBJECTIVES FOR BLENDING

Blending is designed to meet certain overall objectives: economic, environmental, and technical. The economic objectives are always tied to producing the useful energy product at the lowest cost. This may be process steam, where cost is expressed in $ per 10^3 lb of useful steam; electricity, where cost is expressed in $ per MWh, or process heat, where cost is expressed in $ per 10^6 Btu (or $ per GJ). This may involve blending for the cheapest usable fuel product or for the most cost-effective fuel product given the constraints of the existing boiler, kiln, or process heat generator.

Different coals exhibit significantly different price structures, as shown in Figure 1.3. Biomass and waste fuels also exhibit very different cost structures. Blending provides a means for managing total fuel costs in $ per unit of energy produced, recognizing that different fuels have different efficiency and heat transfer characteristics, different slagging and fouling characteristics, and different capacity implications, as is discussed in subsequent chapters.

It is within the economic constraint that the environmental objectives exist. Blending received significant impetus for environmental reasons; using low-sulfur PRB coal blended with Eastern or Interior Province bituminous coal

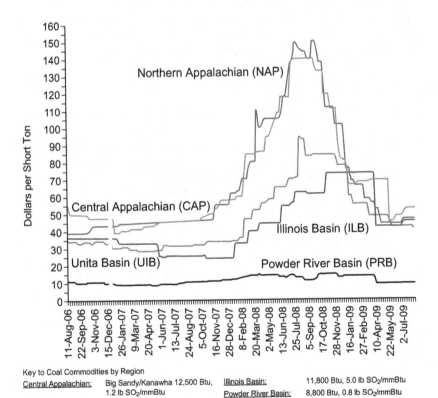

Key to Coal Commodities by Region

Central Appalachian:	Big Sandy/Kanawha 12,500 Btu, 1.2 lb SO₂/mmBtu	Illinois Basin:	11,800 Btu, 5.0 lb SO₂/mmBtu
		Powder River Basin:	8,800 Btu, 0.8 lb SO₂/mmBtu
Northern Appalachian:	Pittsburgh Seam 13,000 Btu, <3.0 lb SO₂/mmBtu	Unita Basin in Colo.:	11,700 Btu, 0.8 lb SO₂/mmBtu

FIGURE 1.3 Recent spot prices for various coals ($/ton). Notice the dramatically lower cost of Powder River Basin coal on a per-ton basis; it is also lower on a $/106 Btu basis. This promotes blending for economic reasons. *Source: [12].*

provided a means for meeting the SO_2 requirements of the Clean Air Act without substantial capital investment [13]. Blending has provided a least cost approach to meeting sulfur dioxide (SO_2) or oxides of nitrogen (NO_x) regulations. Reducing the sulfur content of the fuel is the direct approach to SO_2 emissions, and this formed a priority in blending PRB coal with bituminous coal during the 1970s and 1980s; this approach is still used today.

Reducing fuel nitrogen concentrations and increasing fuel (and fuel nitrogen) reactivity help to reduce NO_x emissions and were additional benefits for blending subbituminous coal with bituminous coal. Blending can provide mechanisms for reducing the ash content and, consequently, the generation of particulates. Blending may provide a means for reducing the concentration of hazardous air pollutants (HAPs) in the fuel as fired; alternatively, blending can make such HAP emissions more manageable. For example, bituminous coals with minor concentrations of chlorine can provide a means for oxidizing mercury and making that metal more easily captured. Cofiring blending, with biomass, provides a means for reducing the carbon footprint of any installation. These subjects are discussed extensively in Chapters 2 through 6.

Strategically, blending provides a means for increasing the usable fuel supply for any given power plant. While U.S. and world coal reserves are vast, the coal appropriate for any given utility or industrial boiler is limited by certain parameters, including moisture content, calorific values, heteroatoms causing pollution (e.g., sulfur, nitrogen), ash content, and so on. Limitations are also in the form of boiler dimensions, heat release requirements, and related physical constraints. Blending can be used to expand the available supply of coal, thereby increasing plant reliability while creating a competitive basis for providing fuel to the plant. Strategically blending biomass or waste-based fuels also increases fuel supply while at the same time addressing environmental concerns.

1.3.1. Economic Considerations with Fuel Blending

The blending of different fuels can have significant economic benefits to a plant and/or utility. Even though the main driver for utilizing subbituminous Powder River Basin (PRB) coal was to decrease the amount of sulfur dioxide emission, it can have a significant impact on economic savings, as shown in Figure 1.3. In 2009, the average sale price of PRB coal was $12.41 per short ton. The average price for bituminous coals in the United States was $55.44 per short ton [12]. For Monroe Power Plant, the 3200-MW station located in Monroe, Michigan, and shown in Figure 1.1, the incremental savings are significant for increases of PRB within the blend.

Blending at Monroe Power Plant is just one example of the economic benefits of blending. It must be recognized that blending is not economically beneficial in all situations. The blending of many biomass fuels typically results in increased fuel costs to a plant; blending or cofiring with biomass is done, most commonly, to meet environmental and regulatory demands. Further, as is discussed in subsequent chapters, there is a difference between fuel price and fuel cost. Fuel cost must take into account additional outages for furnace cleaning, additional investments in capital equipment, and additional operating costs associated with the blending process. These are discussed in Chapter 3.

1.3.2. Environmental Considerations with Fuel Blending

The blending of different solid fuels over the years has mostly been driven by emissions and regulatory drivers. The constituents that are of most concern are SO_2, NO_x, particulates, selected trace metals such as mercury and arsenic, and, more currently, fossil CO_2.

SO_2 Considerations with Fuel Blending

One of the biggest drivers for blending low-sulfur PRB coal with higher-sulfur bituminous coals was the need to reduce SO_2 emissions to meet the requirements of the Clean Air Act as amended. The technique of blending PRB coal with bituminous coals for SO_2 reduction has been used by plants such as Monroe Power Plant, B.L. England Station, and many others. This is a primary driver in the use of PRB coals, and these coals are utilized to such an extent that they support the generation of 20% of the electricity consumed in the United States.

SO$_2$ emission reduction has also been realized when blending PRB coal with lignite coals; many lignites contain more sulfur than PRB coals, when measured on either a percentage basis or a lb/10^6 Btu basis. The drive for low-sulfur-blend coals has also led to the increased use of Adaro coal and other very low-sulfur Indonesian coals throughout the world. Cofiring biomass fuels with coal will typically result in SO_2 emission reduction, since most biomass fuels are inherently low in sulfur. SO_2 has been and will continue to be a key aspect that promotes the blending of various solid fuels. This is discussed more substantially in Chapters 3 through 6.

NO_x Considerations with Fuel Blending

The reduction in NO_x when blending certain solid fuels is typically realized as a function of both fuel bound nitrogen and fuel reactivity. As discussed in Chapter 2 and then further in Chapters 3 through 5, due to the highly reactive nature of certain fuels such as PRB coal and particular biomass fuels, the nitrogen can be released much more rapidly within the furnace. If the nitrogen can be released within a fuel-rich environment, the formation of N_2 is favored, and consequently the formation of NO_x constituents is reduced. Alternatively, if fuel nitrogen is released in a fuel-lean environment, NO_x production is more prominent [3, 14]. Blending of coals and alternative fuels such as petroleum coke with highly reactive coals, biomass fuels, and wastes can be used to influence NO_x production, as will be discussed in subsequent chapters.

Particulates Considerations with Fuel Blending

The blending of solid fuels can have either beneficial or adverse effects for particulate emission. These effects are fuel and blend dependent. The use of lower-ash fuels does not always correspond to beneficial results. When blends of PRB coal with petroleum coke are used, due to the low-ash contents in both fuels, dilution effects are not present, consequently affecting the unburned carbon concentration and salability of the flyash. This phenomenon is discussed extensively in Chapter 2.

Greenhouse Gas Considerations with Fuel Blending

The release of fossil CO_2 into the atmosphere has been of increased concern over the years. This has led to major initiatives and demonstrations for biomass cofiring in the United States [3, 15]. Equally, if not even more significant, this has led many European countries to experiment with and then commercialize biomass cofiring: blending biomass fuels with coal for power generation. Significant efforts have been made in England, Denmark, and other countries, which have led to the export of biomass fuels—typically wood pellets—from the United States and Canada to the European continent.

Biomass is considered a CO_2 neutral fuel (see Hughes [16]). During the life of the plant, photosynthetic processes occur, thus consuming CO_2 from the atmosphere. When combusted, the same amount of CO_2 is considered to be released, so the net CO_2 production is essentially zero (Figure 1.4). As a result, fossil CO_2 is decreased when biomass cofiring is employed. This discussion is amplified in Chapters 4 and 6.

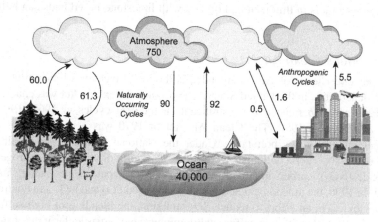

FIGURE 1.4 The global carbon cycle showing carbon flux in billions of metric tons. *Source: [3]; supplied by the Federal Energy Technology Center.*

Trace Metal Emissions

The blending of fuels can have either beneficial or adverse effects on hazardous air pollutants (HAPs). Mercury concentrations in coals and petroleum cokes vary dramatically as a function of their origin, the coalification process, or the coking process. Mercury concentrations in biomass vary as a function of where the biomass is grown and the economic activity in the immediate vicinity [17]. Speciation of mercury for the different fuels has an impact on the amount of mercury that can be captured in postcombustion equipment; the different forms of mercury can be found in elemental form or oxidized form, or they can be oxidized during the combustion process.

Typically, oxidized forms of mercury can be more easily captured in postcombustion equipment. Mercury in PRB coals is typically in elemental form. Blending of PRB coals with Eastern or Interior Province bituminous coal helps to oxidize the mercury because of the chlorine concentration present in these coals, particularly if $\geq$30% bituminous coal is used. Conversely, the interaction between chlorine and lead has adverse effects on lead emissions. Lead chloride compounds can be formed in the gaseous phase (e.g., $PbCl_4$) and do not condense readily; consequently, they cannot be captured.

Arsenic (As) exists in solid fuels in varying concentrations. In blending, As can be reacted with calcium compounds or calcium and aluminum compounds to form Calcium arsenates and calcium-aluminum arsenates. These are readily managed. Alternatively the arsenic may be produced in both the +5 oxidation state and the +3 oxidation state. It is in the +3 oxidation state, as As_2O_3, that the arsenic is toxic and carcinogenic. Blending fuels with significant concentrations of arsenic with fuels containing high concentrations of calcium (e.g., southern PRB coals, biomass fuels) and/or firing these fuels in fluidized bed boilers with limestone-based beds can help address this issue.

The Federal Clean Air Act

The Federal Clean Air Act as amended is the key driver for all air pollution control activities in the United States. The original Clean Air Act was enacted in 1963 and experienced five significant amendment cycles in 1965, 1967, 1970, 1977, and 1990. The Clean Air Act of 1970 was a major driver in addressing the major pollutants. Under the National Ambient Air Quality Standards, six criteria pollutants were addressed: sulfur dioxide (SO_2), nitrogen dioxide (NO_2), carbon monoxide (CO), ozone (O_3), particulate matter, and lead. In 1990, the amendment additionally addressed SO_2 and NO_x and control of air toxics. In an effort to reduce SO_2 emission, the blending of high-sulfur bituminous coals with low-sulfur subbituminous coals—particularly that from PRB coals—became common practice.

1.3.3. Historical and Technical Considerations for Fuel Blending

For decades, power plants were designed to burn a single "design fuel," typically supplied by a single coal mine or coal tipple. This design fuel had limited variability in terms of its analytical properties expressed as proximate analysis, ultimate analysis, calorific value, and ash elemental analysis. The variability at the time could be characterized as "in-seam" or "in-mine" variability—the variation in fuel properties caused by variable coal characteristics that exist in a single coal seam or coal mine. Most plants, typically built in the Northeast, Southeast, and Midwest, had 30+-year coal supplies when designed, constructed, and operated.

Similarly, industrial boilers for the pulp and paper industry, the sugar industry, and other industries had steady supplies of wood waste, or bagasse (the forms of biomass used to generate process heat and electricity in the forest products and food industries), coal, petroleum coke, and other fuels. Similar experiences have occurred for those industries firing process kilns, including the cement industry, the expanded aggregate industry, selected mineral processing industries, and more.

Steps have been taken to introduce technical fuel blending concepts to the energy industry. Fuel blending with a scientific basis emerged in the steel industry—in coke making—in order to ensure appropriate properties of the feedstock and consistent, useful properties in the resulting cokes produced in the slot ovens and beehive ovens of U.S. Steel, Bethlehem Steel, Inland Steel, and other such industries in North America and throughout the world. Blending based on concepts of macerals became common. In addition to the typical fuel analyses just identified, coals became characterized in terms of the free swelling index (FSI), caking properties, agglomeration properties, Giesler plasticity, and the like [18–20].

In a second phenomenon, many utility boilers and industrial kilns have outlived their original design fuel supplies. Utility boilers now have an average current life of more than 35 years. Many have lives exceeding 50 years. For example, the River Rouge Power Plant of DTE Energy is, as of this writing, 55 years old. The power block at the Trenton Channel Power Plant of DTE Energy, referred to as "The High Side," is well over 60 years old. Examples like this abound in virtually all eastern and midwestern utilities.

Many other power plants in the United States and around the world are as old as or older than these examples. Rotary kilns used in the pulp and paper industry, the cement industry, and similar process applications and industrial boilers used in manufacturing plants from sawmills to pulp mills, sugar mills, and other industries also can be quite old. It is not unusual to find utility and industrial boilers that are 50 to 75 years old. Clearly these boilers have outlived their original fuel supply. Blending provides a means for fueling these systems while maintaining the performance parameters of these units.

A third phenomenon has generated significant interest in blending: the growth of PRB subbituminous coal, which now is used to generate about 20% of the electricity consumed in the United States. This coal, which is also exported to other nations as a steam coal, is a low-cost fuel. Further, it is relatively low in calorific value at a nominal 8800 Btu/lb (4900 kcal/kg). Its high reactivity, measured by a volatile/fixed carbon ratio or by atomic hydrogen/carbon and oxygen/carbon ratios, contributes to problems with spontaneous combustion when stored in silos or bunkers.

For the most part, blending of subbituminous coals involves their use with high-Btu/lb bituminous coals; however, sometimes PRB coals are blended with lignites to increase the calorific value of the fuel fired in a given boiler. As discussed previously, sulfur content and fuel volatility are primary drivers in blending with Powder River Basin coals. At the same time, its inorganic constituents or mineral matter (e.g., calcium, sodium) can cause significant problems with respect to slagging and fouling. Blending can address, or exacerbate, these concerns.

Internationally, the Indonesian subbituminous coals (e.g., Adaro coal), which are extremely low in sulfur, present opportunities similar to those associated with PRB coals. Alaskan coal—Usibelli coal—presents opportunities similar to those of Adaro coal. These coals were introduced as blending coals and are still used frequently with that approach. They have additional blending potential throughout the world, as shown by the increased international trade in these subbituminous coals.

The PRB subbituminous coals initially were used only in western power plants; however, they became attractive to midwestern utilities due to their low cost and low sulfur content. Now they are used throughout the United States and Canada and are exported to other nations such as China. They are used more in blend applications than as 100% of the fuel fed to any given boiler. As will be discussed in Chapters 2 and 3, the inorganic constituents in these coals initially presented severe problems.

Further, the relatively low calorific content (typically ~8800 Btu/lb or ~4900 kcal/kg) and high moisture content (typically ~18%) meant use in existing boilers designed for bituminous coals could result in significant derates. The inorganic content, and its behavior in the boiler, also contributed significantly to derates. The consequence was the need to blend this low-sulfur, low-cost coal with bituminous coals in order to operate units successfully. The massive demand for PRB coal—which has grown significantly in recent years—has depended on moving the coal east to boilers designed for bituminous coals and blending it with bituminous coals in order to use both successfully.

PRB blending with bituminous coal has been well documented as a consequence of its economic and environmental benefits. But it is an intricate and complex issue. Further, the industry at times approaches PRB as a single coal or as a southern (calcium based) and northern (sodium based) pair of coals. In reality, there is significant variation among the PRB coals, requiring considerable attention. Variability comes in calorific value, sulfur content, and, most significantly, ash chemistry. Chapter 3 details this variability and its impact on fuel blending.

In the past 15 to 20 years, emphasis has been placed on renewable and sustainable energy. In the United States, more than half of the states have renewable portfolio standards (RPS) mandating use of renewable energy in the mix of power supplied to consumers. Some European nations, such as England, Denmark, and many others, also have RPS laws on the books, and these have led to the use of biomass in fuel blends. While RPS legislation has become increasingly popular throughout the advanced economies of Europe and North America, the demand for electricity has grown; use of electricity has become almost synonymous with standard of living.

While policy makers seek higher percentages of renewable power, utilities continue to pursue technologies of reliable and dispatchable power. The consequence has been continued interest and activity in pursuing biomass cofiring in coal-fired power plants. Cofiring typically involves firing 10% to 30% of opportunity fuels—most commonly various forms of biomass—with coal (or heavy oil) in a given boiler. This approach capitalizes on the higher efficiency of large-scale power generating installations equipped with higher-pressure and higher-temperature boilers (e.g., 2400 psig/1025°F steam) and supercritical boilers (e.g., 3500 psig/1025°F steam). These boilers increase the efficiency of power generation by employing reheat cycles. Consequently, more efficient dispatchable renewable power can be obtained by blending biomass with coal in boilers—cofiring [3, 15]. At times this requires blending biomass fuels together to manage their fuel properties and then cofiring the resulting blend with coal. Cofiring is simply a name for firing dissimilar fuels such as biomass and coal. It is not a new strategy; it has been used in industrial boilers such as power boilers in the pulp and paper industry for many decades (see [11]).

1.4. BLENDING FOR THE STEEL INDUSTRY—THE DEVELOPMENT OF PETROLOGY

It can be argued that the scientific blending of solid fuels began in the steel industry rather than in the electric industry. Coke making originally was done with very little understanding of the science and chemistry involved in the process. Consequently, accidents and unexpected problems occurred. The development of petrography was instrumental and necessary to further the understanding of coke making for the steel industry.

Metallurgical coke is the primary fuel source and is the reductant in blast furnaces. In addition to being consumed as a fuel, the coke must have sufficient strength to support the weight of the blast furnace burden while also being porous enough to permit the passage of hot air or a "blast" for combustion.

Metallurgical coke quality is measured by composition, size, cold and/or hot strength, and reactivity to carbon dioxide at elevated temperatures. The quality and properties of the coke are inherited from the selected coals used, in addition to how they are handled and carbonized in the coke plant operations.

The qualities are influenced by coal rank, the reactive and inert macerals and minerals in the coal, and the ability to become plastic and resolidify into a coherent mass. Bituminous coals of high-volatile A, medium-volatile, and low-volatile rank possess the necessary properties. However, not all coals in these ranks produce desirable quality, and at times they may even be detrimental to the coke ovens. Consequently, the lack of individual coals possessing all the necessary properties resulted in the need to blend different coals. Blends can consist of anywhere from 2 to 20 different coals; the United States typically uses a blend of 3 to 5 coals [21].

1.4.1. Basics of Macerals

The *Glossary of Geology* [22] defines *macerals* as "the microscopically recognizable constituents of coals that evolved from the different organs and tissues of plants." Macerals can be ranked into three basic groups based on similar origin or mode of conservation, as well as chemical composition: vitrinite, liptinite (sometimes referred to as exinite), and inertinite [21]. Table 1.1 details the maceral groups and the macerals that can be found in each group.

TABLE 1.1 ASTM Maceral Group and Terminology

Maceral Group	Maceral
Vitrinite	Vitrinite (various)
Liptinite (or Exinite)	Alginite Cutinite Resinite Sporinite
Inertinite	Fusinite Inertodetrinite Macrinite Micrinite Funginite Secretrinite Semifusinite

Vitrinite group macerals are typically derived from the humification of woody tissues and can possess remnant cell structures or may be structureless. They tend to have more oxygen than other macerals for any given rank. The group of liptinites are derived from plant resins, spores, cuticles, and algal remains that are resistant to bacterial and fungal decay. They are generally characterized as having higher hydrogen content than other macerals; this is particularly true for the lower-rank coals. At the transition from subbituminous to bituminous coal, there is a marked decrease in volatile content and an increase in carbon content. A further decrease in hydrogen and volatile matter content occurs by the medium volatile rank, which makes liptinite macerals difficult to distinguish from vitrinite macerals. Inertinite macerals are derived mainly from woody tissues, degraded plant products, or fungal remains. They typically are associated with high–inherent carbon content that is a result of thermal or biological oxidation [21].

Vitrinite identification and quantification is essential to the steel-making industry, but it is also useful in evaluating coals for combustion and gasification as well (see, for example, Bryers [23], Zhang et al. [24], and Su et al. [25]). Fuel blending and fuel analysis are based on certain analyses: proximate and ultimate analysis, calorific values, ash elemental analysis, fuel reactivity, ash reactivity, and the Hardgrove grindability index (HGI). The linkage between petrology and combustion/gasification analysis is considered by Bryers, Zhang et al., and Su et al. Further, it is explored more completely in Chapter 3.

1.4.2. Petrography Applied to the Steel Industry

In the coke making process, an understanding of the composition of coal is important to evaluating the quality and value of a coking coal. Volatile matter and maximum vitrinite reflectance are some of the most important values used to determine the worth of coking coals. Vitrinite macerals are the principal reactive components of a coking coal. During the heating process in a reducing atmosphere, vitrinite becomes plastic, devolatilizes, and then solidifies to form the porous, carbonaceous matrix of a metallurgical coke. Liptinite macerals are also highly reactive, but because of their higher volatile content, they contribute more to the by-products.

On the other hand, inertinite macerals are inert during the carbonization process because they have limited thermoplastic properties and volatile contents. Inertinite macerals function as filler for the other reactive macerals. Reactive and inert macerals are derived from the concept of whether a particular maceral becomes fluid during the coking process and the degree by which it can be recognized in the coke product. Vitrinite and liptinite, which become completely fluid during the coking process, are the main reactive macerals. Conversely, inert macerals such as fusinite, micrinite, macrinite, and inertodetrinite will maintain their shape and morphology after they have been heated [21]. The use of these factors in combustion/gasification analysis is discussed more completely elsewhere [23–25] and is explored in Chapter 3.

The understanding of clay minerals is also important in coke quality. Illite and sericite contain potassium, contributing to alkali loading. Alkali loading in blast furnaces has a negative influence on coke reactivity. Besides petrography, x-ray diffraction is necessary to further the understanding of clay minerals [21]. These minerals are also of significance in analyzing fuels for combustion and gasification, as is considered in Chapters 2 through 5.

1.4.3. Conclusions on Blending for the Steel Industry

The concept of blending different coals in order to achieve more desirable qualities has been in use within the steel industry for many years. It can be argued that the scientific approach to blending really began within the steel industry. The use of petrography, along with other advanced techniques, enabled the process of coke making to be optimized. The techniques developed for the steel industry have since been "borrowed" and extended in the analysis of coals, petroleum cokes, and other fuels used in combustion and gasification applications.

1.5. TYPICAL FUEL BLENDS

The blending of solid fuels is typically accomplished with two different fuels and commonly three different fuels in order to achieve the desired fuel properties. For example, Monroe Power Plant commonly blends 60% to 70% PRB subbituminous coal with one or two types of Central Appalachian bituminous coal, depending on the conditions at the time of blending [10]. At both Allegheny's Willow Island cyclone boiler and the Allen Fossil Plant of TVA, trifiring of coal, sawdust, and tire-derived fuel (TDF) was conducted [3, 26]. The blending of the particular parent fuels was driven by technical, economical, and environmental factors. The principles of blending coal and alternative fuels are discussed in Chapter 2, and subsequent chapters discuss specific fuel blends.

1.5.1. Coal–Coal Blends

Coal/coal blending (see Chapter 3) is the most commonly performed solid fuel blending. In addition to the well-recognized PRB blending, coal/coal blending has been used with eastern and midwestern coals. Such blending is often used to reduce the sulfur content of the total fuel mass or to manage the inorganic constituents in the total supply of fossil fuel. The blending of PRB coals with bituminous coals is commonly practiced in the United States. Key drivers that initiated the blending of PRB coal with bituminous coals included the low sulfur content of PRB coal and the high reactivity of this material. Other technical and economical factors also became important as well.

The blending of PRB coal with bituminous coal can modify the chemical characteristics of the fuel. Most of the PRB coal is highly reactive, so it greatly influences the behavior and characteristics of the blended fuel. This has been observed at utility boilers and in lab scale experiments [28]. The blending of a highly reactive PRB coal with Central Appalachian bituminous coal at Monroe Power Plant has shown that the reactivity of the blends exceed that of the parents. This was further studied and detailed by looking at reactivity during the pyrolysis and char oxidation stages for both the parent fuels and the blends [29]. This has significant technical influence on the combustion process and emission constituents such as SO_2 and NO_x. These issues are all discussed in detail in Chapter 3.

The blending of PRB coal with lignite coal has been demonstrated and can be very successful. Blending at the Big Brown Plant of Luminant and Texas Genco's Limestone Generating Station showed significant technical and environmental benefits. By blending Powder River Basin coal with a lower-quality fuel, the plant achieved improved fuel characteristics and reduced emissions. The blending of PRB coal with lignite improves fuel quality and can reduce emissions. PRB coal is an inherently low-ash fuel with lower concentrations of sulfur compared to lignite. Blending PRB coal with lignite reduces the ash concentration, and slagging and fouling potentials are less than when blending PRB coal with most bituminous coals.

Fuel qualities such as volatile matter content, sulfur content, and calorific value are improved. PRB coal is a lower-sulfur fuel compared to lignite, so when it is blended, SO_2 emission is reduced. In addition, most PRB coals are very reactive fuels, promoting the early release of nitrogen in a fuel-rich region. NO_x emissions can be reduced with the optimal blend of PRB coal with lignite coal. This is also discussed in Chapter 3.

The blending of Eastern or Interior Province bituminous coals with offshore coals is common practice. Offshore coals will include those from the "Ring of Fire" or from nations such as Russia, South Africa, Australia, Germany, Poland, and others. Adaro coal comes from Indonesia and is part of the Ring of Fire as well. The Ring of Fire is a 40,000-km-long zone encircling the Pacific Ocean Basin that frequently experiences earthquakes and volcanic activity. Deposits from this region are influenced by volcanic activity, impacting the coalification process. Adaro coal is a very low-sulfur coal (e.g., 0.1–0.2% sulfur) that when blended with Eastern or Interior Province bituminous coals lowers sulfur dioxide emissions. Similar to the use of PRB coal, the need to meet the Clean Air Act was a driving force behind blending with Adaro coal.

International coals are also of significance in blending. Coals from locations such as Australia, Colombia, Venezuela, Brazil, South Africa, Russia, China,

Poland, and Germany are of considerable technical and economic importance. Northern European power plants commonly import such coals. China is particularly important as a market for Australian and South African coals. Blending with these coals depends on the technical, economic, and environmental drivers. Each parent fuel and blend must be evaluated on a case-by-case basis. This is examined further in Chapter 3.

1.5.2. Coal–Biomass Blends

Biomass/coal blends are the focus of Chapter 4. In an effort to increase biomass usage in the U.S. electric utility sector, cofiring development was accelerated in the 1990s [30]. Cofiring of various biomass fuels and coals has been evaluated and demonstrated successfully in numerous electric utilities. By 2003, more than 30 individual cofiring tests had been performed in over 19 states of the United States. Among these U.S. demonstrations were those shown in Table 1.2 and the example shown in Figure 1.5.

At the same time, significant demonstrations and commercial installations took place in Denmark, the United Kingdom, and other European countries. The experience over the years has been significant and successful, but for biomass cofiring to be heavily implemented in the United States, there must exist positive economics through regulation and incentives. Biomass firing is now extensively practiced in the European community and has been demonstrated successfully in many installations.

Woody biomass, particularly mill or forest residue, is considered a premium biomass fuel. Woody material is typically low in sulfur, nitrogen, and ash, which are highly reactive and have significant volatility. Despite these generalizations, the fuels are also complex and variable. Fuel characteristics depend on the source, the type of processing facility, the type of processing, and other factors. The cofiring of woody biomass with coal has been successfully demonstrated in many installations and has significant potential [3, 31, 32]. Woody biomass is cofired commercially at Plant Gadsden and will be cofired at up to 20% (heat input basis) at the new 600-MWe generating station being built by Dominion Power based on two Foster Wheeler circulating fluidized bed boilers.

Switchgrass, straw, corn stover, vineyard and orchard prunings, rice hulls, olive pits, and animal manures are among the other biomass fuels that have been cofired with coal in the United States and Europe. Switchgrass was demonstrated at Plant Gadsden of Alabama Power (Figure 1.6), Southern Company, Blount St. Station of Madison Gas & Electric, and Ottumwa Generating Station of Alliant Energy. Straw was demonstrated at Studstrup Generating Station of Midkraft (now Dong Energy) in Denmark and in other stations as well. Like woody biomass cofiring, cofiring of agricultural materials with coal is discussed in Chapter 4.

TABLE 1.2 Representative Biomass and Opportunity Fuel Cofiring Demonstrations and Tests

Power Plant	Owner	Boiler Type and Capacity of Tested Boiler	Opportunity Fuels Fired	Maximum of Biomass Fuel (%)
Allen Fossil Plant	TVA	Cyclone; 280 MW/unit	Sawdust, tire chips	20
Michigan City	NIPSCO (NiSources)	Cyclone; 468 MW	Wood waste	10
Bailly (Unit #6)	NIPSCO (NiSources)	Cyclone; 165 MW	Wood waste, petroleum coke	15
Willow Island Generation Station	Allegheny Energy Supply Company	Cyclone; 195 MW	Wood waste, tire chips	10
King Station	Northern States Power	Cyclone; 600 MW	Dry sanderdust	10
Albright Generation Station	Allegheny Energy Supply Company	Pulverized coal (PC); 140 MW	Sawdust	1
Seward Generation Station	GPU Genco	PC; 32 MW	Sawdust	20
Kingston Fossil Plant	TVA	PC; 190 MW	Sawdust	5
Colbert Fossil Plant	TVA	PC; 190 MW	Sawdust	5
Greenidge Station	NYSEG	PC; 108 MW	Sawdust; wood waste	15
Plant Gadsden	Southern Company	PC; 70 MW	Switchgrass; other	10
Blount Street Station	Madison Gas & Electric	PC; 50 MW	Switchgrass	20
Ottumwa Generation Station	Alliant Energy	PC; 700 MW	Switchgrass	3
Shawville Generation Station	GPU Genco	PC; 138 MW	Wood waste	3
Plant Hammond	Southern Company	PC; 125 MW	Sawdust	10
Plant Kraft	Southern Company	PC; 50 MW	Sawdust	30

Note: This does not include MSW cofiring plant tests and demonstrations.
Source: [3].

FIGURE 1.5 The cofiring demonstration facility at Albright Generating Station, Allegheny Energy Supply Co., LLC (now FirstEnergy).

FIGURE 1.6 Switchgrass cofiring demonstration in operation at Plant Gadsden.

1.5.3. Coal–Opportunity Fuel Blends

Opportunity fuels can be defined as "combustible resources that are outside of the mainstream of fuels of commerce but can be used productively in the generation of electricity or the raising of process and space heat in industrial and commercial applications" [30]. Opportunity fuels are the focus of Chapter 5.

The most common types of opportunity fuels are residues or low-value products from other processes. These can include tire-derived fuel, petroleum coke, plastics, paper products, and others. Blending opportunity fuel with coal provides a significant opportunity to reduce annual operating costs with decreased fuel costs. Using opportunity fuels with coal has been shown to be successful economically and environmentally in the industrial and power generation sectors.

Petroleum coke is a by-product of petroleum refining and is dependent on the source of crude oil. Petroleum cokes are typically classified as delayed coke (sponge), shot coke, fluid coke, and flexicoke. Cofiring petroleum coke with coal is a common practice in the United States, where the vast majority of the petroleum coke is fired in PC boilers. Petroleum coke is used in many electricity generating boilers and industrial plants.

Traditionally, petroleum coke is an excellent fuel to blend with coal because of its high calorific value. The cofiring of petroleum coke with coal is highly favorable in cyclone boilers. Requirements to have sufficient ash and volatile matter concentration in cyclone firing, typically >5% ash, limit the quantity of petroleum coke in the blend. Similar to cyclone firing, PC boilers can cofire petroleum coke with coal if an SO_2 scrubber exists. PC boilers that fire petroleum coke are common in the United States. Again, volatility of the petroleum coke limits its use, typically 20% to 30% (heat input basis). In addition, ASTM C-618 further limits the percentage of petroleum coke due to the limitation of unburned carbon content in the flyash. Combusting petroleum coke in fluidized bed boilers favors boiler efficiencies, availability, and control of airborne emissions.

The cost of petroleum coke is typically lower than coal. The economic benefits of cofiring petroleum coke have traditionally been a major factor in its use. However, driven by demand, spikes in pricing are common; as of the writing of this text, petroleum coke fuel prices are higher than those of coal. Petroleum coke firing is currently favored, however, thus increasing its demand.

Tire-derived fuel is another popular opportunity fuel discussed in Chapter 5. The use of TDF from automobiles, trucks, and other mobile equipment is of significant interest to the energy industry. TDF can be classified into three basic types—TDF with steel, TDF without steel, and crumb rubber—which have high calorific values (typically >15,000 Btu/lb), so they can replace coal with less mass when fed to the boiler. Tire-derived fuel is also lower in sulfur than most Eastern bituminous coals and comparable to medium-sulfur coals on a lb/10^6 Btu basis.

TDF cofiring is commonplace and can be very attractive because a sufficient supply of scrap tires is available in most areas. Because of its advantages, TDF has been fired with coal, with and without biomass, at the Willow Island Generating Station of Allegheny Energy Supply Co., LLC (now FirstEnergy); at Allen Fossil Plant of TVA; and at numerous other installations. These cofiring locations involve commercial installations, going beyond the demonstration phase.

Plastics-based fuels, also discussed in Chapter 5, have also been fired with coal at such locations as Blount St. Station of Madison Gas & Electric. Plastic-based fuels are promising resources because they have high combustion temperatures (typically >17,000 Btu/lb) with low inherent moisture content. Optimally the fuel comes from low-density polyethylene (LDPE) or high-density polyethylene (HDPE) but not polyvinyl chloride (PVC). Proper fuel selection and combustion of plastics are important. When properly equipped, operated, and maintained, the cofiring of plastic-based fuels can meet U.S. emissions standards. The elimination of dioxin and furan can be achieved through base plastic selection and high-temperature combustion. Cofiring of plastics with coals has significant potential, but care must given to the design and combustion of the fuel.

Waste-based fuels (discussed in Chapter 5) also include such opportunity fuels as reclaimed and reprocessed coal wastes from energy industries and other industries. These fuels include anthracite culm, gob, slack, and other coal mining wastes. Coal fines and impoundments are found throughout the coal industry. Substantial quantities of coal products have promoted the development of several opportunity fuels such as coal-water slurries (CWS) and direct combustion of waste coal. Consequently, the blending of waste fuels was developed to address concerns for environmental issues, cleanup of waste coal ponds and impoundments, and control of airborne emissions.

1.6. BLENDS AND FIRING SYSTEMS

Blending is a process by which the fuel chemistry of the mass of solid material is designed to obtain certain chemical parameters. One of the most common solid fuel blends is bituminous coal combined with PRB subbituminous coal; this was driven mostly by the need to decrease sulfur emissions without installing scrubbers. Fuel blending goes beyond combining various coals. Blending and cofiring have been conducted with biomass and coal, petroleum coke and coal, and other combinations. Blending has been conducted in different types of firing systems, including pulverized coal boilers, cyclones, stokers, bubbling bed boilers, and circulating fluidized bed boilers. These technologies are focused on in Chapter 2.

1.6.1. Types of Firing Systems

Pulverized coal (PC) systems are the most commonly employed methods of fuel combustion for power generation. Pulverized coal combustion includes wall-fired, tangentially fired, arch-fired, and roof-fired systems. Wall-fired and tangentially fired systems are the most common type of boilers. PC firing uses pulverizers to grind the fuel to particle sizes of typically >70% passing through 200 mesh or 74 µm [33]. PC firing typically occurs at temperatures of 2500°F to 2900°F (1370–1590°C).

Optimal fuel chemistry can be designed for PC firing through fuel blending, as is discussed in detail in subsequent chapters of this book. Properties such as high calorific value, low moisture, low sulfur, and low fuel nitrogen are important parameters. Fuel blending can achieve desired characteristics that can be effectively and successfully fired in PC boilers. Furthermore, the use of lower-quality fuels such as PRB coals or opportunity fuels such as petroleum coke or biomass fuels has been successfully demonstrated. In PC firing, attention must be given to particle size and its consequence on the combustion process. Therefore, using oversized fuel particles, such as certain biomass fuels (e.g., wood chips and TDF), has a significant negative influence on the overall system. Comingling of the fuels when blending will typically result in decreased performance of the pulverizers, resulting in larger particle sizes.

Cyclone firing is fundamentally slagging combustion, occurring at 3300°F to 3550°F (1815–1950°C). It uses centrifugal force to combust the crushed fuel particles. As the spiraled fuel particles are thrown outward, they build up against the outer wall of the cyclone barrels. Because the fuel particles are only crushed and not pulverized, the particle sizes typically are larger: usually $3/8$-inch by 0-inch. The technology originally was developed to combust fuels with high slagging propensities, removing approximately 70% of the inorganic material as tapped slag or bottom ash [33, 34]. Cyclones are fuel-flexible boilers that typically burn a variety of fuel and fuel blends—high slagging bituminous coal, petroleum coke, biomass, subbituminous coals, lignites, and others. High-ash, high-sulfur, and high-chlorine fuels are commonly encountered. Fuels that are high in iron and calcium are also very effectively combusted in cyclone boilers (Figure 1.7).

Despite the fact that the fuel is not pulverized, oversized fuel particles must still be managed in cyclone firing. Large fuel particles will not stick to the slag layer and will experience minimal combustion in the cyclone barrel. Particle sizing is a significant parameter. Due to high combustion temperatures and the nature of the fuels typically combusted in cyclone firing, NO_x and SO_2 emissions can be extremely high compared to other firing mechanisms. The need for postcombustion capture systems exists.

In stoker firing, the fuel is burned on a grate where the range of fuel types can be from coal to biomass. In this type of firing, as shown earlier in Table 1.1,

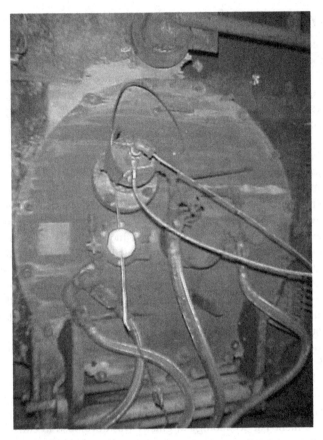

FIGURE 1.7 A cyclone burner at Paradise Fossil Plant Unit #3, the largest one that has been built. This unit has 23 cyclone burners, all fed by radial feeders, and generates 1150 MWe. The plant has been the location of numerous fuel-blending experiments, including petroleum coke/ Interior Province slagging coal blend tests.

combustion can be more easily achieved compared to other conventional firing systems. Depending on the manner in which the fuel is loaded onto the grate, there are two main types of stoker systems: chain grate stokers and spreader stokers [33, 34].

Stoker grate designs can be traveling pinhole grates, chain grates, hydro-grates, sloping grates, and roller grates. The design depends on the material being burned. Fuel feeding is accomplished through several mechanisms. For chain and sloping grate systems, fuel is fed into a feed chute and dragged onto the grate. For units burning wood, a typical design uses a "windswept spout," where the combustion air injects the fuel above the bed and the fuel falls onto the bed. For units burning coal, a "paddle wheel" type of feeder puts the fuel onto the grate.

Another type is the spreader style, where feeders spread the fuel onto the grates [33, 34]. From a fuel blending standpoint, stoker boilers are very flexible. Stokers have high combustion capacities and are capable of burning a large variety of fuels, including wood wastes, tire chips, municipal wastes, and many others. Mixing of fuels is easily achieved. Their limitation is total capacity and the consequent steam conditions that are appropriate. Since stokers are limited to a maximum of about 70 to 75 MW_e, there is a practical limitation of about 1800 psig/1000°F for steam conditions. Reheat cycles are not practical. Further, the quantity of fines in the fuel should be minimized.

Fluidized bed boilers offer an alternative to conventional firing systems, as discussed in Chapter 2 and detailed in Miller and Miller [35]. The distinguishing feature of all types of fluidized bed boilers is the air velocity traveling through the boiler. Bubbling beds will have lower fluidization velocities, thus preventing solids elutriation from the bed into the convection pass. They are operated at gas velocities that are several times greater than the minimum fluidizing velocity, occupying 20% to 50% of the bed volume. The bed material experiences intense agitation and mixing, while maintaining relatively close contact with a well-defined upper surface [35].

Circulating fluidized bed boilers utilize higher velocities in order to promote solids elutriation. Bed particles are entrained and are removed from the combustor. The entrained solids are separated from the gas stream by utilizing a cyclone and are recycled to the bed. Bed inventory is maintained by the recirculation of solids separated by the off-gas. Effective combustion of the fuel and maximized sulfur capture with the sorbent can be achieved [35]. Typically, the feed to fluidized bed systems consists of the fuel (or fuels), an inert material such as sand or ash, and usually limestone suspended by combustion air introduced below the combustor floor.

Inert materials have several key functions. These include dispersing the fuel particles throughout the bed, heating the fuel particles, acting as a thermal flywheel for the combustion process, and providing residence time for combustion to take place. The increased turbulence of fluidized bed boilers is an attractive feature; increased turbulence permits the generation of heat at a lower and more uniformly distributed temperature. High thermal energy allows for a variety of fuels to be combusted; blending within fluidized bed boilers is a common practice.

The chemistry of fluidized bed boilers is fundamentally different from those of PC boilers [35]. Fluidized bed boilers can combust low-calorific-value fuels with higher-moisture and higher-sulfur contents. Operating temperatures for fluidized bed boilers are lower than those of PC boilers, so slagging is insignificant and agglomeration potential still exists. Fouling and corrosion are still of concern, particularly when fuels or blend of fuels with high concentrations of alkali metals and chlorine are combusted. Blending in fluidized bed boilers is commonly practiced and is successful.

Fluidized bed boilers operate normally in a temperature range of 1450°F to 1650°F (780–900°C). At this temperature range, most inorganic components will not melt and form slag. Thermally induced NO_x is also not a major concern at the operating temperatures. The reactions of sulfur dioxide with sorbents (commonly limestone) are thermodynamically and kinetically balanced; sulfur capture decreases outside of the temperature range. Fluidized bed boilers enable the use of a wide variety of fuels. Fuels that are lower in calorific value and have higher concentrations of ash and moisture are commonly burned.

Blending of fuels is easily achieved and maintained in these boilers. Fuels can be added through the front wall or return leg(s), enabling multiple locations where fuel can be introduced. By dropping the fuel into the return leg(s), increased residence time in the furnace can be obtained. The practice of blending in fluidized bed boilers is used throughout the industry. Even though slagging is not a concern with fluidized bed boilers, fouling remains an issue. High concentrations of alkali metals such as sodium or potassium will cause fouling deposits.

Corrosion is also another area of concern; again, it is contributed by alkali metals. In the presence of chlorine, corrosion potential is exasperated. Agglomeration in the loopseal has been experienced where particle-to-particle bonding occurs when in close contact. With increasing regulations on emissions, polishing scrubbers and selective noncatalytic reduction (SNCR) systems are required on fluidized bed boilers in order to reduce SO_2 and NO_x. The advantage of traditional fluidized bed boilers compared to PC and cyclone firing—not requiring postcombustion systems—has decreased to some extent in certain applications and installations.

1.6.2. Types of Boilers

Each boiler type has different characteristics, such as temperature and residence time. These characteristics determine the type of fuel blends and blending capacities that can be achieved. Modifications are commonly employed throughout the industry to achieve increased blending capabilities. The inorganic behavior of fuels and fuel blends has significant consequences on slagging, fouling, and corrosion. The interaction between fuels when they are blended does not necessarily behave as the weighted average of the parent fuels; reactivity must be explored and understood when blending.

Slagging characteristics when different fuels and fuel types are blended have been studied by numerous researchers (see Chapters 2–5; see also [36–38]). Depending on the parent fuels used in the blends, some of the studies have shown linearity of the slagging properties, while other studies have shown nonlinearity. When the parent fuel characteristics are relatively similar in nature, linearity in the slagging properties is observed. However, when the parent fuel characteristics are vastly different—for example,

Powder River Basin coal with bituminous coals—weighted average calculations do not represent the slagging behaviors. The study of inorganic reactivity becomes important.

Fouling differs from slagging in that deposition occurs in the convective pass, the superheater and reheater sections, and the economizer sections of the boiler. The consequences of blending mirror those of slagging. Nonlinear behaviors are typically observed, and the interaction between the parent fuels exists. The characteristics of corrosion are different from those of deposition and are different from those relating to low-NO_x firing or waterwall corrosion. The blending of fuels can either reduce or accelerate the effects of corrosion. Blending will affect concentrations of constituents like sodium, potassium, chlorine, or sulfur. The interaction and loading of these parameters determine the behavior and extent of corrosion experienced by a system.

1.7. CONCLUSIONS

This book discusses the concepts associated with fuel blending, focusing on the power generation system. It discusses the fundamentals and chemistry of blending different fuels; it looks at the benefits, issues, and different management techniques utilized. Modeling techniques used in studying fuel blends are covered. Finally, environmental aspects and institutional issues with fuel blending are discussed where blending typically is conducted because of environmental and economic drivers.

REFERENCES

[1] McGowin CR, Wiltsee GA. Strategic analysis of biomass and waste fuels for electric power generation. Proceedings EPRI conference on strategic benefits of biomass and waste fuels. Washington, DC; 1993, March 30–April 1.

[2] Prinzingt DE, Hunt EF, Battista JJ. Impacts of wood cofiring on coal pulverization at the Shawville Generating Station. Proceedings Engineering Foundation conference on biomass usage for utility and industrial power. Snowbird, UT; 1996, April 28–May 3.

[3] Tillman DA. Final report: EPRI-USDOE cooperative agreement, vol. 1: cofiring. Palo Alto, CA: Electric Power Research Institute; 2001.

[4] Tillman DA. Petroleum coke as a supplementary fuel for cyclone boilers. Proceedings ASME international joint power generation conference. Phoenix; 2002, June 24–27.

[5] Dobrzanski A. Opportunity fuels: a plant perspective. Proceedings electric power conference. Chicago; 2005.

[6] Tillman D. Opportunity fuels: combustion characteristics. Proceedings electric power conference, Chicago; 2005.

[7] Hus PJ, Tillman DA. Cofiring multiple opportunity fuels with coal at Bailly Generating Station. Biomass & Bioenergy 2000;19(6):385–94.

[8] Tillman DA, Hus PJ. Blending opportunity fuels with coal for efficiency and environmental benefit. Proceedings 25th international technical conference on coal utilization and fuel systems. Clearwater, FL; 2000, March 6–9. p. 659–70.

[9] Boylan, DM. Southern company tests of wood/coal cofiring in pulverized coal units. Proceedings EPRI conference on strategic benefits of biomass and waste fuels. Washington, DC; 1993, March 30–April 1.

[10] Tillman DA, Dobrzanski A, Duong D, Dezsi P. Fuel blending with PRB coals for combustion optimization: a tutorial. Proceedings 31st international technical conference on coal utilization and fuel systems. Clearwater, FL; 2006, May 21.

[11] Villesvik G, Tillman DA. Cofiring of dissimilar solid fuels: a review of some fundamental and design considerations. Proceedings American Power Conference; April 1983.

[12] U.S. Energy Information Administration. Annual Energy Review 2009. Washington, DC: US Department of Energy; 2010.

[13] Bryers RW, Harding NS, editors. Coal-blending and switching of low-sulfur Western coals. New York: American Society of Mechanical Engineers; 1994.

[14] Baxter LL, Mitchell RE, Fletcher TH, Hurt RH. Nitrogen release during coal combustion. Energy & Fuels 1996;10(1):188–96.

[15] Tillman D, Evan H. Issues associated with cofiring biomass in coal-fired boilers. Proceedings EPRI conference on effects of coal quality on power plants. Kansas City, MO; 1997, May 20–22.

[16] Hughes E, Tillman D. Biomass cofiring: status and prospects 1996. Proceedings Engineering Foundation conference on biomass usage for utility and industrial power. Snowbird, UT; 1996, April 28–May 3.

[17] Tillman DA. Trace metals from combustion systems. San Diego: Academic Press; 1994.

[18] Speight J. The chemistry and technology of coal. New York: Marcel Dekker; 1983.

[19] Miller BG, Tillman DA. Coal characteristics. In: Miller BG, Tillman DA, editors. Combustion engineering issues for solid fuel systems. Boston: Academic Press; 2008. p. 33–80.

[20] Miller BG. Clean coal engineering technology. Boston: Academic Press; 2011.

[21] Mitchell GD. Selecting coals for quality coke: a ten-part series from March–December. Iron and Steelmaker Magazine; 1999.

[22] Neuendorf KKE, Mehl Jr JP, Jackson JA, editors. Glossary of geology, 5th ed. Alexandria, VA: American Geological Institute; 2005.

[23] Bryers R. Investigation of the reactivity of macerals using thermal analysis. Fuel Processing Technology 1995;44:25–54.

[24] Zhang J, Yuan J-W, Sheng C-D, Xu Y-Q. Characterization of coals utilized in power stations in China. Fuel 2000;79:95–102.

[25] Su S, Pohl JH, Holcombe D, Hart JA. A proposed maceral index to predict combustion behavior of coal. Fuel 2001;80:699–706.

[26] Tillman DA. Opportunity fuel cofiring at Allegheny Energy: final report. EPRI Report #1004811, Palo Alto, CA; 2004.

[27] Duong D, Kerry M. The application of advanced fuel characterization to power plant operations. Proceedings electric power conference. Baltimore; 2008, May 6–8.

[28] Duong D, Miller B, Tillman D. Characterizing blends of PRB and Central Appalachian coals for fuel optimization purposes. Proceedings 31st international technical conference on coal utilization and fuel systems. Clearwater, FL; 2006, May 21–25.

[29] Johnson DK, Miller BG, Wasco RS. Pyrolysis and char oxidation kinetics of selected coals: final report. University Park: The Energy Institute, Pennsylvania State University; 2005.

[30] Tillman DA, Harding SN. Fuels of opportunity: characteristics and uses in combustion systems. Amsterdam: Elsevier Ltd; 2004.

[31] Duong D, Tillman D, Widenman A. Fuel blending for combustion management. In: Miller BG, Tillman DA, editors. Combustion engineering issues for solid fuel systems. Boston: Academic Press; 2008. p. 171–98.

[32] Tillman DA. Biomass cofiring: The technology, the experience, the combustion consequences. Biomass & Bioenergy 2000;19:365–84.

[33] Marx P, Morin J. Conventional firing systems. In: Miller BG, Tillman DA, editors. Combustion engineering issues for solid fuel systems. Boston: Academic Press; 2008. p. 241–74.

[34] Tillman DA. The combustion of solid fuels and wastes. San Diego: Academic Press; 1991.

[35] Miller BG, Miller SF. Fluidized-bed firing systems. In: Miller BG, Tillman DA, editors. Combustion engineering issues for solid fuel systems. Boston: Academic Press; 2008. p. 275–340.

[36] Kitto JB, Stultz SC, editors. Steam: its generation and use, 42nd ed. Barberton, OH: Babcock & Wilcox; 2005.

[37] Widenman A. Fuel characterization: a tutorial. Proceedings 35th international technical conference on coal systems and fuel utilization. Clearwater, FL; 2010, June 6–10.

[38] Widenman T. Ash fusion characteristics for binary blends of selected PRB and Pittsburgh seam coals. Proceedings 28th international technical conference on coal systems and fuel utilization. Clearwater, FL; 2003, March 8–13.

Chapter 2

Principles of Solid Fuel Blending

2.1. INTRODUCTION: BLENDING FOR DOLLARS

Blending is a physical process designed largely to impact the fuel chemistry of the mass of solid material being fed to the boiler. Because it is a physical process designed to impact chemical parameters, it may or may not result in a product that is the weighted average of the component of parent fuels being blended. The most common blends are between two or more coals. Most frequently the blends involve bituminous and Powder River Basin (PRB) or comparable subbituminous coals. However, blending also applies to coal/noncoal fuels being combined in an engineering fashion. There are some cases where blends involve alternative solid fuels and do not include coals. However, in all cases the basic principles of solid fuel blending apply [1].

All fuel blending is designed to provide the fuel user with an economic advantage. To this end DTE Energy invested some $400 million in its blending facility at the Monroe Power Plant, as shown in Figure 2.1, and it continues to invest in its blending processes [2]. Other utilities also have made significant investments in controlled blending systems, and still more are considering doing the same. The economic advantages of blending may come from fuel price reductions—purchasing lower-cost fuel—or they may come from achieving economically beneficial technical advantages, of which there are many, or they may come from both avenues simultaneously [1–4].

The purchase of lower-cost fuel may involve fuels with little or no cost—or negative cost—to the great advantage of the power plant, cement kiln, or industrial boiler. Such "fuels" can include the traditional petroleum coke. They also include everything from waste coals to hazardous wastes. Creativity, boiler or kiln design, and permitting are the only limitations to the selection and use of such fuels in commercial applications [5].

Fuel blending, from a technical perspective, is performed largely to modify the chemical characteristics of the fuel as fired. In modifying the fuel chemistry, blending may achieve economic benefits by mitigating problems—blending away from difficulties such as slagging, fouling, and corrosion, or the formation of one or more air pollutants (e.g., particulates, sulfur dioxide, or oxides of nitrogen) [1]. More recently fuel blending has been used to generate

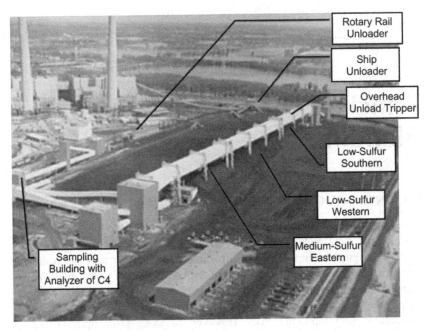

FIGURE 2.1 The coal handling and blending facility at Monroe Power Plant, showing key aspects. Blending occurs on the conveyor in the tunnel under the overhead tripper. Low-sulfur Southern, low-sulfur Western, and medium-sulfur Eastern refer to coal types. *(Photo and facility identification courtesy of ECG Consultants)*

a component of "green power"—power from renewable and sustainable resources; this typically is proposed and/or implemented in the form of "cofiring" biomass fuels with coal [6, 7]. Generation of "green power" is commonly pursued as a consequence of current or expected regulations. Cofiring biomass with coal is simply another form of fuel blending.

2.2. DESIGNING THE MOST FAVORABLE FUEL

There is an extensive body of literature relating to fuel blending (see, for example, [1–20]), including fuel databases [21], research studies (see, for example, [17, 18, 20, 22]), case studies (see, for example, [7, 23]), and more. While much of this literature deals with coal-coal blends, a significant portion deals with firing opportunity fuels such as petroleum coke (see, for example, [24–28] or biomass fuels from wood waste to switchgrass to straw and corn stover (see, for example, [6, 7, 21, 29–31]). This body of literature is international in scope, with major efforts conducted in the United States, Europe, Asia, and Australia. This body of literature provides the engineering community with a sound basis for optimizing the fuel being fed to utility and industrial boilers, kilns, and other combustion systems.

Virtually all solid fuel–fired plants have a most favorable fuel. It has a relatively low price in $/10^6 Btu ($/GJ) and has sufficient calorific value to support achieving full capacity of the boiler or kiln. Further, it has a low impact on airborne emissions—low in fuel nitrogen and sulfur; potentially low in ash, depending on the firing method; and low in mercury and other hazardous air pollutant trace minerals; has a low impact on deposition—slagging and fouling; and produces little if any corrosion. Deposition is governed by such constituents in the ash as iron, calcium, potassium, and sodium. Corrosion is impacted by deposition constituents and by such chemical elements as chlorine and the halogens and by sulfur (see, for example, [32]).

Influences on corrosion may be subtle, and they include the interactions between the fuel and the specific firing system. Ideally, a fuel or fuel blend should be low in moisture content to promote a thermodynamically efficient boiler or kiln. It should be sufficiently reactive, as exemplified by reaction kinetics, to promote a stable and vigorous flame front. Blending can significantly impact all of these properties.

While blending is performed principally to modify the fuel chemistry, the physical properties of the blend are not to be ignored in designing for an ideal fuel mix. These properties include bulk density (hence the ability of the conveyor to deliver Btu/ft^3; the conveyors are volumetric devices); Hardgrove grindability index (HGI) and the total concept of comminution or crushing, grinding, and pulverizing; the particle size distribution, which is critical to some firing methods such as bubbling and circulating fluidized bed boilers; the potential for dust generation and associated spontaneous combustion hazards; and more.

Some engineers are now specifying free swelling index (FSI) and even the Geisler index measuring agglomeration tendencies of bituminous coals; these indices are virtually linear in blending considerations and are sometimes specified for fluidized bed applications. Again, blending has the potential to impact all of these properties. The ideal fuel should have a reasonably high bulk density, an acceptable HGI, a particle size distribution consistent with the firing method, and minimal dusting, and it must be easily handled. Moisture also is a physical property of importance as well as a chemical property. High moisture significantly impacts fuel handling on conveyors, in silos, in pipes, and in chutes. For pulverized coal (PC) boilers, it impacts pulverizer outlet temperatures and, in severe conditions, the potential for pneumatic coal pipes to operate below dew point temperatures. This can cause severe plugging. These issues all need to be considered.

Cost issues also must be considered. Is the system, and the particular blend, labor intensive? Is it maintenance intensive? Are the capital costs such that they can readily be recovered by the economics of the blend achieved? Is the system designed for maximum flexibility and, simultaneously, maximum control through information management? Some blends, particularly those including high percentages of PRB coal, can enter the system with a relatively fine particle size (e.g., $\frac{3}{8} \times 0$ in. or 0.95×0 cm), and the coal yard can be designed such that this fuel bypasses the crusher; maintenance is reduced, and the time between crusher rebuilds is extended. All of these issues must be addressed in seeking the optimal blend.

2.3. INFLUENCES ON THE MOST FAVORABLE FUEL BLEND

The most favorable fuel blend will vary as a function of the fuels available for blending, such as various types of coal, petroleum coke, various types and forms of biomass, and so on. The most favorable blend will also vary as a function of the firing method. Permit conditions also will influence optimal blend. This is explored more completely in subsequent chapters.

2.3.1. Firing Method Considerations

The firing methods that influence blending of fuels are many and diverse. Some examples of the influence of firing methods on optimal blend properties are as follows (from [33–36]):

Pulverized Coal Boilers

The optimal fuel chemistry creates a fuel blend with the highest calorific value expressed as Btu/lb, the lowest moisture, the lowest sulfur and fuel nitrogen expressed in $lb/10^6$ Btu (or kJ/kg) sulfur as SO_2 and N, the lowest ash content, also in $lb/10^6$ Btu, and as high a set of ash fusion temperatures as possible Note that there are eight ash fusion temperature measurements: initial-oxidizing, softening-oxidizing, hemispherical-oxidizing, fluid-oxidizing, initial-reducing, softening-reducing, hemispherical-reducing, and fluid-reducing. The blend ash fusion temperatures are typically measured in terms of initial-oxidizing and hemispherical-reducing. The blend is also considered optimal if it has as high an HGI as possible.

Fines in coal supply are acceptable at practical percentages. HGI determines the ease with which pulverizers can reduce particles to the desired particle sizes (e.g., <0.5% retained on 50 mesh and >70% passing 200 mesh). In addition to these parameters, a certain minimal fuel reactivity is required to achieve fuel stability; this varies depending on the boiler type and design. In general, higher reactivity is preferred. Reactivity depends on the interaction between the fuels. Ash reactivity is also significant for PC boilers due to issues of deposition—slagging and fouling—and corrosion. Again, interaction between the fuels is significant because this may influence the chemical mechanisms of ash behavior.

Fluidized Bed Boilers

The chemistry of fuels for fluidized bed boilers is quite different from that of PC boilers. Bubbling and circulating fluidized bed boilers (BFB and CFB) can accept low-calorific-value fuels and relatively high-moisture fuels. Sulfur content may be high, particularly because limestone can be added to the bed, to drive high-temperature calcium-sulfur capture reactions, but chlorine content must be minimized. Ash content should be sufficient to generate bed material if possible. Alternatively, limestone or sand must be added for bed makeup. Fuel fines can be tolerated but only at low percentages. Pulverized coal boiler flyash or bottom ash can also be used as bed makeup material. The chemistry of the entire bed of solids, including the fuel, is of significant concern and can be influenced substantially by fuel blending and the consequent fuel composition.

Fluidized bed boilers are commonly operated at relatively low temperatures (e.g., 1500–1700°F or 815–926°C). Consequently, slagging is of minor consideration. Further, ash fusion temperatures are of less concern. However, fouling remains an issue, since this can be caused by sodium (Na) or other alkali metals. Fouling deposition is of critical concern. Corrosion, which can also be a consequence of alkali metals with or without the presence of chlorine, also remains a serious concern.

Agglomeration is also of concern, particularly between bed particles either in the furnace or in the loopseal that links the cyclone with the primary furnace. The measures of agglomeration in fuels, typically FSI and Geisler index, were originally developed for coking processes and later were applied to fixed bed gasifiers; they have limited application to fluidized bed firing. Because fluidized bed boilers burn crushed but not pulverized fuel, HGI is of lesser concern. Reactivity of both fuel and ash remain considerations.

Cyclone Boilers

Cyclone boilers burn crushed (but not pulverized) coal in large (e.g., 10-ft dia.) cyclone barrels and send hot (e.g., 3200–3550°F or 1760–1950°C) gases into the furnace and then convective pass. Cyclone boilers are designed to maximize the production of slag from the inorganic constituents in fuel; inorganics are managed in this manner. Typically 70% to 80% of the inorganic constituents are removed as slag or bottom ash, and only 20% to 30% of these materials are removed as flyash. Combustion occurs at very high temperatures: 3200°F to 3550°F (1760–1950°C). The fuel for cyclone boilers is typically a bituminous coal prone to slagging; it may be blended with other coals and/or petroleum coke, biomass, or other fuel materials. Some cyclone boilers are fired with subbituminous coals, while others are fired with lignites.

While cyclone units are fuel-flexible boilers, the fuel chemistry is optimized toward the highest calorific value and lowest moisture content possible. Sulfur and chlorine can both be relatively high (particularly if the unit is scrubbed), and the sulfur will bring with it the desired iron as a fluxing agent. Fuels that bring in calcium such as Western bituminous coals and PRB coals are also desired, and blends that produce appropriate ratios of Fe and Ca can be very effective.

Alliant Energy uses sodium-based PRB coal—typically Spring Creek coal—at its Nelson Dewey station, with favorable results. In cyclone firing, ash should be sufficient in quantity to coat cyclone barrels, and it should contain sufficient fluxing agents (Fe, Ca) to depress ash fusion temperatures and T_{250} temperatures (the temperature at which the slag has a viscosity of 250 poise). Fuel fines are acceptable but only in limited percentages.

Fuel reactivity is highly desirable; fuels with low reactivity measures can be quite problematic. Ash chemistry favors reactivity. Lower ash fusion temperatures, lower T_{250} values, and slag forming mechanisms are favored to a point. HGI values are not particularly important, since the cyclone boiler burns crushed fuel and not pulverized fuel.

Cyclone boilers are, to some extent, the inverse of pulverized coal boilers. Highly reactive ash is desired; unreactive ash with high ash fusion temperatures causes serious problems. This can be addressed by adding fluxing agents to the fuel feed—either a reactive calcium-based material (e.g., hydrated lime, limestone) or an iron-based material (e.g., blast furnace dust). Alternatively, and preferably, this can be addressed by fuel blending.

Stoker Boilers

The fuel chemistry is essentially the same as pulverized coal boilers, with the exceptions that moisture can be high and that sufficient ash to cover and insulate the grate bars is desirable. Ash chemistry is essentially the same as pulverized coal boilers. Fuel reactivity is particularly significant, but ash reactivity is not desirable. HGI is irrelevant. Particle size is of importance; fuel fines should be minimized.

2.3.2. Market Considerations

Beyond the firing method, access to various fuels is not consistent across the entire market. PRB coal is now essentially universally available in the United States and can be exported to Europe and Asia. However, access to the very reactive low-sulfur and low-ash Indonesian coals (e.g., Adaro coal) requires some access to ocean transport. Similarly, the very low-sulfur Alaskan coals (e.g., Usibelli coal) require ocean transport. Access to petroleum coke depends on either proximity to oil refineries or an appropriate infrastructure.

Access to woody biomass in the United States requires proximity to forests and forest products industries, although wood pellets can move significant distances. Access to herbaceous biomass is a local phenomenon. Many of the waste fuels from paint sludges to off-spec products and selected hazardous wastes are available on a local basis. All of these issues must be considered in developing and implementing a blending strategy.

2.4. DEVELOPING A FUEL BLENDING STRATEGY

Blending of dissimilar coals, cofiring of biomass or petroleum coke with one or another of the range of coals, or blending of nonfossil fuels among themselves provides the opportunity to develop a fuel supply strategy (see [2, 3, 8]). Blending can be used to address many issues, specifically:

- Fuel price ($/10^6 Btu or $/GJ)
- Fuel energy content (Btu/lb or GJ/kg; Btu/ft^3 or GJ/m^3)
- Carbon (and volatile carbon) concentration and boiler or kiln thermal efficiency to address carbon footprint
- Pollutant considerations (sulfur, nitrogen, halogens such as chlorine, HAPS)
- Ash concentrations
- Ash chemistry, particularly iron, calcium, magnesium, potassium, sodium, silica, and alumina and their influence on slagging, fouling, and corrosion
- Fuel reactivity
- Fuel physical properties

Extensions of these properties and issues are certain consequences of firing one or more fuels or fuel blends. These consequences include system reliability, maintenance issues, capacity achievement (or derates), and operational flexibility. All of these considerations are influenced by the blending process [2].

2.4.1. Blend Fuel Considerations

Since fuel blending involves combining different fuels—two or more coals, one or two coals plus a biomass (e.g., wood waste, corn stover), opportunity fuel (e.g., petroleum coke), or industrial or municipal waste fuel (e.g., refuse-derived fuel, or RDF, hazardous waste)—it becomes essential to understand the fuel characteristics and principles involved. Properly managed, and under the right circumstances, fuel blending can be used to "design" a fuel to optimize certain beneficial properties while minimizing deleterious effects. Properly managed, the blend can achieve more beneficial properties than its parent fuels when the blend is managed along with firing conditions [2]. To consider this further, data are presented in Tables 2.1 through 2.4.

Table 2.1 summarizes chemical characteristics of a few bituminous coals. Table 2.2 summarizes chemical characteristics of some subbituminous and lignite coals, and Table 2.3 summarizes characteristics of a few biomass fuels. Table 2.4 summarizes chemical characteristics of a few petroleum cokes. Note that there is considerable variability among coals, even coals from the same general deposit and seam [37]. Similarly there is significant variability in types of biomass fuels (e.g., wood waste) [2–4]. As shown in Table 2.3, there are several types of petroleum coke—with varying chemistries and physical properties [38]. These considerations will be covered in detail in subsequent chapters.

2.4.2. Combustion Characteristics of Binary and Ternary Blends

Some fuel combustion properties are essentially the weighted averages of the fuels being blended. These properties include heating value (Btu/lb or MJ/kg) and

TABLE 2.1 Representative Characteristics of Some Bituminous Coals

Parameter	Central Appalachian	Illinois Basin	Utah
Proximate (wt%)			
Moisture	6.14	2.5	8.40
Ash	8.23	12.0	6.57
Volatile Matter	31.40	38.4	40.59
Fixed Carbon	54.23	47.1	44.44
Ultimate (wt%)			
Moisture	6.14	2.5	8.40
Carbon	72.31	67.66	67.50
Hydrogen	4.86	4.68	4.82
Nitrogen	1.32	1.17	1.15
Sulfur	0.85	3.27	0.52
Chlorine	0.20	0.30	0.01
Ash	8.23	12.0	6.57
Oxygen (by diff.)	6.37	8.39	11.04
Higher Heating Value			
Btu/lb	13,050	11,940	11,867
MJ/kg	30.28	27.7	27.53
Ash Elemental (wt% ash)			
SiO_2	56.39	48.9	47.53
Al_2O_3	20.58	18.3	10.89
TiO_2	1.12	0.9	0.48
Fe_2O_3	6.89	18.1	6.60
CaO	1.04	4.8	19.30
MgO	0.99	1.0	2.68
Na_2O	0.19	1.0	0.34
K_2O	2.25	2.1	0.67
P_2O_5	0.48	0.2	0.13
SO_3	0.78	4.7	11.20

Note: These are representative values only. Measures of variability are presented in Chapter 3, which discusses blending of coal in more detail. **Source:** [39].

TABLE 2.2 Characteristics of Some Subbituminous and Lignite Coals

Parameter	Montana Subbit	Wyoming Subbit	Texas Lignite
Proximate (wt%)			
Moisture	23.53	26.98	31.40
Ash	4.21	5.49	16.40
Volatile Matter	32.68	33.03	27.60
Fixed Carbon	39.66	34.49	24.60
Ultimate (wt%)			
Moisture	23.53	26.98	31.40
Carbon	55.36	50.33	36.81
Hydrogen	3.90	3.73	2.67
Nitrogen	0.82	0.63	0.69
Sulfur	0.35	0.32	0.96
Chlorine	N/A	<0.01	0.07
Ash	4.21	5.49	16.40
Oxygen (by diff.)	11.79	12.52	11.04
Higher Heating Value			
Btu/lb	9586	8852	6681
MJ/kg	22.24	20.54	15.5
Ash Elemental (wt% ash)			
SiO_2	33.42	32.50	51.2
Al_2O_3	17.93	15.20	16.1
TiO_2	1.17	1.09	1.2
Fe_2O_3	6.08	4.66	7.6
CaO	14.22	18.90	8.9
MgO	4.01	3.88	2.0
Na_2O	6.84	1.55	0.2
K_2O	0.99	1.01	0.9
P_2O_5	14.59	1.18	0.4
SO_3	14.59	13.15	8.4

Note: These are representative values only. Measures of variability are presented in Chapter 3, which discusses blending of coal in more detail. **Sources:** [23, 39].

TABLE 2.3 Characteristics of Some Representative Biomass Fuels

Parameter	Wood Waste	Switchgrass	Corn Stover
Proximate (wt%)			
Moisture	39.79	8.12	6.23
Ash	1.57	3.08	7.05
Volatile Matter	47.99	76.12	71.08
Fixed Carbon	10.65	12.68	15.64
Ultimate (wt%)			
Moisture	39.79	8.12	6.23
Carbon	32.67	43.01	43.39
Hydrogen	3.61	5.04	5.15
Nitrogen	0.12	0.19	0.83
Sulfur	0.03	0.06	0.10
Chlorine	0.011	<0.01	0.23
Ash	1.57	3.08	7.05
Oxygen (by diff.)	22.21	40.50	37.25
Higher Heating Value			
Btu/lb	5497	7320	7088
MJ/kg	12.76	16.99	16.45
Ash Elemental (wt% ash)			
SiO_2	1.48	71.98	52.13
Al_2O_3	0.45	1.43	3.77
TiO_2	0.04	0.14	0.13
Fe_2O_3	0.09	0.60	1.79
CaO	59.80	9.65	8.75
MgO	4.45	3.12	3.36
Na_2O	0.45	0.14	1.52
K_2O	8.84	3.57	17.30
P_2O_5	2.46	25.55	2.68
SO_3	0.25	1.24	3.71

Note: *These are typical values only; many values have been measured for these and other biomass* *fuels.* **Sources:** *[23]; Tillman, D.A. personal files.*

TABLE 2.4 Characteristics of Some Petroleum Cokes

Parameter	Sponge Coke	Shot Coke	Fluid Coke
Proximate (wt%)			
Moisture	7.60	6.29	2.24
Ash	0.72	1.06	1.32
Volatile Matter	4.48	3.07	4.94
Fixed Carbon	80.2	89.59	91.50
Ultimate (wt%)			
Moisture	7.60	6.29	2.24
Carbon	81.12	81.29	84.41
Hydrogen	3.60	3.17	2.12
Nitrogen	2.55	1.60	2.35
Sulfur	4.37	5.96	6.74
Ash	0.72	1.06	1.32
Oxygen (by diff)	0.04	0.93	0.82
Higher Heating Value			
Btu/lb	14,298	14,364	14,017
MJ/kg	33.18	33.34	32.53
Ash Elemental (wt% ash)			
SiO_2	10.1	13.8	23.6
Al_2O_3	6.9	5.9	9.4
TiO_2	0.2	0.3	0.4
Fe_2O_3	5.3	4.5	31.6
CaO	2.2	3.6	8.9
MgO	0.3	0.6	0.4
Na_2O	1.8	0.4	0.1
K_2O	0.3	0.3	1.2
SO_3	0.8	1.6	2.0
V_2O_5	58.2	57.0	19.7
NiO	12.0	10.2	2.9

Note: These are typical values without regard to the source of the crude oil. More values are presented in Chapter 5. **Sources:** [5, 38].

TABLE 2.5 Example Blending Impacts

Parameter	Blend		
Percent PRB Coal	60	70	80
Percent Bituminous Coal	40	30	20
Calorific Value (Btu/lb)	10456	10037	9618
Ash Percentage	5.96	5.70	5.44
Ash Loading (lb/10^6 Btu)	5.68	5.66	5.64
Moisture Percentage	18.75	20.83	22.92
Base/Acid Ratio	0.522	0.569	0.616

Note: These are linear impacts only. Fuels blended are similar to the Central Appalachian and Wyoming PRB coals in Tables 2.1 and 2.2.

percentages of components of the ultimate analysis (C, H, N, S, H_2O, ash, O). Table 2.5 shows such properties, based on blending calculations for binary blends of Central Appalachian bituminous coal and southern PRB coal, which are common coals used in binary blends [6].

2.4.3. Reactivity, Ignition, and Flame Characteristics of Fuel Blends

Certain critical combustion properties of blended fuels do not reflect the weighted average of the two or more fuels being blended. These properties include fuel reactivity as measured by ignition temperature and the kinetics of solid particle pyrolysis. They also include fuel reactivity as measured by the kinetics of char oxidation. The kinetics as measured are in the form of an Arrhenius equation; the variables measured are the preexponential constant (A) and the activation energy (E). Because the reactivity of both the fuels and the fuel blends is so critical to the assessment of fuel blending, the methodologies are discussed here, along with other measures of fuel and fuel blend reactivity.

These critical parameters and related reactivity measurements (e.g., the temperature at which pyrolysis commences) are significantly influenced by the blending process, as has been shown by Duong et al. [40–42] and Johnson et al. [43]. Figures 2.2 through 2.4 illustrate some of these reactivity considerations of fuel blending [2, 3, 41, 42]. Note that reactivity is shown by measured activation energy for coal blends and parent fuels and by burning profiles of coal-biomass blends with both bituminous coal and PRB subbituminous coal. These are among the several measurement techniques available to evaluate fuel reactivity. The reactivity measurements, including activation energy, for coal-coal blends are discussed more extensively in Chapter 3. Burning profiles have

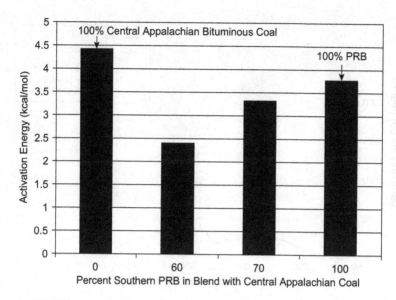

FIGURE 2.2   Activation energy for coal blends and parent coals. *Source: [2].*

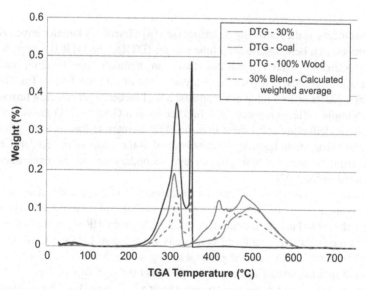

FIGURE 2.3   Burning profile for bituminous coal and wood, and 30% wood/70% coal blend.
Source: [40].

also been developed for coal-switchgrass blends and coal-corn stover blends, as discussed in Chapter 4.

The methods used to develop the activation energy data—and the entire devolatilization or pyrolysis kinetic data set—rely heavily on drop tube furnace

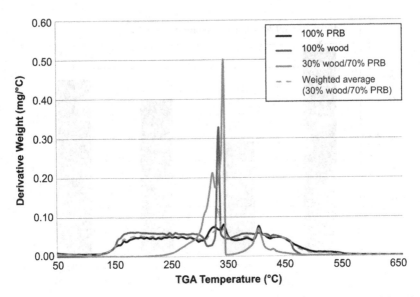

FIGURE 2.4 Burning profile for subbituminous coal and wood, and 30% wood/70% PRB coal blend. *Source: [42].*

measurements as the following describes (see [8]). Pyrolysis kinetics have been determined in a bench-scale drop tube reactor (DTR). The DTR is a vertically fed, electrically heated apparatus that can simulate fuel heating rates, temperature profiles, and particle residence time of a utility boiler. The DTR system consists of four main components: a fuel feeder, a preheater, a furnace, and a sample collection probe. The fuel feeder is a Gerike GMD microfeeder capable of delivering solid fuels into the DTR at rates as low as <0.1 g/min. Fuel and entrainment (primary) gas are mixed at the outlet of the feeder. UHP grade argon is used for both primary and secondary gas for all experiments conducted in the DTR.

The preheater designed by Penn State is cylindrical in shape, with a diameter of 22.9 cm and a height of 28.0 cm. A total flow rate of 10 L/min of UHP argon is fed into the DTR. Fuel and entrainment gas, 2.5 L/min of UHP argon, are delivered through the preheater at ambient conditions via a 3-mm I.D. × 41-cm water-cooled brass tube. The remaining (secondary) gas, at a flow rate of 7.5 L/min, is delivered into the heated section of the preheater through two 12.7-mm I.D. × 8.0-cm alumina cylinders. Secondary gases enter a 7.62-cm I.D. × 18-cm cylinder and are heated to 850°C by a Kanthal A-1 braided heating element, as measured by a type S thermocouple. The secondary gas exits the preheater through a 3-cm-thick Mullite honeycomb flow straightener with 1.6-mm × 1.6-mm square cells. The remaining volume of the preheater contains insulation.

The DTR has both an upper and a lower furnace to heat the sample gas to the desired temperature. For the proposed study, only the upper furnace is used, while the lower furnace remains open to gain access to the collection probe. The upper furnace is electrically heated and contains a 6.35-cm I.D. × 95.25-cm alumina muffle tube. The furnace can maintain a maximum temperature of 1700°C using 6 Kanthal silicon carbide heating elements. For the proposed experiments, the furnace temperatures used are 600, 800, 1000, 1500, 1600, and 1700°C. Solid samples are collected from a filter housing located at the end of a 1.25-cm I.D. × 53.5-cm refractory-lined, water-cooled copper tube.

The kinetic parameters A and E are calculated from the rate of weight loss data as a function of temperature. The DTR is operated in an inert UHP argon environment. The fuel is carefully weighed from the feeder before and after the test to determine the total amount of fuel feed to the DTR. Likewise, the char is carefully collected and weighed from three sources: the char deposited on top of the filter paper, the char trapped in the filter paper, and the char deposited on the probe wall. The sum of these three yields the total char collected for a given test time. From the weight of the fuel fed (w_f) and the weight of the char collected (w_c), the % weight loss (V) can be calculated by equation (2.1):

$$V = \left[\frac{w_f - w_c}{w_f}\right] \times 100 \tag{2.1}$$

The reactivity R, at a given DTR temperature, can then be calculated by equation (2.2):

$$R = \frac{\dfrac{V}{V_\infty}}{t_r} \tag{2.2}$$

where V_∞ is the maximum % weight loss that occurs at any DTR temperature, and t_r is the residence time of the fuel particle in the DTR. The particle residence time is calculated by dividing the length (L) that the particle travels through the region of the DTR that is at the desired temperature by the particle velocity. The particle velocity is the sum of two components: the gas stream velocity (V_g) and the terminal velocity (V_t).

$$t_r = \frac{L}{V_g + V_t} \tag{2.3}$$

The centerline gas velocity is approximated by doubling the bulk gas velocity. The bulk gas velocity is simply the volumetric gas flow rate divided by the cross-sectional area of the tube inside the DTR. The terminal velocity can be calculated by Stokes's law:

$$V_t = \frac{g d_p^2 \Delta\rho}{18\mu} \tag{2.4}$$

where d_p is the average particle diameter, μ is the gas viscosity, $\Delta\rho$ is the density difference between the fuel particle and the carrier gas, and g is the gravitational constant. At a temperature of 1700°C, where the char is generated for the char oxidation studies, the calculated residence time for the coal particles is 186 ms.

There are other measures of combustion reactivity as well. These include thermogravimetric analysis (TGA) and the temperature at which devolatilization is initiated (T_{init}). Ignition temperature is a related measure of reactivity. Chi et al. [20], using drop tube furnace measurements, have demonstrated that this, too, is a nonlinear phenomenon. Blending two coals, a high-volatile subbituminous coal and a lower-volatile bituminous coal, their measurements of ignition temperature lead to equation (2.5):

$$T(°F) = 1048.2 + 0.0132(\%S)^2 - 2.5976(\%S) \qquad (2.5)$$

where T is ignition temperature and %S is the percentage of subbituminous coal in the blend. The metric equivalent equation is:

$$T(°C) = 564.35 + 0.0073(\%S)^2 - 1.4431(\%S) \qquad (2.6)$$

The data from Chi et al. can be shown graphically; this is presented in Figure 2.5. Note that the regression curve, previously shown in equation (2.5), has a very strong coefficient of determination (R^2). This is consistent with the

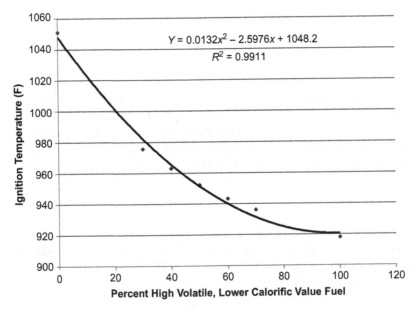

FIGURE 2.5   Ignition temperature as a function of reactive subbituminous coal in a coal blend. *Source: Adapted from [20].*

TGA/DTG work of Duong et al. [41, 42] blending both bituminous coals with various forms of biomass and PRB coals with biofuels.

TGA has been used to evaluate char oxidation as well [3]. The use of TGA to determine char oxidation kinetics was developed at the Energy Institute of Pennsylvania State University [44], as summarized by Duong, Tillman, and Miller [8]. To determine the char oxidation kinetics, char generated at high temperatures of 3092°F (1700°C) is collected and evaluated by thermogravimetric analyzer techniques because this temperature most realistically represents the heating rates in an industrial boiler.

The thermogravimetric analyses were conducted in air at atmospheric pressure with sample weights of approximately 5 mg. Operating temperatures were typically 350, 375, 400, and 425°C. The char sample was heated to the desired reaction temperature at a rate of 20°C/min. Once the sample reached the desired temperature, the gas flow was changed from nitrogen to air. The sample was then held at temperature on the order of 10 hours, depending on the reaction rate. Weight change data were recorded as a function of time. From this data, instantaneous reactivity (R_u) was calculated by equation (2.7):

$$R_u = \frac{1}{W_u} \cdot \frac{dW}{dt} \tag{2.7}$$

where W_u is the dry ash free weight of unreacted char at time t, and dW/dt is the slope of the burn-off curve at corresponding time t.

Reaction rates were determined from this data using a method developed by Tsai and Scaroni as discussed in Duong, Tillman, and Miller [8]; Miller and Miller [36]; and Scaroni et al. [44]. They showed that the reactivity of several chars was a function of burn-off and that the value of reactivity can vary by a factor of 2, depending on the extent of burn-off at which the reactivity was calculated. Therefore, a more meaningful value than the instantaneous reactivity at a given weight loss for predicting char burn-out rates would be a reactivity value averaged over the entire char burn-out range. An average reactivity can then be calculated by:

$$R_{a,x\%} = \frac{\sum(R_u \cdot \Delta W)}{\sum \Delta W} \tag{2.8}$$

where R_u is the instantaneous reaction rate with respect to unreacted char calculated from equation (2.7), and ΔW is the weight loss occurring within each time interval.

The summation of the ΔW values is equal to x, which can be taken anywhere in the range from 0% to 100%, depending on the range of char burn-off interest. The results of such testing show that when a high-volatile subbituminous coal is introduced to a bituminous coal blend, the kinetics of the char oxidation, and particularly the activation energy of char oxidation, are governed by the kinetics and activation energy associated with the subbituminous coal, as shown in Figure 2.6.

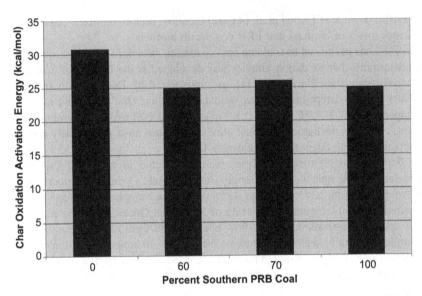

FIGURE 2.6 Char oxidation activation energy for bituminous coal, PRB coal, and blends. *Source: [3].*

Many researchers have found similar results. For example Haas, Tamura, and Weber [19] found similar linear or additive properties—and nonlinear or nonadditive properties—in a study of 6 coals and 15 coal blends, including Polish coal, German coal, brown coal (lignite), and South American coals. Using an isothermal plug flow reactor, these researchers measured devolatilization and char burn-out under conditions common to utility and industrial boilers. Devolatilization curves followed a traditional inverted S shape for virtually all coals and blends.

Faundez et al. [9] also made similar investigations of blends containing high-volatile bituminous coals, low-volatile bituminous coals, and subbituminous coals. Like other researchers, they found that the more volatile coals enhanced lower-temperature ignition of the fuel being fired. Taniguchi et al. [26] made measurements of ignition and flame propagation, including flame velocity, with high- and low-volatile bituminous coals, anthracite, and petroleum coke. Like other researchers, they found the high-volatile coals and other high-volatile solid fuels drive the reactivity of the fuel blend.

2.5. FORMATION OF POLLUTANTS

Blending of coals, or cofiring biomass with coal and/or petroleum coke, has long been used as a strategy for managing the formation of emissions—particularly airborne emissions. The initial push for opening up the Powder River Basin was to meet the Clean Air Act requirements for sulfur dioxide (SO_2) emissions (see, for example, [45, 46]). This force has also promoted the use of Adaro coal from Indonesia and other Pacific Rim coals that are very low in sulfur. Usibelli coal, from the Nenana coal field in Alaska, is a similar coal. Reduction of sulfur emissions also was one consideration (of many) that promoted the interest of EPRI, several major utilities, and the U.S. Department of Energy in the process of cofiring biomass with coal in utility boilers [47]. The data in Tables 2.1 through 2.4 illustrate this consideration, since SO_2 emissions are governed by the sulfur content of the fuel mix.

Fuel blending also influences particulate emissions—particularly PM_{10} emissions. Total suspended particulates (TSP) are a function of the percentage of ash, or mineral matter, in the fuel, coupled with the unburned carbon (UBC) resulting from combustion and the firing method. PC firing, for example, typically yields 80%+ solid products of combustion as flyash, while cyclone firing typically yields on the order of 30% solid products of combustion as flyash. In all firing system technologies, fuels can be blended to reduce TSP by seeking low-ash fuels and high-volatile fuels. These can be blended with high-ash bituminous coals to reduce TSP. The volatility is used to reduce UBC in the products of combustion. Fluidized bed combustion does not conform to such relationships; typically there is only 2% to 5% fuel in the total flow of solids in a bubbling or circulating fluidized bed, and both the fuel and the bed contribute to flyash.

Environmental regulations now govern more than TSP; they govern PM_{10} and $PM_{2.5}$, as well. PM_{10} refers to particles smaller than 10 μm in diameter, and $PM_{2.5}$ refers to particles with diameters less than 2.5 μm. Extensive research by Wang et al. [48, 49] shows that particles less than 1 μm are formed by vaporization and subsequent condensation and nucleation of inorganic matter. Particles in the range of 1 to 10 μm range are formed either by liberation of fine mineral particles during combustion or by collision and coalescence of mineral particles in both the solid and liquid condensed phases. These particles have consistent compositions of calcium (Ca), iron (Fe), aluminum (Al), and silica (Si) and are considered to be formed by the same mechanisms [48].

Research by Wang et al. [48, 49] involved blends of bituminous coals with dissimilar compositions—particularly dissimilar calcium concentrations in the fuel and ash. Blends ranged in numerous increments from single coals through 40%/60% and 60%/40% high-calcium coals. The results showed that formation of PM_{10} was not linear and that by adding calcium, finer particles could be "glued" together by calcium in the liquid phase. This research showed that higher concentrations of calcium increase the amount of liquid phase material resulting from combustion, and this results in less PM_{10} and more PM_{10+}.

Oxides of nitrogen (NO_x) management also can be influenced by solid fuel blending. It is recognized that there are two basic forms of NO_x: thermal- and fuel nitrogen–based NO_x. Pershing and coworkers [50, cited in 51] have shown that thermal NO_x typically occurs at temperatures >2700°F (1480°C) and is most commonly a problem with cyclone boilers or large, opposed wall–fired

PC boilers—particularly those using cell burners. Fuel NO_x is more prevalent and more ubiquitous. Further, it is more readily influenced by fuel blending.

Like all pollutants, NO_x calculations are best made on a $lb/10^6$ Btu (kg/GJ or kg/kcal) basis. Fuels can then be evaluated in terms of total fuel nitrogen content and, using drop tube furnace data, volatile fuel nitrogen content. Further, fuel volatility—critical in managing NO_x—can be expressed from the proximate analysis in terms of fixed carbon/volatile matter (or its inverse) ratio.

Alternatively, because volatile yield is somewhat a function of pyrolysis temperature (see Miller and Tillman [52]), the fixed carbon/volatile matter or inverse ratio can be calculated based on maximum volatile yield in a drop tube reactor operated at up to 3090°F (1700°C). The volatility is further a function of fuel aromaticity [52]. Figure 2.7 shows fuel nitrogen content and volatile nitrogen content for a suite of solid fuels including biomass, various ranks of coal, and petroleum coke. Figure 2.8 shows fuel volatility as a function of drop tube reactor maximum temperature measurements.

Baxter et al. [54] have shown that controlling fuel nitrogen evolution to NO_x can be best managed by evolving volatile nitrogen compounds in a fuel-rich environment. The more rapidly the nitrogen volatiles evolve, the more likely they are to evolve in a fuel-rich region and the less likely they are to oxidize to NO, NO_2, and related species. Baxter et al. [54] have shown that, typically, there is a lag between the initial devolatilization of the fuel mass and the

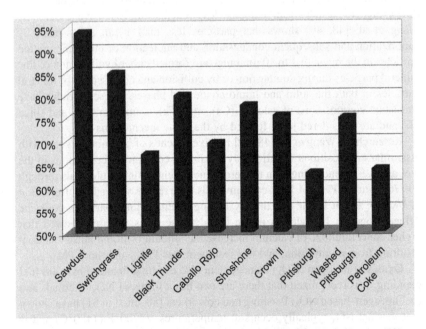

FIGURE 2.7   Maximum nitrogen volatile yield content for a suite of fuels. *Source: [53].*

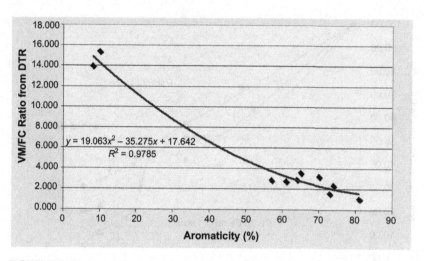

FIGURE 2.8 Fuel reactivity for a suite of fuels. Reactivity is reflected by the volatile matter (VM)/fixed carbon (FC) ratio and is a function of fuel aromaticity. The two data points with the very high VM/FC ratios are biomass fuels, while the cluster of fuels with low ratios is composed of various ranks of coal and petroleum coke. *Source: [53].*

devolatilization of nitrogen from the fuel mass. This lag causes many of the volatile nitrogen compounds and species to evolve in a fuel-lean environment; hence, many are converted into NO_x compounds. Virtually all of the char-bound nitrogenous species evolve in a fuel-lean environment as well.

Blending can be used to increase the volatile fuel nitrogen component, as can be seen in Figure 2.7. Depending on the two or three fuels being blended, this can provide a significant means for reducing NO_x. Further, there are some blends that promote more rapid and complete nitrogen volatile evolution than either parent fuels. Figure 2.9 shows one such blend—70% southern PRB coal and 30% Texas lignite—where the nitrogen evolution was more rapid and more complete than either parent fuel [2]. This blend, when burned in a large utility boiler, produced a very low NO_x emission rate. The same principle was used to achieve NO_x reduction through biomass cofiring [6, 7]. As is shown in Figure 2.10, most biomass cofiring demonstrations produced a disproportionate level of NO_x reduction [6]. Again, the consequences of blending either reactive coals or coals with biomass are not linear but exceed the expectations of linearity.

Trace metals emissions such as mercury (Hg), lead (Pb), arsenic (As), nickel (Ni), and more can be considered essentially linear with fuel blends; however, it is essential to calculate their concentrations on a mass basis rather than in $lb/10^6$ Btu (g/GJ). At the same time, the management of these emissions must utilize the concentrations on a heat input basis. Blending can address certain issues with trace metal emissions control. With respect to Hg, control of these emissions can be facilitated by oxidizing the metal with halogens such as

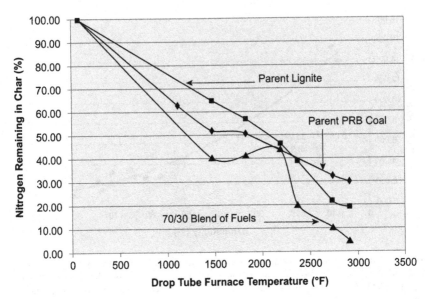

FIGURE 2.9   Nitrogen evolution from a Texas Lignite A, a PRB coal—Rawhide—and a 70% PRB/30% lignite blend. Note that the blend yields more nitrogen in volatile form, facilitating NO_x control, than either parent coal. *Source: [2].*

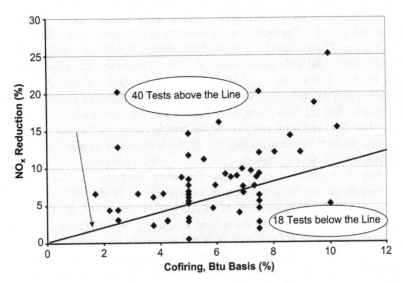

FIGURE 2.10 NO_x reduction as a function of cofiring. The diagonal line represents the NO_x reduction that would be achieved as a function of reduced nitrogen in the fuel. *Source: [55].*

bromine or chlorine. Blending a modest amount of chlorine-bearing coal in the total fuel blend can oxidize the Hg and facilitate the capture of a portion of this metal with an electrostatic precipitator or baghouse [32]. Experience showed that a blend of 30% Central Appalachian coal/70% PRB coal facilitated such control. The Hg in the PRB coal, in reduced form, is oxidized by the Cl in the bituminous coal.

2.6. FUEL BLENDING CHARACTERISTICS INFLUENCING DEPOSITION

The body of literature concerning deposition is particularly large (see, for example, [10–20, 56, 57]). This literature deals with both negative and positive impacts of solid fuel blending, with emphasis on the inorganic constituents in the fuel. Deposition—slagging and fouling—and corrosion are largely a function of inorganic constituents and the way they interact during the combustion process. It is recognized that in coal, biomass, and other solid fuels, these constituents most commonly do not occur—in the unburned fuel—in oxide form [30, 31]. These minerals may exist in a wide variety of forms, as shown in Table 2.6, which is adapted from Baxter et al. [58]. However, for the most part, they are reported in oxide form (e.g., SiO_2, Al_2O_3, Fe_2O_3, CaO). The

TABLE 2.6 Representative Selected Minerals in Coal

Group	Mineral/Inorganic Material	Formula
Clay	Montmorillonite	$(NaCa)_{0.3}(Al,Mg)_2Si_4O_{10}(OH)_2nH_2O$
	Illite	$KAl_2Si_3AlO_{10} \cdot H_2O$ $KMg_2Si_3AlO_{10} \cdot H_2O$ $KFe_2Si_2O_{10} \cdot H_2O$
	Kaolinite	$Al_2O_3 \cdot 2SiO_2 \cdot {}_2H_2O$
Feldspars	Orthoclase	$KAlSi_2O_2$
	Albite	$NaAlSi_2O_2$
	Anorthite	$CaAl_2Si_2O_2$
Silicates	Mullite	$_3Al_2O_3 \cdot 2SiO_2$
	Gehlenite	$CaAlSiO_2$
	Meta-kaolinite	$Al_2O_3 \cdot 2SiO_2$
Oxides	Quartz/Tridymite/Cristobalite	SiO_2
	Rutile	TiO_2
	Magnetite	Fe_2O_3

Source: Adapted from [58].

interactions of these constituents, such as those reported earlier in Tables 2.1 through 2.4, are many and complex, and they also result in blend results that are not necessarily the weighted average of the fuels.

Deposition—slagging and fouling—is largely governed by fuel chemistry, combustion temperatures, and whether the localized conditions are oxidizing or reducing. Slagging is deposition in the furnace area, while fouling is deposition in the convective pass, the superheater and reheater sections, and the economizer sections of the boiler.

Several of the studies have focused on ash fusion temperatures as a means for understanding deposition and corrosion [13, 14, 22]. There is conflicting evidence in these studies concerning the mechanisms and consequences of blending. Widenman [17], in a detailed study of Pittsburgh Seam-PRB coals, has shown that ash fusion temperatures—particularly initial deformation and ash softening temperatures—are somewhat nonlinear with respect to blends. The PRB coals have a disproportionate impact on ash fusion temperatures. At the same time, the bituminous coals hold ash fusion temperatures at high levels when the percentage of PRB coal is <20%. For the most part, however, the Widenman data show reasonable linearity.

Sakurova, Lynch, and Maher [12] demonstrate essential linearity in the thermoplastic properties of coal blends, as measured by Gieseler plastometry. As a consequence, they state that fusibility of coal blends can be predicted from the weighted average of the two or more coals. The research by Sakurova [18] on blends of Australian coals, however, shows distinct nonlinearity of thermoplastic properties. This research shows that the logarithm of maximum fluidity (LMF) as determined by Gieseler plastometry is linear or additive only when coals of similar or equal rank are blended. Using proton magnetic resonance thermal analysis (PMRTA), Sakurova shows that nonlinearity is a consequence of interaction between inorganic components of the coals being blended. This research shows that as the percentage of subbituminous coal with reactive inorganic constituents increases, the temperature at which the blend exhibits mineral fluidity decreases, and the extent of inorganic fluidity—ash fusion—increases.

The research shows that this nonlinear increase in fluidity or ash fusion is a consequence of the transfer of mineral matter between coals or solid fuels. The interactions are caused by transfer of materials from coal that is volatile at or below the temperatures at which the coal has softened. Nonlinearity is also caused by the ability of mineral matter to be adsorbed by low-volatile and high-rank coals, thereby reducing the overall fusibility of a blend of more reactive subbituminous coals and lower-volatile, higher-rank coals.

Analytically, a common approach to the analysis of the ash fusion and the deposition behavior of fuel blends is to evaluate them with respect to ternary diagrams (see, for example, Qiu, Zheng, Zheng, and Zhou [14]). Such diagrams, along with binary diagrams, have been published by the American

Ceramic Society [59] and by others as well. They can also be created using FactSage thermodynamic modeling [private communication by Sharon Falcone Miller, 2010]. Common systems for analysis are shown in Figures 2.11 through 2.13 and include calcium oxide–silica–sodium oxide systems, calcium oxide–silica–iron oxide systems, and potassium oxide–sodium oxide–silica systems. Binary diagrams are also used as shown in Figures 2.14 and 2.15 and include calcium oxide–iron oxide systems (see pages 59 and 60). Temperatures in these diagrams are reported in °C.

Kondratiev and Jak [57] used similar analytical techniques to model the alumina–calcium oxide–iron oxide–silica system. The use of these diagrams involves establishing the ratio of the components in the blend (e.g., CaO, SiO_2, Fe_2O_3) and determining where they are on the ternary diagram. Note the very low temperatures on the Ca-Si-Na system. This indicates the problems associated with blends containing high silica content.

The eutectic effect of Fe_2O_3 and CaO—both fluxing agents—has been very pronounced and well reported [34]. Experience at the Monroe Power Plant was most instructive, showing that a blend with an Fe_2O_3/CaO blend of about 0.8 to 1.0 had the highest potential for slagging. Including analyses of slag deposits, it also showed that the total blend of fuel did not have to have the ratio of 0.8 to 1.0 ratio in order to get deposits—sometimes quite large deposits—with such ratios. Miller [60] has reported this phenomenon as well, indicating that the eutectic region has an Fe_2O_3/CaO ratio of 0.3 to 3.0.

When the blends include coal and petroleum coke, then the ternary diagrams of significance include vanadium and, at times, nickel. One such binary diagram is shown in Figure 2.14. For reasons of flame stability, petroleum coke is typically limited to 20% of a fuel blend in pulverized coal firing. With petroleum coke/PRB coal blends, this may push the ash fusion temperature to reasonably high levels; however, when blends are petroleum coke/bituminous coal, there may be more potential for some slag deposition.

In summary, the interaction of inorganic components, particularly in high-temperature PC and cyclone firing, can be dramatic, and the consequences for slagging and fouling can be substantial. At the same time, developing blends to capitalize on the slagging and fouling characteristics is quite possible. For cyclone firing, this means developing blends to facilitate slag formation. For PC firing, this means exhibiting higher-temperature slagging and fouling characteristics.

2.7. FUEL BLENDING AND CORROSION

Corrosion characteristics of fuel blends are distinct from deposition characteristics, although they also are based on inorganic chemistry. This corrosion is distinct and separate from the corrosion caused by such firing systems as

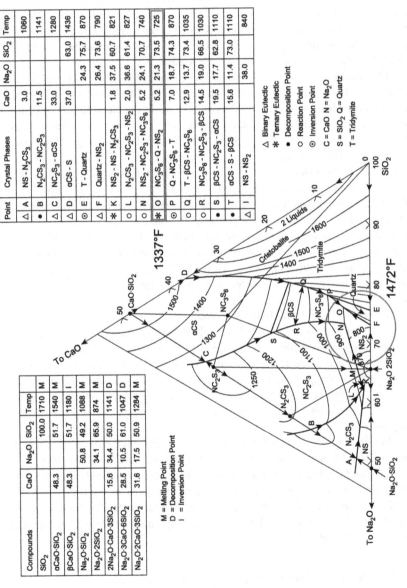

Point		Crystal Phases	CaO	Na2O	SiO2	Temp
△	A	NS - N2CS3	3.0			1060
●	B	N2CS3 - NC2S3	11.5			1141
△	C	NC2S3 - αCS	33.0			1280
△	D	αCS - S	37.0		63.0	1436
⊙	E	T - Quartz		24.3	75.7	870
△	F	Quartz - NS2		26.4	73.6	790
*	K	NS2 - NS - N2CS3	1.8	37.5	60.7	821
○	L	N2CS3 - NC2S3 - NS2	2.0	36.6	61.4	827
○	N	NS2 - NC2S3 - NC3S6	5.2	24.1	70.7	740
*	O	NC3S6 - Q - NS2	5.2	21.3	73.5	725
⊙	P	Q - NC3S6 - T	7.0	18.7	74.3	870
○	Q	T - βCS - NC3S6	12.9	13.7	73.4	1035
○	R	NC3S6 - NC2S3 - βCS	14.5	19.0	66.5	1030
●	S	βCS - NC2S3 - αCS	19.5	17.7	62.8	1110
●	T	αCS - S - βCS	15.6	11.4	73.0	1110
△	I	NS - NS2		38.0		840

△ Binary Eutectic
* Ternary Eutectic
● Decomposition Point
○ Reaction Point
⊙ Inversion Point

C = CaO N = Na2O
S = SiO2 Q = Quartz
T = Tridymite

Compounds	CaO	Na2O	SiO2	Temp	
SiO2			100.0	1710	M
αCaO·SiO2	48.3		51.7	1540	M
βCaO·SiO2	48.3		51.7	1180	I
Na2O·SiO2		50.8	49.2	1088	M
Na2O·2SiO2		34.1	65.9	874	M
2Na2O·CaO·3SiO2	15.6	34.4	50.0	1141	D
Na2O·3CaO·6SiO2	28.5	10.5	61.0	1047	D
Na2O·2CaO·3SiO2	31.6	17.5	50.9	1284	M

M = Melting Point
D = Decomposition Point
I = Inversion Point

FIGURE 2.11 The ternary diagram for CaO-Na2O-SiO2. Note the potential for a very low-temperature melt (or condensation) of one compound along the Na2O-SiO2 line. *Source: [59]. Reprinted with permission of the American Ceramic Society, www.ceramics.org. All rights reserved.*

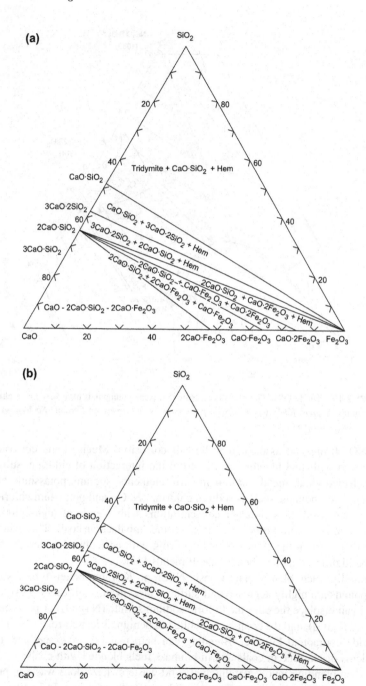

FIGURE 2.12   The CaO-Fe$_2$O$_3$-SiO$_2$ system at (a) above 1155°C (2110°F) or (b) below 1155°C (2110°F). *Source: [59]. Reprinted with permission of the American Ceramic Society, www. ceramics.org. All rights reserved.*

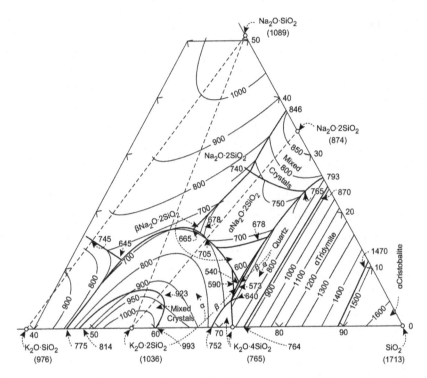

FIGURE 2.13　The K_2O-Na_2O-SiO_2 system. Note that some compounds have very low melting temperatures. *Source: [59]. Reprinted with permission of the American Ceramic Society, www. ceramics.org. All rights reserved.*

low-NO_x firing and associated waterwall corrosion. Much of the corrosion associated with fuel blending results from the interaction of chlorine, sulfur, and selected alkali metals and alkali earth elements: sodium, potassium, and calcium. Compounds, such as sodium chloride (NaCl) and potassium chloride (KCl), can result in significant concentrations with blends of high-chlorine coals, such as Interior Province coals and some Appalachian coals. They can be exacerbated when the blends include cofiring with agricultural materials that can be high in both chlorine and alkali metals [32].

Initially such blends can deposit NaCl or KCl on boiler tubes; such compounds are highly aggressive. If sulfur is present in significant excess, the sulfur can displace the chlorine to form sodium sulfate (Na_2SO_4) or potassium sulfate (K_2SO_4), and the compound, while not benign, is less corrosive than the chloride compound [30, 32]. Blending can help manage, or exacerbate, the problems of chlorine in coals and in biomass fuels blended with coals.

When blending involves municipal solid waste or hazardous waste in processed form (e.g., cofiring refuse-derived fuel with coal in a PC boiler, or chlorinated solvents with coal in a rotary kiln), chlorine-based corrosion must be given substantial attention. RDF can contain upward of 0.5% chlorine on

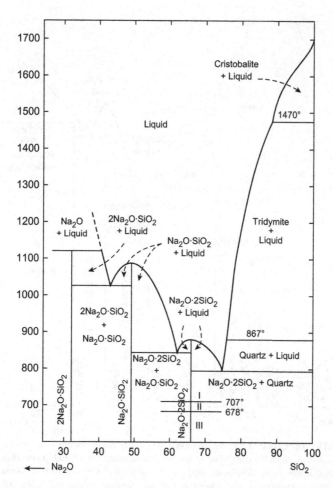

FIGURE 2.14 The Na_2O-SiO_2 system. Note liquids forming as low as 800°C or 1470°F. *Source: [59]. Reprinted with permission of the American Ceramic Society, www.ceramics.org. All rights reserved.*

a total basis; this creates a substantial problem. Burning chlorinated solvents either with coal in a rotary kiln or with other hazardous wastes in an incinerator also provides fuel with substantial chlorine content. Other halogens such as fluorine can also create similar problems.

2.8. BLENDING'S IMPACT ON THE PHYSICAL CHARACTERISTICS OF SOLID FUELS

Blending impacts the ease of grinding or grindability of the fuel, measured by the HGI. It impacts the handling and storage characteristics of the fuel as well. Vuthaluru et al. [61], citing the work of Conroy and Bennette [62], report

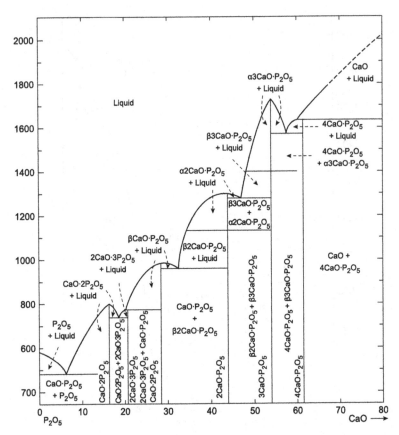

FIGURE 2.15 The Cao-P_2O_5 system. Note that the phosphorus drives down to very low levels the temperatures at which the compound undergoes a phase change to a liquid. As the phosphorus pentoxide is at levels above 65%, liquid temperatures are below 1000°C or 1800°F. *Source: [59]. Reprinted with permission of the American Ceramic Society, www.ceramics.org. All rights reserved.*

that studies of Australian and Indonesian coals—both bituminous and subbituminous—indicate that blending lowers the HGI for the blend, sometimes lower than the HGI of the parent coals. Vuthaluru et al. [61], however, found that both binary and ternary blends could have higher HGI values than the parent coals. The characteristics of the parent coal heavily influence this property. Ural and Akyildtz [63] have also performed similar research.

Research by Dosch and Taylor (see [2]) shows that blending fuels with different HGI values results in preferential grinding within the pulverizer. Blends of PRB coals—from the Black Thunder and Antelope mines—and lower HGI Central Appalachian coals were not pulverized equally. The outputs of the mills favored the softer (higher-HGI) coals over the harder bituminous coals. Research by E. Stephens of TVA, cited by Tillman [6] in a major EPRI report, shows that blending biomass with coal has even more dramatic impacts on the pulverizer performance. When wood waste was added to bituminous coal at low mass percentages, the performance of the pulverizer began to degrade. Above 5% biomass on a mass basis, the performance of the mill was totally unacceptable.

This research was performed using a Raymond bowl mill. The choice of mill makes a significant difference. Ball-and-race mills, bowl mills, and similar designs have significant difficulties with dissimilar materials. Atritta mills, which are essentially modified hammer mills, can handle dissimilar materials to a greater extent. Blends of coal on coal have modest impacts on grinding properties of the blend. Blends of biomass on coal have dramatic and deleterious impacts on the grinding properties of the blend. Petroleum coke tends to have relatively low HGI values and behaves, with coal in blends, as the harder material.

The handling ability of coal blends has been given scant attention in the literature. A study by Zhong, Ooi, and Rotter [64], however, indicates that blending can have dramatic—beneficial or deleterious—impacts on the handling characteristics of coal blends. They conclude that dominant influences are the extent to which fines, including clay particles, are included in the blends. Coarse particles improve handling characteristics. Such characteristics can result in coal flow stoppages or blockages if the blend is inappropriate. The blend handling properties are less a function of coal rank or chemistry; rather, they are influenced by whether the coal has been processed by a preparation plant with extensive cleaning cycles—resulting in a high percentage of fine particles in the product. If the blend involves petroleum coke, it may come with an oily surface, and this, too, influences handling characteristics.

The handling characteristics of coal–woody biomass blends have been studied by numerous researchers, including Zulfiquar, Moghtaderi, and Wall [65]. They found that sawdust-coal blends improved the handling characteristics of the fuel blend relative to the coal alone. The sawdust helped reduce the strength of the mass of solids. At the same time, the frictional properties of the parent coal were maintained. No flow stoppage or blockage was expected. Wood chips, however, had the potential to cause blockage and pluggage when blended with coal. The physical form of the biomass is critical to flow properties of the blend. This is consistent with the experience of the cofiring tests and demonstrations reported by Tillman [6, 7] for cyclone boilers, where biomass was added to coal prior to bunkers and silos for final fuel storage and feeding.

Storage is the final physical issue reviewed here, and only briefly. It is recognized that blending PRB coal with bituminous coals increases the moisture content of the blend and that moisture may migrate through the blend either in outside storage or in silos. The blends with the more reactive subbituminous coals have a higher tendency to spontaneous combustion, and this must be managed by plant practices. Storage of wood waste with coal has been accomplished satisfactorily at TVA and at NIPSCO plants (see Tillman [6]). Other agricultural products such as ground switchgrass cannot be stored with coal unless constant bunker or silo pluggage is accepted by the plant owner.

2.9. MANAGEMENT AND CONTROL OF FUEL BLENDING

The preceding data indicate that blending to optimize the fuel product can be a very complex process requiring considerable attention to science and engineering. Parameters managed include moisture, calorific value, volatility and reactivity, pollution potential (e.g., particulates, SO_2, NO_x, HAPS), slagging and fouling, and corrosion. To this, physical properties such as HGI and pulverizer performance can be added. Many other complicating factors can make management of the blends more difficult. One such factor is the inherent variability of coal as shipped to the power plant fleet.

Numerous studies have shown significant inherent coal variability as a function of coal region, coal seam, and coal mine. The days when a single coal sample could be used to represent the output from a given mine are long over. Inherent variability also characterizes noncoal fuels: woody biomass, agricultural biomass, petroleum coke, municipal solid waste and refuse-derived fuel, and the various industrial wastes. Variables include moisture, calorific value, and chemical composition, as well as physical parameters.

Control of the blending process, at its best, requires constant attention to the properties of the fuel being blended. If a consistent blend is being fired by a plant—without change in response to market conditions—this vigilance can be applied by extensive testing of the fuel properties of the two or three coals or coals/opportunity fuels and by determining the acceptable blend. This approach has been commonly applied (see Peltier and Wicker [23]). This is a common approach. When testing coal-biomass blends, comparable approaches have been taken. With extensive fuel characterization before commencing the demonstration, the inherent fuel variability can be accommodated.

A more intricate and sophisticated approach involves adjusting the blend to meet market conditions and inherent fuel variability; this approach has been taken by the Monroe Power Plant of DTE Energy (see Peltier and Wicker [23]; Tillman et al. [2]). The Monroe Power Plant is a 3200-MW_e plant consisting of four supercritical boilers built between 1970 and 1974. The approach taken by the Monroe Power Plant requires extensive fuel characterization. Further, it requires vigilance in knowing blend ratios and the consequences being fed to the power plant. At the Monroe Power Plant, fuel is blended in a $400 million facility using a combination of underpile ploughs and belt scales, with the ability to vary the plough feeder speeds to vary the specific coal inputs (see Figure 2.16). The coals being blended include PRB, low-sulfur Central Appalachian coal, and medium-sulfur Central Appalachian coal.

For the Monroe Power Plant, an x-ray fluorescence (XRF) on-line coal analyzer supplied by QC, Inc., has been installed, and its calibration is

FIGURE 2.16 Under pile belt at the Monroe Power Plant with ploughs and belt scales to manage accurate, reproducible blending that can be controlled by the AccuTrack® system and the on-line analyzer.

constantly maintained. This analyzer provides the plant with an analysis of the blended fuel, including a full proximate analysis, a calorific value analysis, and an ash elemental analysis. These data are obtained every 90 seconds. Further, a complete fuel tracking software system called AccuTrack®, which was designed and implemented by Engineering Consultants Group, incorporates fuel properties into the total information package. AccuTrack® runs on the PI platform. Several other vendors supply on-line analyzers and fuel tracking software, and many such combinations can be established.

The combination of the XRF on-line analyzer and AccuTrack® provides the control room supervising operator and the shift supervisor with data concerning the fuel being burned, the fuel that will arrive at the burners 1 hour and 2 hours from the current time, and the quality of the fuel being put up in the silos. Figures 2.17 and 2.18 show the AccuTrack® system and the advisory screen providing fuels data to the operators and shift supervisors. Additional information has also been derived from these data, including an approximate heat and material balance about each boiler, identifying controllable losses; a temperature profile for the unit; and an analysis of flyash resistivity and conditioning needs.

One key to the system has been training operators and shift supervisors to obtain the most value from this fuel blend management system. The fuel blend can be adjusted periodically to optimize the combination of plant load and fuel cost. At the same time, the blend can be determined with attention to the potential for pollution formation, slagging and fouling, and other phenomena.

The Monroe Power Plant system represents perhaps the greatest degree of sophistication, and it is being duplicated in modified form by other utilities in

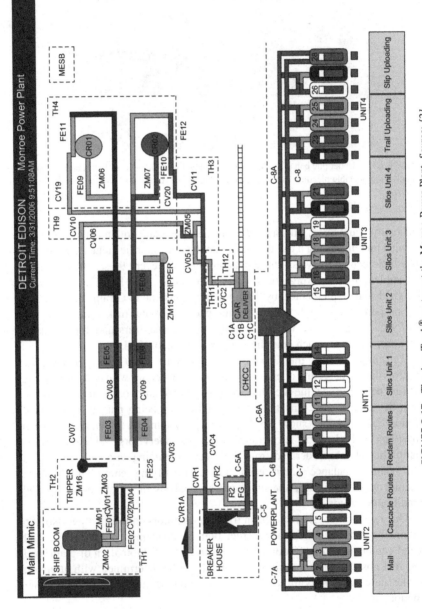

FIGURE 2.17 The AccuTrack® system at the Monroe Power Plant. *Source: [2].*

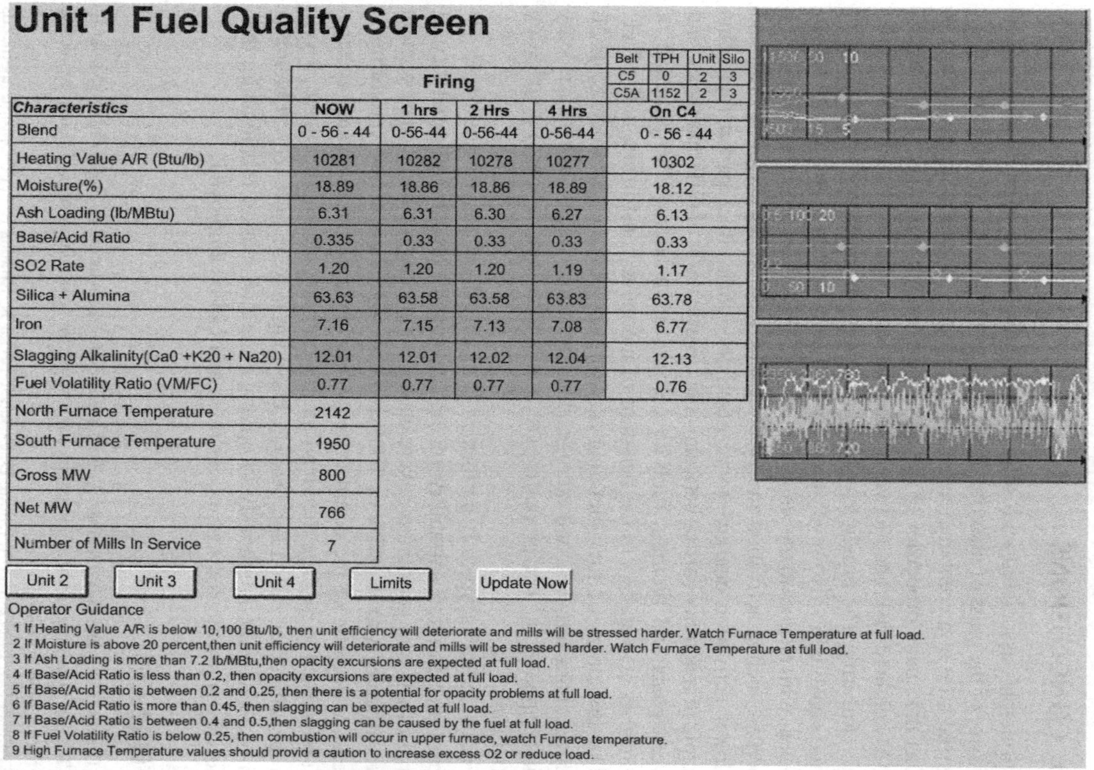

Unit 1 Fuel Quality Screen

Belt	TPH	Unit	Silo
C5	0	2	3
C5A	1152	2	3

	Firing				
Characteristics	NOW	1 hrs	2 Hrs	4 Hrs	On C4
Blend	0 - 56 - 44	0-56-44	0-56-44	0-56-44	0 - 56 - 44
Heating Value A/R (Btu/lb)	10281	10282	10278	10277	10302
Moisture(%)	18.89	18.86	18.86	18.89	18.12
Ash Loading (lb/MBtu)	6.31	6.31	6.30	6.27	6.13
Base/Acid Ratio	0.335	0.33	0.33	0.33	0.33
SO2 Rate	1.20	1.20	1.20	1.19	1.17
Silica + Alumina	63.63	63.58	63.58	63.83	63.78
Iron	7.16	7.15	7.13	7.08	6.77
Slagging Alkalinity(Ca0 +K20 + Na20)	12.01	12.01	12.02	12.04	12.13
Fuel Volatility Ratio (VM/FC)	0.77	0.77	0.77	0.77	0.76
North Furnace Temperature	2142				
South Furnace Temperature	1950				
Gross MW	800				
Net MW	766				
Number of Mills In Service	7				

Unit 2 | Unit 3 | Unit 4 | Limits | Update Now

Operator Guidance

1 If Heating Value A/R is below 10,100 Btu/lb, then unit efficiency will deteriorate and mills will be stressed harder. Watch Furnace Temperature at full load.
2 If Moisture is above 20 percent,then unit efficiency will deteriorate and mills will be stressed harder. Watch Furnace Temperature at full load.
3 If Ash Loading is more than 7.2 lb/MBtu,then opacity excursions are expected at full load.
4 If Base/Acid Ratio is less than 0.2, then opacity excursions are expected at full load.
5 If Base/Acid Ratio is between 0.2 and 0.25, then there is a potential for opacity problems at full load.
6 If Base/Acid Ratio is more than 0.45, then slagging can be expected at full load.
7 If Base/Acid Ratio is between 0.4 and 0.5,then slagging can be caused by the fuel at full load.
8 If Fuel Volatility Ratio is below 0.25, then combustion will occur in upper furnace, watch Furnace temperature.
9 High Furnace Temperature values should provid a caution to increase excess O2 or reduce load.

FIGURE 2.18 The fuel quality screen available to supervising operators and shift supervisors at the Monroe Power Plant. *Source: [3].*

response to increasing regulatory requirements and economic constraints on solid fuel power plants. Many more intermediate control systems are being implemented that use some, but not all, of the blending principles that the Monroe Power Plant uses.

2.10. CONCLUSIONS

The principles associated with solid fuel blending are based on modifying the chemistry of the mass of material being fired in an industrial or utility boiler, an industrial kiln, or a combination boiler/incinerator. Attention must be given to the fact that many properties of the fuel cannot be predicted as a consequence of the weighted average between the two or more fuels included in the blend—linearity. The principles of linearity apply to calorific value and many specific constituents. However, the very critical concerns of fuel reactivity are not linear with respect to the weighted average of fuel components. Many of the properties, such as activation energy of devolatilization, are such that the blend has lower activation energy—more reactivity—than either parent fuel. This principle of reactivity carries between coals, particularly of different ranks, and between coals and biomass fuels in cofiring applications. The principle of nonlinearity also applies to the behavior of inorganic constituents in the fuel.

The principles of fuel blending also impact physical parameters: HGI, handling characteristics, and storage characteristics. Again, linearity is only partially applicable. The properties are as much a function of fuel preparation as coal rank.

Fuel blending is typically performed to modify the chemistry of the fuel, yet in-seam and in-mine variability of all coals, and comparable variability of biomass fuels and petroleum cokes, makes the process of blend management more difficult. And if MSW or RDF is used, the inherent fuel variability is increased. Because of the complexity of fuel blending, sophisticated systems must be created to manage the blending. If a constant blend is being employed, then fuel characterization and periodic checking may be sufficient. If blending is to be managed continuously to optimize the fuel-market relationship, then some very sophisticated techniques must and can be employed.

REFERENCES

[1] Duong D, Tillman D, Widenman A. Fuel blending for combustion management. In: Miller B, Tillman D, editors. Combustion engineering issues for solid fuel systems. Boston: Academic Press; 2008. p. 171–98.

[2] Tillman DA, Dobrzanski A, Duong D, Dosch J, Taylor K, Kinnick R, et al. Optimizing blends of Powder River Basin subbituminous coal and bituminous coal. Proceedings PRB users group annual meeting. Baltimore; 2006.

[3] Tillman DA, Dobrzanski A, Duong D, Dezsi P. Fuel blending with PRB coals for combustion optimization: a tutorial. Proceedings 31st international technical conference on coal utilization and fuel systems. Clearwater, FL; 2006, May 21.

[4] Tillman DA. Biomass cofiring: the challenge before us. Proceedings EPRI workshop on biomass energy. Atlanta; 2009, Oct. 2.

[5] Tillman DA, Stanley Harding N. Fuels of opportunity. Amsterdam: Elsevier; 2004.

[6] Tillman DA. Cofiring. In: Final report: EPRI-USDOE cooperative agreement, vol. 1. Palo Alto, CA: Electric Power Research Institute; 2001.

[7] Tillman DA. Opportunity fuel cofiring at Allegheny Energy: final report. Palo Alto, CA: EPRI; 2004. Report #1004811.

[8] Duong D, Tillman D, Miller B. Characterizing blends of PRB and Central Appalachian coals for fuel optimization purposes. Proceedings 31st international technical conference on coal utilization and fuel systems. Clearwater, FL; 2006, May.

[9] Faundez J, Arias B, Rubiera F, Arenillas A, Garcia X, Gordon AL, Pis JJ. Ignition characteristics of coal blends in an entrained flow furnace. Fuel 2007;86:2076–80.

[10] Shi S, Pohl JH, Holcombe D, Hart JA. Slagging propensities of blended coals. Fuel 2001;80:1351–60.

[11] Shi S, Pohl JH, Holcombe D. Fouling propensities of blended coals in pulverized coal-fired power station boilers. Fuel 2003;82:1653–67.

[12] Sakurova R, Lynch LJ, Maher TP. The prediction of the fusibility of coal blends. Fuel Processing Technology 1994;37:255–69.

[13] Rushdi A, Sharma A, Gupta R. An experimental study of the effect of coal blending on ash deposition. Fuel 2004;83:495–506.

[14] Qiu JR, Li F, Zheng Y, Zheng CG, Zhou HC. The influences of mineral behavior on blended coal ash fusion characteristics. Fuel 1999;78:963–9.

[15] Shih J-S, Frey HC. Theory and methodology: coal blending optimization under uncertainty. European Journal of Operational Research 1995;83:452–65.

[16] Smith DJ. Blending of opportunity fuels with coal can reduce emission and generating costs. Coal Power 2006;2(2):24–6.

[17] Widenman T. Ash fusion characteristics for binary blends of selected PRB and Pittsburgh seam coals. Proceedings 28th international technical conference on coal systems and fuel utilization. Clearwater, FL; 2003, March 8–13.

[18] Sakurova Richard. Direct evidence that the thermoplastic properties of blends are modified by interactions between the component coals. Fuel 1997;76(7):615–21.

[19] Haas J, Tamura M, Weber R. Characterisation of coal blends for pulverized fuel combustion. Fuel 2001;80:1317–23.

[20] Chi T, Zhang H, Yan Y, Zhou H, Zheng H. Investigations into the ignition behaviors of pulverized coals and coal blends in a drop tube furnace using flame monitoring techniques. Fuel 2010;89:743–51.

[21] Prinzing D. EPRI alternate fuels database. Palo Alto, CA (prepared by Foster Wheeler Environmental Corporation, Sacramento); 1996. EPRI TR 107602.

[22] Barroso J, Ballester J, Ferrer LM, Jimenez S. Study of coal ash deposition in an entrained flow reactor: influence of coal type, blend composition, and operating conditions. Fuel Processing Technology 2006;87:737–52.

[23] Peltier R, Wicker K. PRB coal makes the grade. Power 2003;147(8):28–36.

[24] Anthony EJ, Iribarne AP, Iribarne JV, Talbot R, Jia L, Granatstein DL. Fouling in a 160 MW$_e$ FBC boiler firing coal and petroleum coke. Fuel 2001;80:1009–14.

[25] Wang J, Anthony EJ, Abanades JC. Clean and efficient use of petroleum coke for combustion and power generation. Fuel 2004;83:1341–8.

[26] Taniguchi M, Kobayashi H, Kiyama K, Shimogori Y. Comparison of flame propagation properties of petroleum coke and coals of different rank. Fuel 2009;88:1478–84.

[27] Bryers RW. Utilization of petroleum coke and petroleum coke/coal blends as a means of steam raising. Fuel Processing Technology 1995;44:121–41.

[28] Tillman DA. Petroleum coke as a supplementary fuel for cyclone boilers. Proceedings ASME international joint power generation conference. Phoenix; 2002, June 24–27.

[29] Weick-Hansen K, Overgaard P, Larsen OH. Cofiring coal and straw in a 150 MW$_e$ power boiler experiences. Biomass & Bioenergy 2000;19(6):395–410.

[30] Baxter LL, Miles TR, Miles Jr TR, Jenkins BM, Dayton DC, Milne TA, et al. The behavior of inorganic material in biomass-fired power boilers: field and laboratory experiences. Proceedings Engineering Foundation conference on biomass usage for utility and industrial power. Snowbird, UT; 1996, April 28–May 3.

[31] Baxter LL, Miles TR, Miles Jr TR, Jenkins BM, Dayton DC, Milne TA, et al. The behavior of inorganic material in biomass-fired power boilers—field and laboratory experiences: vol. 2, of alkali deposits found in biomass power plants. Sandia National Laboratories, USDOE; 1996. Report NREL/TP-433–8142.

[32] Tillman DA, Duong D, Miller B. Chlorine in solid fuels fired in pulverized fuel boilers—sources, forms, reactions and consequences: a literature review. Energy & Fuels 2009;23:3379–91.

[33] Kitto JB, Stultz SC, editors. Steam: its generation and use, 41st ed. Barberton, OH: Babcock and Wilcox; 2005.

[34] Singer JG. Combustion: fossil power systems, 3rd ed. Windsor, CT: Combustion Engineering; 1981.

[35] Marx P, Morin J. Conventional firing systems. In: Miller BG, Tillman DA, editors. Combustion engineering issues for solid fuel systems. Boston: Academic Press; 2008. p. 241–74.

[36] Miller BG, Miller SF. Fluidized-bed firing systems. In: Miller BG, Tillman DA, editors. Combustion engineering issues for solid fuel systems. Boston: Academic Press; 2008. p. 275–340.

[37] Widenman A. Fuel characterization: a tutorial. Proceedings 35th international technical conference on coal systems and fuel utilization. Clearwater, FL; 2010, June 6–10.

[38] Bryers RW. Utilization of petroleum coke and petroleum coke/coal blends as a means of steam raising. In: Bryers RW, Harding NS, editors. Coal blending and switching of low-sulfur Western coals. New York: the American Society of Mechanical Engineers (for the Engineering Foundation); 1994. p. 185–206.

[39] Pierce BS, Dennen KO, editors. The National Coal Resource Assessment overview. US Geological Survey, Washington, DC.

[40] Duong D. Biomass Cofiring and its effect on the combustion process. M. Eng. Thesis. Bethlehem, PA: Lehigh University; 2010.

[41] Duong D, Lantos G, Tillman D, Kawecki D, Coleman A. Biomass cofiring and its effects on the combustion process. Proceedings 35th international technical conference on coal systems and fuel utilization. Clearwater, FL; 2010, June 6–10.

[42] Duong D, Lantos G, Tillman D. Biomass cofiring: update on TGA analysis of biomass with PRB coal. Proceedings 13th annual electric power conference. Rosemont, IL; 2011, May 10–12.

[43] Johnson D, Miller B, Pisupati S, Clifford D, Badger M, Tillman D. Characterizing biomass fuels for cofiring applications. Kauai, Hawaii: AFRC International Symposium; 2001.

[44] Scaroni AW, Khan MR, Eser S, Radovic LR. Ullmann's encyclopedia of industrial chemistry, vol. A7. New York: VCH Publishers; 1986.

[45] Gunderson JR, Selle SJ, Harding NS. Utility experience blending Western and Eastern coals: survey results. In: Bryers RW, Harding NS, editors. Coal blending and switching of low-sulfur Western coals. New York: American Society of Mechanical Engineers (for the Engineering Foundation); 1994. p. 111–30.

[46] Morgan JI, Boesen RN. Powder River Basin coal: do we really want to burn this stuff? In: Bryers RW, Harding NS, editors. Coal blending and switching of low-sulfur Western coals. New York: American Society of Mechanical Engineers (for the Engineering Foundation); 1994. p. 225–30.

[47] Tillman DA. Biomass cofiring: the technology, the experience, the combustion consequences. Biomass & Bioenergy 2000;19:365–84.

[48] Wang Q, Zhang L, Sato A, Ninomiya Y, Yamashita T. Effects of coal blending on the reduction of PM_{10} during high-temperature combustion: 1. Mineral transformations. Fuel 2008;87:2997–3005.

[49] Wang Q, Zhang L, Sato A, Ninomiya Y, Yamashita T. Effects of coal blending on the reduction of PM_{10} during high-temperature combustion: 2. A coalescence-fragmentation model. Fuel 2009;88:150–7.

[50] Pershing DW, Wendt J. Pulverized coal combustion. The influence of flame temperature and coal composition on thermal and fuel NO_x. Proceedings Combustion Institute 16th international symposium; 1976.

[51] Tillman DA. The combustion of solid fuels and wastes. San Diego: Academic Press; 1991.

[52] Miller BG, Tillman DA. Coal characteristics. In: Miller BG, Tillman DA, editors. Combustion engineering issues for solid fuel systems. Boston: Academic Press; 2008.

[53] Tillman DA, Miller BG, Johnson DK, Clifford DJ. Structure, reactivity, and nitrogen evolution characteristics of a suite of solid fuels. Proceedings 28th international technical conference on coal utilization and fuel systems. Clearwater, FL; 2003, March 10–13.

[54] Baxter LL, Mitchell RE, Fletcher TH, Hurt RH. Nitrogen release during coal combustion. Energy & Fuels 1996;10(1):188–96.

[55] Plasynski SI, Costello R, Hughes E, Tillman D. Biomass cofiring in full-sized coal-fired boilers. Proceedings 24th international technical conference on coal utilization and fuel systems. Clearwater, FL; 1999, March 8–11. p. 281–92.

[56] Wiedong L, Ming L, Weifeng L, Haifeng L. Study on the ash fusion temperatures of coal and sewage sludge mixtures. Fuel 2010;89:1566–72.

[57] Kondratiev A, Jak E. Predicting coal ash slag flow characteristics (viscosity model for the Al_2O_3-CaO-'FeO'-SiO_2 system). Fuel 2001;80:1989–2000.

[58] Baxter LL. Ash deposit formation and deposit properties: a comprehensive summary of research conducted at Sandia's combustion research facility. Final report. Livermore, CA: Sandia National Laboratories; 2000. SAND2000-8253.

[59] Levin EM, Carl RR, McMurdle HF. Phase equilibrium diagrams, vol 1. Columbus, OH: American Ceramic Society; 1964.

[60] Miller BG. Coal energy systems. Boston: Academic Press; 2005.

[61] Vuthaluru HB, Brooke RJ, Zhang DK, Yan HM. Effects of moisture and coal blending on Hardgrove Grindability Index of Western Australian coal. Fuel Processing Technology 2003;81:67–76.

[62] Conroy A, Bennette C. The combustion behavior of Australian export and overseas low rank coal blends. ACARP final report C3097; 1997.

[63] Ural S, Akyildiz M. Studies of the relationship between mineral matter and grinding properties for low-rank coals. International Journal of Coal Geology 2004;60:81–4.

[64] Zhong Z, Ooi JY, Rotter JM. Predicting the handlability of a coal blend from measurements on the source coals. Fuel 2005;84:2267–74.

[65] Zulfiqar M, Moghtaden B, Wall TF. Flow properties of biomass and coal blends. Fuel Processing Technology 2006;87:281–8.

Chapter 3

Blending Coal on Coal

3.1. INTRODUCTION AND BASIC PRINCIPLES

Blending of alternative coals is the backbone of the fuel blending practice. It is estimated that blends of coals are more commonly combusted or gasified than single coals in commercial applications. It dwarfs the practices of blending coals with biomass or waste-based fuels. Initially the practice of blending— with an engineering and scientific base—was dominated by environmental issues. Control of SO_2, NO_x, and particulates was of significance, as discussed in Chapters 2 and 6. Today the practice is dominated by economic and technical considerations.

The U.S. electricity generating fleet is fueled, to about 50%, with various ranks and types of coal (Figure 3.1). At the same time, coal is used extensively in the pulp and paper industry, the minerals processing industry, cement kilns, and similar locations. Coal is emerging as one of the fuels of choice for the rapidly growing Chinese economy and for other emerging economies in Asia. Coal is also used to fuel economies in Europe, particularly Germany, Poland, Russia and the former USSR nations, and others. It is the energy backbone of the Republic of South Africa, where it is not only used to generate electricity but also converted into liquid fuels by SASOL, Ltd. All of these economies and applications utilize fuel blending because it permits a broader fuel supply and at the same time permits individual plants and processes to manipulate the technical parameters of the fuel.

The United States burns about 1 billion tons of coal annually—with variations caused by economic activity. The Chinese economy burns about 2 billion tons of coal annually, and it is on a growth trajectory. It both mines significant quantities of coal and imports coal from other nations. Other nations with significant use of coal, as of 2009, include Germany, Russia, the Ukraine, Poland, the United Kingdom and the Republic of South Africa (Table 3.1). International trade in coal is, consequently, quite significant. Major coal exporting countries include the United States, Germany, the Republic of South Africa, Australia, Poland, Indonesia, Colombia, and more, as shown in Table 3.2. With the quantities of coal used in the worldwide economy, it is understandable that coal-coal blending is a major fact of life.

Solid Fuel Blending: Principles, Practices, and Problems

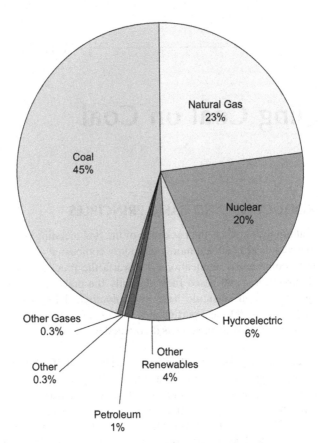

FIGURE 3.1   Production of U.S. electricity by fuel/energy source in 2010. *Source: [1].*

The science of solid fuel blending began with coal, although not with the combustion of coal used to generate electricity or process heat. Rather, the science began with the production of coke for the steel industry. Inconsistencies in the properties of coke caused a major emphasis on petrography and the science of coal properties, as alluded to in Chapter 1. The science was based on a careful study of macerals—defined by Thrush et al. [2] as "organic units composing the coal mass, being the descriptive equivalent of the inorganic units composing rock masses and universally called minerals. Individual macerals have the termination *-inite*—that is, vitrinite—as the organic unit making up the lithologic specimen vitrain."

In 1995, Bryers [3] provided an insightful linkage between blending based on macerals and the combustion/gasification process, evaluating individual macerals and the combustion process. His research showed that coal vitrinite reflectance in oil is directly correlated to fuel reactivity. High-volatile coal vitrinite has a reflectance <1, medium-volatile bituminous coal has

TABLE 3.1 World Energy Consumption of Coal by Region and Major Country

Region/Country	Consumption (in short tons)
Total World Coal Consumption	**7,577,379,000**
North America	1,072,224,000
Canada	51,910,000
United States	1,000,424,000
Mexico	19,890,110
Central and South America	46,607,220
Brazil	25,364,180
Chile	7,378,872
Colombia	6,687,315
Europe	943,710,800
Bulgaria	33,440,820
Czech Republic	57,928,870
Germany	249,696,700
Greece	71,083,650
Poland	141,119,000
Turkey	102,527,100
United Kingdom	53,826,970
Eurasia	395,561,800
Kazakhstan	87,355,970
Russia	222,630,500
Ukraine	63,991,380
Middle East	15,807,140
Africa	213,486,800
South Africa	199,069,700
Asia and Oceania	4,889,982,000
Australia	149,671,800
China	3,474,665
Japan	181,498,900
South Korea	113,293,400

Source: [1].

TABLE 3.2 Total Coal Exports by Major Exporting Region and Country

Country	Total Steam Coal Exports (in short tons)
World	**1,090,776,000**
North America	92,351,000
Canada	31,947,000
United States	60,404,000
South America	85,209,000
Colombia	75,740,000
Europe	40,028,000
Germany	500,000
Poland	14,606,000
Eurasia	163,829,000
Kazakhstan	25,694,000
Russia	130,863,000
Africa	74,548,000
Southern Africa	73,768,000
Asia and Oceania	634,710,000
Australia	288,524,000
China	38,354,000
Indonesia	261,419,000

Source: [1]

a reflectance of 1.0 to 1.3, and low-volatile bituminous coal has a reflectance >1.3. His research further related various macerals to thermogravimetric analysis and burning profiles for a range of U.S. and international coals. Zhang et al. [4] provided a detailed analysis of Chinese coals using macerals analysis. Petrographic analysis provides a general outline of coal properties for combustion and gasification; more detailed chemical analyses follow.

A basic linkage between petrography and combustion concerns was proposed by Su et al. [5], based on both individual coals and blends and using tests conducted both in the United States by EER Corporation, and in Australia by ACIRL. This linkage is a maceral index (MI) of coal burn-out percentage (96% to 99.9%) calculated as follows [5]:

$$
\text{MI} = \left(\frac{L + \dfrac{V}{R^2}}{I^{1.25}} \right) \left(\frac{HV}{30} \right)^{1.25}
\tag{3.1}
$$

where L is the percent liptinite that is normalized to 100% on a mineral-free basis; V is the percent vitrinite that is normalized to 100% on a mineral-free basis; I is the percent inertinite that is normalized to 100% on a mineral-free basis; R is the mean maximum vitrinite reflectance; HV is the heating value of the coal or coal blend on an air-dried basis, expressed in MJ/kg; and 30 is the typical heating value of bituminous coal in MJ/kg. Embedded in this equation is a reactivity factor (RF) calculated as follows:

$$
\text{RF} = \left(\frac{L + \dfrac{V}{R^2}}{I^{1.25}} \right)
\tag{3.2}
$$

The basics of this linkage are that liptinite burns rapidly and contains relatively high concentrations of hydrogen, while vitrinite combustion is a function of its reflectance. Inertinite burns most slowly among the macerals. The HV/30 is based on the fact that burn-out depends, to a significant extent, on heat release of the coal or coal blend. This macerals index works for both individual coals and coal blends. It illustrates the significant linkage between macerals analysis and combustion/gasification. It also demonstrates that the evaluation of blending through macerals, from a combustion or gasification perspective, is general in nature. More detailed assessments are necessary if blending is to be optimized from technical and economic perspectives. This maceral index is consistent with the work of Bryers [3] correlating coal rank to vitrain reflectance.

3.2. BLENDING OF COAL FOR COMBUSTION AND/OR GASIFICATION PURPOSES

Blending for combustion, or gasification, purposes takes a different tack from blending for coke production. As indicated in Chapter 2, the data sets are different to accommodate the different purpose of the combustion and gasification processes. The starting point is the same—the raw (or washed) coal in unreacted form. The differences are defined by purpose. Combustion is accomplished to convert the coal into CO_2, H_2O, SO_2 (for capture), and useful heat within the regulatory constraints of environmental laws. With air as the oxidant, which is virtually 100% of the cases, combustion produces a gas stream including N_2, O_2, and some NO_x. Solid products of combustion include ash.

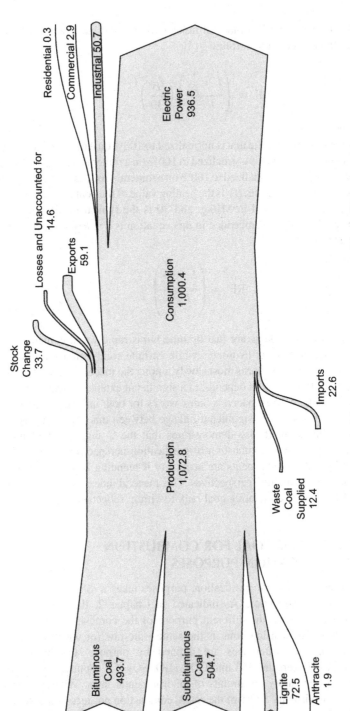

FIGURE 3.2 Consumption of coal in the United States (in tons). *Source: [1].*

FIGURE 3.3 Polk County IGCC gasifier and gas cleanup system. The synthesis gas produced by this system is fired in a combustion turbine-generator. Steam from the radiant cooler is combined with steam produced in a heat recovery steam generator (HRSG) to generate additional electricity in a steam turbine-generator system.

Ignoring, for the moment, the oxidant selected—which may be either air or pure oxygen, or some combination of the two—coal gasification is accomplished to convert the coal into CO, H_2, and occasionally CH_4, C_2H_{2-6}, and other hydrocarbons. The synthesis gas—CO and H_2—can be combusted directly to produce heat or electricity or can be further reacted to produce CH_4 as substitute natural gas; alternatively, synthesis gas can be reacted to produce a wide range of liquid fuels and chemical feedstocks. Today well over 90% of the coal used in the United States is employed in the generation of electricity, as shown in Figure 3.2. While almost all of this is consumed directly in direct combustion applications—coal-fired boilers—some is consumed in integrated gasification combined cycle (IGCC) systems, as shown in Figure 3.3. Worldwide electricity generation is the dominant market for coal, although IGCC technology appears to be more popular outside of North America.

3.3. COMBUSTION AND GASIFICATION PROCESSES

Blending can be used to facilitate and improve the processes of direct combustion and gasification—largely for electricity generation but also for industrial applications such as steel making, cement kiln operation, and other process industries.

3.3.1. Combustion Processes and Fuel Blending

For purposes of this discussion, since coal combustion is the dominant process, it is the focus of this analysis and presentation. In doing so, it is useful to review its basic reactions as well as the gasification process. While recognizing that combustion is a complex process, the global, simplistic combustion reactions are summarized as follows:

$$C + O_2 \rightarrow CO_2 \tag{3.3}$$

$$H_2 + \frac{1}{2}O_2 \rightarrow H_2O \tag{3.4}$$

$$S + O_2 \rightarrow SO_2 \tag{3.5}$$

or (and disregarding NO_x formation):

$$C_aH_bO_dN_eS_f\text{Ash} + XcsO_2(\text{with } N_2) \rightarrow aCO_2 + \frac{1}{2}bH_2O + fSO_2 \\ + gO_2 + hN_2 \tag{3.6}$$

The actual combustion process is far more complex than equations (3.3) through (3.6) indicate. While this is not a review of the combustion process in detail, it is useful to provide an overview of the coal combustion mechanisms in order to facilitate an understanding of the impacts and potentials of coal blending. More detailed discussions of the combustion mechanisms can be found in Tillman [6, 7], Sarofim and Bartok [8], Palmer and Beer [9], and numerous other texts.

The general mechanism is shown in Figure 3.4. The coal particle is first heated, and then surface and inherent moisture is removed during the heating and drying process. Subsequent to heating and drying, the particle undergoes pyrolysis or devolatilization, as summarized in Figure 3.4 and equation (3.7). Products of devolatilization include both noncondensable volatiles and radical fragments, condensable volatiles or tars, and solid products or chars composed of carbon, some residual hydrogen and heteroatoms such as nitrogen and sulfur, and inorganic matter. The tars further pyrolyze to noncondensable volatile compounds and radical fragments and fine solid char-rich products.

$$\text{Coal} + \text{heat} \rightarrow \text{volatiles, fragments, tars, chars} \tag{3.7}$$

The volatile matter consists of hydrocarbons, including CO, CH_4, C_2H_{2-6}, C_3H_{6-8}, H_2, H_2O, and so on. It also includes volatile nitrogen compounds, volatile sulfur, and other volatiles, including some halogenated compounds. Volatile fragments include $\cdot CH_3$, $\cdot C_2H_5$, $\cdot OH$, and many more radicals. Since combustion is, ultimately, a free radical series of reactions, these fragments play a significant role in chain initiation and chain propagation.

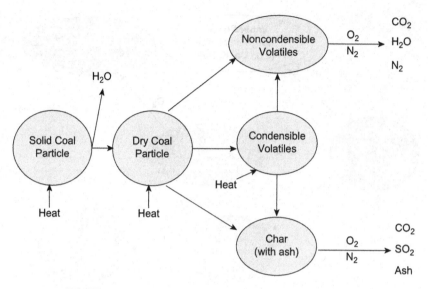

FIGURE 3.4 The combustion mechanism. *Source: Adapted from [6].*

Volatile matter, once released from the solid material, oxidizes rapidly in the presence of oxygen and heat. Char oxidation proceeds more slowly and is generally considered a diffusion-controlled process. It proceeds by radical attack on exposed, reactive carbon sites. The radical attack releases the carbon atom into a volatile compound or fragment (e.g., CO, HCO); this volatile compound then completes the combustion process. Sulfur can be oxidized to SO_2, as shown in equation (3.5), either from volatile matter or, more commonly, from char. The SO_2 can then be captured by combustion reactions in a fluidized bed reactor carrying calcium as part of the bed, or it can be captured in post-combustion scrubbers of various types.

The combustion processes applied to the inorganic matter in coal are reviewed in detail by Baxter [10], Raask [11], and numerous other authors (e.g., [12–17]). These processes are depicted, in summary form, in Figure 3.5. Inorganic matter may exist in coal either as mineral matter bound to the coal or as extraneous or adventitious matter brought in with the coal. Adventitious matter may be material from the roof or floor of underground mines, as overburden brought in with the coal from surface mines, or as dust blown into coal piles. It may be released during the combustion process by vaporization, by oxidation of the inorganic matter with subsequent phase change to vapor, or by fragmentation and carryover. It may return to solid phase material by condensation on other particles or by homogeneous condensation into fine fume particles.

Baxter [10] observes that sulfates, carbonates, and sulfides thermally decompose on heating and thus produce minerals in the vapor phase. Alkali earth elements decompose prior to melting; in some cases alkali metals exhibit

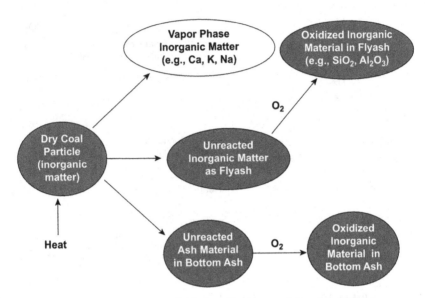

FIGURE 3.5   Inorganics in combustion. *Source: Adapted from [6].*

similar behavior, while in other cases the alkali metals melt, vaporize, and then decompose in the gas phase [10].

3.3.2. Coal Blending and the Combustion Process

The combustion process provides a clear lead into a discussion of coal blending. Chapter 2 introduced the different approaches to combustion: pulverized coal boilers and burners (e.g., for rotary kilns in the cement and pulp and paper industries), which represent the dominant technology today; cyclone burners for large utility boilers; spreader-stokers used largely in industrial boilers; and fluidized beds of either bubbling or circulating design. Each of these designs requires fuel of varying characteristics.

In all of the conventional firing systems, combustion temperature control is from load, excess air or stoichiometric ratio, temperature of combustion air, and manipulation of the combustion process through burner settings as well as fuel composition. Fuel composition variables include calorific value (Btu/lb or kcal/kg), moisture percentage, ash percentage, and other aspects of fuel chemistry. Reactivity also significantly influences both flame temperature and NO_x formation, with more rapid release of volatile fuel nitrogen in a fuel-rich zone promoting NO_x control. Typical flame temperatures for pulverized coal-fired boilers are on the order of 2700°F to 2900°F (1480–1590°C), and for cyclone boilers such temperatures can be 3300°F to 3550°F (1810–1950°C) [13]. On the other hand, bubbling and circulating fluidized bed boilers fired with coal typically have furnace temperatures of 1500°F to 1700°F (815–930°C) [14].

Blending can be used to influence chemical composition, as was introduced in Chapter 2. Further, it can be used, through chemical composition and calorific value, to influence flame temperature. This chapter explores such phenomena in some detail. Equally significantly, blending can be used to influence fuel reactivity and ash reactivity—concepts also introduced in Chapter 2. Coal pyrolysis or devolatilization reactivity—kinetics—was discussed briefly in Chapter 2, and it will be explored in more detail in subsequent sections of this chapter.

Briefly, the reactivity appears to be increased disproportionately as volatile compounds, radical fragments, and atoms released from the more reactive coal particles attack previously devolatilized particles of the less reactive coal; the consequence is that more gaseous radicals (e.g., OH, $H, \cdot CH_3$) and compounds are generated, increasing the reactivity of the fuel blend. Reactivity is generally a function of coal rank (see Table 3.3 and Figure 3.6) and inferentially a function of hydrogen/carbon (H/C) and oxygen/carbon (O/C) atomic ratios [6]. Coal blending, particularly with the more reactive Powder River Basin (PRB) coals, capitalizes on the availability of hydrogen and oxygen atoms in the PRB coals to promote increased reactivity of the blend [18].

The influence of blending on inorganic species in coal combustion comes from the interaction of gas-phase molecules and fragments, the interaction of inorganic and organic compounds in the gas phase, and heterogeneous gas-solids reactions. Examples of these interactions include the following:

- Mercury from both Eastern and Western coal, oxidized by chlorine from Eastern (e.g., Central Appalachian) bituminous coals to generate a form of mercury that is more easily captured.

- Interaction between calcium from PRB coal and iron from Eastern bituminous coals, forming lower-temperature eutectics, to increase the slagging potential beyond what each individually would cause; this is expressed as the Fe_2O_3/CaO ratio, and the eutectics are most active when the ratio approaches 1.0 from either side [19, 20].

- Modifications to the base/acid ratios of the coal blend, causing a decrease in ash fusion temperatures measured either in an oxidizing or reducing environment and measured as initial deformation, softening, hemispherical temperature, or fluid temperature (Figure 3.7).

- Modifications of other ratios, including the dolomite ratio, the SiO_2/Al_2O_3 ratio, and related measures of inorganic behavior [12].

- Ratios of chlorine and alkali metals (K, Na), and ratios of Cl and S, driving or inhibiting corrosion mechanisms [21–23].

Like the reactivity of the organic or combustible fraction of coal blends, the reactivity of inorganic matter in coal blends does not behave in a linear or weighted average fashion.

TABLE 3.3 Classification of Coal by Rank (ASTM D 388)

Class	Group	Fixed Carbon Limits, % dry, MMF(*)	Volatile Matter Limits, % dry, MMF(*)	Calorific Value Limits (10^3 Btu/lb as received)	Agglomerating Character
Anthracite	Meta-anthracite	$\geq$98	$\leq$2	—	N(**)
	Anthracite	$\geq$92, $\leq$98	$\geq$2, $\leq$8	—	N(**)
	Semianthracite	$\geq$86, $\leq$92	$\geq$8, $\leq$14	—	N(**)
Bituminous	Low Vol	$\geq$78, $\leq$86	$\geq$14, $\leq$22	—	CA(***)
	Medium Vol	$\geq$69, $\leq$78	$\geq$22, $\leq$31	—	CA(***)
	High Vol A	$\leq$69	$\geq$31	$\geq$14	CA(***)
	High Vol B	—	—	$\geq$13, $\leq$14	CA(***)
	High Vol C	—	—	$\geq$11, $\leq$13	A(#)
				$\geq$10.5, $\leq$11.5	N(**)
Subbituminous	Subbit A	—	—	$\geq$10.5, $\leq$11.5	N(**)
	Subbit B	—	—	$\geq$9.5, $\leq$10.5	N(**)
	Subbit C	—	—	$\geq$8.3, $\leq$9.5	N(**)
Lignite	Lignite A	—	—	$\geq$6.3, $\leq$8.3	N(**)
	Lignite B	—	—	$\leq$6.3	N(**)

Notes: (*) MMF = mineral matter free; (**) N = nonagglomerating; (***) CA = commonly agglomerating; (#) A = agglomerating.
Source: [2].

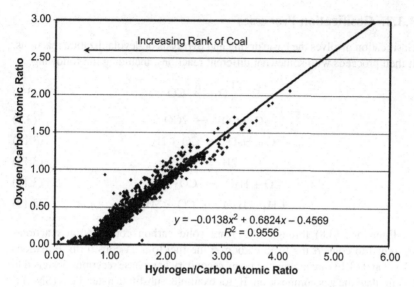

FIGURE 3.6 Influence of H/C and O/C atomic ratios on coal rank and reactivity. *Source: Adapted from [18].*

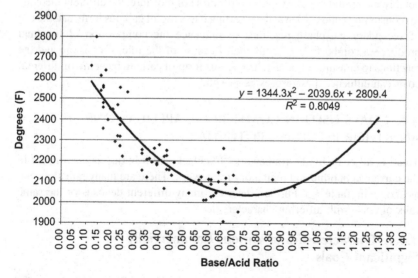

FIGURE 3.7 Initial deformation (reducing) temperature versus base/acid ratio; similar curves exist for all ash fusion measures. *Source: [15].*

3.3.3. Gasification Processes

Gasification involves the basic drying and pyrolysis or devolatilization reactions. It then proceeds with somewhat different reactions, including the following:

$$C + \frac{1}{2}O_2 \ \rightarrow \ CO \tag{3.8}$$

$$C + CO_2 \ \rightarrow \ 2CO \tag{3.9}$$

$$C + H_2O \ \rightarrow \ CO + H_2 \tag{3.10}$$

$$C + 2H_2 \ \rightarrow \ CH_4 \tag{3.11}$$

$$CO + H_2O \ \rightarrow \ CO_2 + H_2 \tag{3.12}$$

$$CH_4 + H_2O \ \rightarrow \ CO + 3H_2 \tag{3.13}$$

Equations (3.8) through (3.11) are solid carbon consumption reactions, designed to convert the fixed carbon in the fuel to a gaseous form. Equations (3.12) and (3.13) are among the many secondary gas-phase reactions designed to manipulate the gas composition. If, for example, substitute natural gas (SNG) is desired, then equation (3.13) will be suppressed. If, however, a synthesis gas is desired, equation (3.13) will be favored. If CO_2 capture and sequestration are desired, the water-gas shift equation (3.12) will be promoted. All of these reactions occur during the combustion of solid fuels containing hydrogen, carbon, oxygen, and moisture; however, they are promoted in gasification systems.

Gasification systems and principles are beyond the scope of this text. It is sufficient to note the basic differences between gasification and combustion—notably the promoting of a more complex role of moisture, the different methods for consuming solid carbon and converting it to a useful gaseous form, and the use of secondary gas-phase reactions to influence the composition of the final product—a combustible gaseous fuel. Because of the role of moisture and the reactions to consume or adjust the gaseous form of carbon, blending of different coals can influence these processes as well.

3.4. COALS USED IN COMMERCIAL APPLICATIONS AND THEIR BLENDING POTENTIAL

Treating all coals by the framework "It's black and it burns" is insufficient in the current economic and regulatory framework. There are many ranks of coal, as shown in Table 3.3. Further, there are many different deposits of the same rank of coal, with differing characteristics.

3.4.1. Characteristics of Various Commercially Significant Coals

Many commercially significant coals exist in the United States and throughout the world. Tables 3.4 through 3.11 summarize representative values for some of

TABLE 3.4 Representative Lignite Values

Parameter	Coal			
Mine	Sandow	San Miguel	So. Beulah	Indianhead
State	Texas	Texas	N. Dakota	N. Dakota
Proximate Analysis (wt% total basis)				
Moisture	20.34	17.99	28.40	30.85
Ash	19.24	22.44	15.90	8.20
Volatile Matter	30.93	32.24	26.50	28.80
Fixed Carbon	29.49	27.31	29.20	33.86
Calorific Value (Btu/lb)	7597	7497	6420	7235
Ultimate Analysis (wt% dry basis)				
Carbon	44.03	42.89	40.20	43.95
Hydrogen	5.52	5.25	5.60	6.25
Oxygen	29.18	26.60	36.50	42.15
Nitrogen	0.79	0.59	0.70	0.70
Sulfur	1.24	2.22	0.73	0.45
Chlorine (ppm)	110	100	0	0
Ash	21.40	24.11	19.10	10.35
Ash Elemental Analysis (wt% dry ash basis)				
SiO_2	44.00	53.00	30.00	22.00
Al_2O_3	25.00	15.00	11.00	9.50
TiO_2	1.40	0.71	0.00	0.00
Fe_2O_3	32.10	2.70	6.50	11.00
CaO	7.10	7.30	21.00	18.00
MgO	3.50	0.41	3.50	5.00
K_2O	0.37	1.50	0.18	0.07
Na_2O	0.20	4.80	3.60	3.40
P_2O_5	0.48	0.04	0.06	0.65
SO_3	13.00	11.00	16.00	25.00
Base/Acid Ratio	0.615	0.243	0.848	1.190
Fe_2O_3/CaO Ratio	4.521	0.370	0.310	0.611

Source: [24].

TABLE 3.5 Some Northern Powder River Basin Coals

Parameter	Coal		
Mine	Decker	Absoloka	Spring Creek
Proximate Analysis (wt% total basis)			
Moisture	22.61	24.17	26.45
Ash	4.22	9.51	4.29
Volatile Matter	31.78	30.31	27.13
Fixed Carbon	41.39	36.01	42.12
Calorific Value (Btu/lb)	9640	8700	9042
Ultimate Analysis (wt% dry basis)			
Carbon	71.57	65.79	71.57
Hydrogen	4.92	4.40	4.92
Oxygen	16.60	15.51	16.11
Nitrogen	1.16	0.87	1.16
Sulfur	0.30	0.88	0.39
Chlorine (ppm)	12	23	90
Ash	5.45	12.54	5.83
Ash Elemental Analysis (wt% dry ash basis)			
SiO_2	36.79	34.81	35.45
Al_2O_3	16.08	18.21	17.70
TiO_2	1.01	0.75	1.23
Fe_2O_3	5.11	4.83	5.29
CaO	14.40	20.48	14.30
MgO	3.79	2.71	3.75
K_2O	0.77	0.67	0.96
Na_2O	8.83	2.17	6.75
P_2O_5	0.28	0.58	0.67
SO_3	12.30	12.97	11.70
Base/Acid Ratio	0.611	0.574	0.571
Fe_2O_3/CaO Ratio	0.355	0.236	0.370

Source: [15].

TABLE 3.6 Representative Values for Some Southern PRB Coals

Parameter	Coal			
Mine	Antelope	Rochelle	Black Thunder	Caballo Rojo
Proximate Analysis (wt% total basis)				
Moisture	27.42	21.39	25.83	30.11
Ash	4.20	4.24	5.05	4.93
Volatile Matter	31.73	32.92	31.56	30.44
Fixed Carbon	36.64	36.45	37.55	34.53
Calorific Value (Btu/lb)	8730	8817	8943	8455
Ultimate Analysis (wt% dry basis)				
Carbon	74.45	74.13	69.96	69.45
Hydrogen	4.85	4.87	4.79	5.12
Oxygen	13.80	14.05	17.22	16.99
Nitrogen	0.85	0.93	0.91	0.94
Sulfur	0.25	0.27	0.31	0.44
Chlorine (ppm)	na	na	na	na
Ash	5.79	5.76	6.81	7.05
Ash Elemental Analysis (wt% dry ash basis)				
SiO_2	31.84	30.70	32.76	30.2
Al_2O_3	15.63	15.43	16.20	15.9
TiO_2	1.35	1.32	1.26	1.40
Fe_2O_3	5.49	5.61	6.02	4.98
CaO	26.16	25.47	22.84	22.30
MgO	6.03	5.80	5.22	4.23
K_2O	0.29	0.30	0.56	1.10
Na_2O	2.54	2.36	2.44	1.62
P_2O_5	1.58	2.15	1.57	0.83
SO_3	7.32	6.49	10.10	14.50
Base/Acid Ratio	0.830	0.833	0.738	0.721
Fe_2O_3/CaO Ratio	0.210	0.220	0.264	0.223

Source: [15].

TABLE 3.7 Representative Values for Selected Northern Appalachian Coals

Parameter	Coal			
Mine	Bailey	Blacksville	Federal 2	Mine 84
Proximate Analysis (wt% total basis)				
Moisture	4.52	5.50	6.55	5.84
Ash	6.78	7.79	7.11	7.06
Volatile Matter	34.31	35.79	33.93	37.72
Fixed Carbon	54.39	50.93	52.41	44.39
Calorific Value (Btu/lb)	13,376	13,192	13,209	13,195
Ultimate Analysis (wt% dry basis)				
Carbon	78.82	77.07	77.41	78.96
Hydrogen	5.24	5.19	5.16	5.19
Oxygen	5.17	54.33	6.11	5.05
Nitrogen	1.52	1.48	1.54	1.66
Sulfur	2.15	2.69	2.17	1.64
Chlorine (ppm)	600	1480	1000	1100
Ash	7.10	8.24	7.61	7.50
Ash Elemental Analysis (wt% dry ash basis)				
SiO_2	50.52	42.31	45.86	48.00
Al_2O_3	23.10	21.94	23.01	27.30
TiO_2	0.93	0.92	1.03	1.00
Fe_2O_3	15.63	21.76	12.98	14.90
CaO	2.64	5.08	5.00	2.54
MgO	0.78	0.89	1.08	0.75
K_2O	1.94	1.39	1.33	0.90
Na_2O	0.56	0.90	1.32	0.52
P_2O_5	0.35	0.44	0.64	0.27
SO_3	1.25	4.65	4.57	0.95
Base/Acid Ratio	0.289	0.461	0.311	0.257
Fe_2O_3/CaO Ratio	5.920	4.283	2.596	5.866

Source: [15].

TABLE 3.8 Representative Values for Selected Central Appalachian Coals

Parameter	Coal			
Mine	Long Fork	Premier Elkhorn	Elk Run	Sequoia
Proximate Analysis (wt% total basis)				
Moisture	7.95	5.12	5.44	2.11
Ash	8.64	7.19	9.15	6.10
Volatile Matter	32.67	34.21	29.42	35.63
Fixed Carbon	50.74	53.48	55.98	56.17
Calorific Value (Btu/lb)	12,525	13,142	12,944	13,859
Ultimate Analysis (wt% dry basis)				
Carbon	74.44	77.23	77.33	77.25
Hydrogen	5.06	3.98	4.83	5.10
Oxygen	8.03	8.03	5.83	8.61
Nitrogen	1.63	1.57	1.45	1.87
Sulfur	1.46	1.49	0.77	0.92
Chlorine (ppm)	NA	1240	1160	230
Ash	9.39	7.58	9.68	6.23
Ash Elemental Analysis (wt% dry ash basis)				
SiO_2	49.20	47.50	53.02	48.58
Al_2O_3	30.21	26.09	32.08	30.30
TiO_2	1.46	1.30	1.35	1.06
Fe_2O_3	10.99	11.54	4.43	8.10
CaO	3.55	2.87	0.86	2.67
MgO	1.13	1.17	0.94	1.04
K_2O	2.59	2.10	3.38	2.41
Na_2O	0.76	0.60	0.36	0.34
P_2O_5	0.29	0.68	0.32	1.30
SO_3	3.16	1.58	0.57	2.07
Base/Acid Ratio	0.235	0.244	0.115	0.182
Fe_2O_3/CaO Ratio	3.096	4.021	5.151	3.034

Source: [15, 24].

TABLE 3.9 Representative Values for Selected Illinois Basin Coals

Parameter	Coal			
Mine	Cottage Grove	Prosperity	Willow Lake	Riola
Proximate Analysis (wt% total basis)				
Moisture	9.00	15.16	9.20	16.35
Ash	6.20	5.38	8.20	8.00
Volatile Matter	35.50	33.79	36.00	33.75
Fixed Carbon	49.30	45.67	46.60	41.90
Calorific Value (Btu/lb)	12,480	11,647	12,171	10,882
Ultimate Analysis (wt% dry basis)				
Carbon	76.40	76.24	73.40	75.05
Hydrogen	5.10	5.33	5.10	4.98
Oxygen	7.11	8.20	7.80	6.81
Nitrogen	1.50	1.72	1.50	1.48
Sulfur	2.43	2.17	3.17	2.12
Chlorine (ppm)	NA	NA	NA	NA
Ash	6.81	6.34	9.03	9.56
Ash Elemental Analysis (wt% dry ash basis)				
SiO_2	49.50	49.14	49.30	49.04
Al_2O_3	20.00	21.70	17.80	22.90
TiO_2	1.00	1.18	1.00	1.10
Fe_2O_3	18.60	18.40	17.70	15.20
CaO	4.00	1.62	4.80	4.46
MgO	0.80	0.80	0.80	0.68
K_2O	2.20	2.38	2.20	1.82
Na_2O	0.40	0.84	0.70	1.40
P_2O_5	0.10	0.42	0.10	0.15
SO_3	3.10	0.09	2.70	2.50
Base/Acid Ratio	0.369	0.334	0.385	0.323
Fe_2O_3/CaO Ratio	4.650	11.358	3.688	3.408

Source: [15, 24].

TABLE 3.10 Representative Values for Some Western Bituminous Coals

Parameter	Coal			
Mine	ColoWyo	Seneca	Sunnyside	Deseret
State	Colorado	Colorado	Utah	Utah
Proximate Analysis (wt% total basis)				
Moisture	13.76	8.24	3.00	5.20
Ash	4.12	7.99	5.50	6.90
Volatile Matter	34.27	39.04	39.00	42.30
Fixed Carbon	47.83	44.71	52.50	45.60
Calorific Value (Btu/lb)	10,947	11,360	13,370	12,520
Ultimate Analysis (wt% dry basis)				
Carbon	63.13	63.09	75.20	70.30
Hydrogen	5.88	5.60	5.50	5.60
Oxygen	24.98	18.89	11.00	15.30
Nitrogen	1.50	1.57	1.60	1.40
Sulfur	0.30	2.82	1.20	0.50
Chlorine (ppm)	70	79	0	0
Ash	4.42	8.43	6.00	7.70
Ash Elemental Analysis (wt% dry ash basis)				
SiO_2	35.00	34.00	44.10	43.00
Al_2O_3	23.00	12.00	20.00	20.00
TiO_2	0.93	0.64	0.99	1.10
Fe_2O_3	4.50	37.00	19.00	3.50
CaO	13.00	4.21	2.50	10.00
MgO	2.00	0.72	0.71	0.59
K_2O	0.24	0.27	0.19	0.54
Na_2O	1.00	0.21	0.79	2.60
P_2O_5	2.70	0.29	0.26	0.65
SO_3	3.90	8.60	3.50	3.10
Base/Acid Ratio	0.352	0.909	0.356	0.269
Fe_2O_3/CaO Ratio	0.346	8.789	7.600	0.350

Source: [24].

TABLE 3.11 Representative Values for Some International Coals

Parameter	Coal			
Country	China	Indonesia	Australia	South Africa
Proximate Analysis (wt% total basis)				
Moisture	0.57	17.84	6.98	4.30
Ash	15.42	3.82	22.34	9.67
Volatile Matter	9.07	37.23	23.09	33.78
Fixed Carbon	70.94	41.11	41.24	48.14
Calorific Value (Btu/lb)	12,067	10,003	9660	12,170
Ultimate Analysis (wt% dry basis)				
Carbon	79.37	71.41	56.00	69.70
Hydrogen	2.75	4.91	3.50	4.50
Oxygen	0.19	17.42	7.43	9.10
Nitrogen	1.11	1.44	1.22	1.60
Sulfur	0.42	0.17	0.35	0.70
Chlorine (ppm)	NA	130	NA	NA
Ash	16.16	4.65	24.00	10.10
Ash Elemental Analysis (wt% dry ash basis)				
SiO_2	47.60	42.84	57.90	44.00
Al_2O_3	33.01	16.30	32.80	32.70
TiO_2	1.26	0.62	1.00	1.20
Fe_2O_3	3.33	12.86	6.20	4.60
CaO	5.96	10.75	0.60	5.70
MgO	1.02	7.34	0.80	1.30
K_2O	0.96	1.59	0.50	0.30
Na_2O	1.13	0.29	0.10	0.10
P_2O_5	0.51	0.37	NA	2.20
SO_3	1.75	5.71	0.80	4.60
Base/Acid Ratio	0.151	0.549	0.089	0.154
Fe_2O_3/CaO Ratio	0.559	1.196	10.333	0.807

Sources: [12, 15].

these coals [12, 15, 18]. Note that these are representative values only. Many values are covered in the literature. There is inherent variability in coal, and that variability is in mines, in seams, and in regions of coal production.

What also becomes significant is that the values shown in Tables 3.4 through 3.11 are representative values that are used to characterize the output of one or another source of coal. However, coal exhibits variability in a coal region, in a given coal seam, and in a given coal mine [24]. Tables 3.12 and 3.13 show the typical variability measured for a selected southern PRB coal used as steam coal for electricity generation by DTE Energy: Black Thunder coal [15, 16, 18]. The coals shown in these tables are only examples. Note that the representative values shown for Black Thunder coal are within the range of but are not identical to the average of the nine Black Thunder samples shown in Tables 3.12 and 3.13. Note also that only four examples of international coals are shown in Table 3.11. In reality, a vast number of such coals could be displayed.

3.4.2. Relationship of Chemical Composition to Petrography

The blending of coals began with coke making for the steel industry, as discussed in Chapter 1. Petrography emerged as the science driving this blending. Petrography classified coals according to concentrations of macerals—vitrinite, exinite, liptinite, and inertinite—and numerous subcategories of macerals [2]. There is significant linkage between the study of macerals and the analysis of steam coals, and it can contribute to the analysis of steam coal blends. This linkage involves general combustion parameters, combustion processes, and environmental consequences of combustion [25–29].

Osorio et al. [25] evaluated a Brazilian subbituminous coal—Faxinal coal, with a reflectance of 0.48%—singly and in blends for pulverized fuel injection into blast furnaces, linking combustion and steel making. In blends with less reactive imported coals (reflectances of 1.62 and 1.68), the Faxinal coal disproportionately improved the reactivity of the blend and facilitated combustion processes. Helle et al. [26] related maceral composition of coals and coal blends to carbon conversion to CO_2 and to unburned carbon in the flyash.

Su, Pohl, Holcombe, and Hart [27] developed an index to predict the combustion behavior of various coals and to evaluate the fouling propensities of various coals and coal blends. The maceral index (MI) predicting the behavior of coals and coal blends calculates burn-out percentage, or carbon conversion to CO_2, as follows:

$$MI = \left(\frac{L + \frac{V}{R^2}}{I^{1.25}} \right) \left(\frac{HV}{30} \right)^{2.5} \tag{3.14}$$

TABLE 3.12 Variability Measured in Black Thunder Coal (typical values)

Parameter	Black Thunder Coal			
Statistical Value	Mean	Std. Dev.	Maximum	Minimum
Proximate Analysis (wt% total basis)				
Moisture	26.87	0.69	28.04	25.78
Ash	5.12	0.31	5.49	4.73
Volatile Matter	33.84	1.19	36.52	32.65
Fixed Carbon	34.36	1.27	35.40	31.61
Calorific Value (Btu/lb)	8844	132	9044	8638
Ultimate Analysis (wt% dry basis)				
Carbon	68.37	1.58	70.79	66.72
Hydrogen	5.12	0.39	5.56	4.34
Oxygen	18.13	1.50	19.94	16.21
Nitrogen	0.94	0.10	1.06	0.79
Sulfur	0.46	0.03	0.53	0.42
Chlorine (ppm)	400	500	1300	100
Ash	6.96	0.46	7.52	6.51
Ash Elemental Analysis (wt% dry ash basis)				
SiO_2	32.85	2.18	34.80	28.90
Al_2O_3	16.69	1.50	19.50	15.20
TiO_2	1.36	0.44	2.20	1.09
Fe_2O_3	4.89	0.50	5.80	4.25
CaO	22.82	4.14	30.00	18.90
MgO	5.56	2.12	9.60	3.88
K_2O	0.70	0.62	2.16	0.10
Na_2O	1.21	0.49	1.95	0.50
P_2O_5	1.11	0.20	1.53	0.91
SO_3	14.00	2.20	16.50	10.50
Base/Acid Ratio	0.691	0.110	0.880	0.560

Source: [15].

TABLE 3.13 Ash Fusion Temperature Variability for Black Thunder Coal

Parameter		Black Thunder Coal		
Statistical Value	Mean	Std. Dev.	Maximum	Minimum
Ash Fusion Temperature (°F, reducing)				
Initial	2080	53	2170	1989
Softening	2151	27	2200	2119
Hemispherical	2171	30	2230	2133
Fluid	2225	68	2373	2143
Ash Fusion Temperature (°F, oxidizing)				
Initial	2133	43	2210	2090
Softening	2190	27	2250	2165
Hemispherical	2204	30	2270	2178
Fluid	2256	27	2310	2229

Source: [15].

or

$$MI = (HVF)^{2.5}(RF) \tag{3.15}$$

$$HVF = \frac{HV}{30} \tag{3.16}$$

$$RF = \frac{L + \dfrac{V}{R^2}}{I^{1.25}} \tag{3.17}$$

where MI is the maceral index number; L is the percent liptinite on a volumetric, mineral-free basis normalized to 100%; V is the percent vitrinite on a volumetric, mineral-free basis normalized to 100%; I is the percent inertinite on a volumetric, mineral-free basis normalized to 100%; R is the maximum vitrinite reflectance; and HV is the heating value of the coal or coal blend on an air-dried basis, measured in MJ/kg. HVF is the heating value factor, and RF is the reflectance factor. The fundamental principles behind this index are that liptinite is very volatile and very reactive, and it combusts extremely rapidly.

Vitrinite ignition and combustion are functions of its reflectance. Inertinite is, by and large, difficult to burn.

Blends of coal can be calculated using the weighted average of the macerals; however, the preceding equations, particularly (3.14) and (3.15), demonstrate that the consequences are not linear. Index values above 3 generally indicate >99% carbon conversion to CO_2, while index values above 4 indicate 99.5% carbon conversion to CO_2. The index does not demonstrate increased carbon conversion above 99.8%. Su et al. [27] utilized extensive data sets from EER Corporation and from ACIRL. The r^2 for the EER data was 0.982, and the r^2 for the ACIRL data was 0.8908. This is a robust set of equations.

Goodarzi [28] used petrography of subbituminous coals as an indicator for the ability of the coal alone to capture mercury in the flyash. Using Canadian coals from the Alberta province, Goodarzi documented that coals and coal blends with relatively high concentrations of inertinite produce more unburned carbon than those coals or blends with low concentrations of inertinite. The unburned carbon caused by the inertinite then captures mercury in a manner essentially identical to activated carbon that can be injected into a boiler.

In conclusion, petrography—and the macerals—provides overall guidance concerning the general combustion behavior of coals and coal blends. The macerals provide some insights into coal reactivity and the completion of the carbon conversion reaction (to CO_2). Further, they provide insights into specific reaction mechanisms such as the capture of mercury. The use of macerals to evaluate behavior of coal blends in combustion and gasification systems, however, provides limited insights into the specifics of coal blend reactivity and behavior. Those insights come more from the assessment of blends through kinetics.

3.4.3. Chemical Composition and Calorific Value

The chemical composition, reflected in the proximate and ultimate analysis, and the calorific value calculated on either a higher heating value (HHV) or a lower heating value (LHV) basis are essentially linear or additive properties. These properties reflect the weighted average of the composition of the fuels being blended. Consequently, blends can be created from the coals shown earlier in Tables 3.4 through 3.11. This linearity or additivity has been shown by numerous researchers, including Haas, Tamura, and Weber [29] and many others (see, for example, [3, 10, 11, 16]). Such additivity or linearity permits the practical blending processes that are installed in many coal yards today. Such processes depend on weigh-belt feeders, variable speed ploughs, and other similar devices. Weigh-belt feeders, connected to controls, provide the mechanisms for controlling the composition of the blend, as will be discussed later in this chapter.

3.5. KINETICS AND THE ANALYSIS OF COAL BLEND REACTIVITY

Miller [30], in an excellent review of combustion chemistry, shows that the overall rate of combustion can be expressed by the following general equation:

$$q = k_c = A\exp\left(\frac{-E}{RT_p}\right)P_s^m \tag{3.18}$$

where A is the true preexponential constant $(kg/C/m^2s(atm\ O_2)^{-m})$, E is the true activation energy (J/mol), R is the universal gas constant, T is the true particle temperature (K), P is the partial pressure of oxygen at the surface of the particle, and m is the true reaction order.

Kinetics introduces overall concepts of coal and coal blend reactivity. Reactivity includes not only pyrolysis or devolatilization but also ignition and combustion performance. This includes char oxidation kinetics. For certain overall combustion mechanisms such as devolatilization and char oxidation, the Arrhenius equation is sufficient for the evaluation of coals and coal blends. This was introduced in Chapter 2, with commentary on activation energies measured for certain coals and associated coal blends. In the subsequent discussion, this concept for determining coal and coal blend reactivity is explored in considerably more detail.

3.5.1. Devolatilization Kinetics

Devolatilization kinetics provides key insights into the reactivity of coals and coal blends and offers key information concerning the fact that blending cannot be considered linear with respect to combustion behavior—particularly with respect to blends of dissimilar coals. Blends involving the highly reactive, oxygenated PRB coals with bituminous coals are particularly susceptible to nonlinear behavior. Further, the reactive coals impart disproportionate reactivity to the blended fuel.

DTE Energy contracted with The Energy Institute of Pennsylvania State University to conduct a study of blended coal reactivity, focusing on both devolatilization and char oxidation [31]. This study resulted in numerous subsequent published analyses (e.g., [17, 32–35]). The blends studied were based on the blended fuels burned at Monroe Power Plant: Antelope (Southern PRB) and Long Fork (Central Appalachian) coals. Blends tested included 60% Antelope coal and 70% Antelope coal on a mass basis—or 54% and 65% Antelope coal on a calorific value basis. The parent coals were also tested. The Antelope coal, and the 54% blend, exhibited 2-stage pyrolysis characteristics with somewhat more reactive activation energies at the low temperature; both the coal and the blend could also be represented by a single Arrhenius-type equation. The 65% blend and the Long Fork coal exhibited single-stage pyrolysis characteristics.

TABLE 3.14 Kinetic Parameters for Devolatilization of Antelope, Long Fork, and Blended Coals

	Temperature Range	A (1/sec)	E (kcal/mol)
Antelope	600–1600°C	6.46	4.72
Antelope Low	600–1200°C	4.08	3.78
Antelope High	1200–1600°C	17.4	7.98
Long Fork	600–1600°C	6.03	4.43
65% Blend	600–1600°C	4.88	3.33
54% Blend	600–1600°C	5.54	3.96
54% Blend Low	600–1000°C	2.52	2.41
54% Blend High	1000–1600°C	9.03	5.47

Source: [31].

Table 3.14 summarizes the preexponential constants and activation energies measured by The Energy Institute, and Figures 3.8 through 3.11 display the kinetics as measured, using single-stage pyrolysis as the basis.

A strong approach to evaluating the relative devolatilization reactivities has been shown by Bradley et al. [36]. That approach involves calculating the reactivity constants as a function of temperature and plotting those constants on a comparative basis. This approach is shown in Figure 3.12 for the Antelope, Long Fork, and blended coals. Note that this figure shows the 65% blend to be more reactive than either parent coal, while the 54% blend conforms to the profile of the more reactive Antelope coal.

It should be noted that this study involved coals that are quite dissimilar from one another, as shown earlier in Tables 3.6 and 3.8. The Antelope coal, further, is highly oxygenated, as shown in Table 3.6. It is believed that the reactivity is caused by radicals and fragments caused by pyrolysis of the Antelope coal (e.g., OH, H·, CH_3) attacking the char structure of the Long Fork coal, creating more volatile matter than would otherwise be created. As was noted in Chapter 2, similar reactivity differences existed in a study of Texas Lignite and Buckskin PRB coal blends [32]. Like the 54% Antelope/Long Fork blend, the Buckskin/Texas Lignite blend shape conformed to the more reactive PRB coal rather than the somewhat less reactive lignite. This reactivity also manifested itself in nitrogen evolution, as noted in Chapter 2. Figure 3.13 shows the relative reactivities of the Buckskin, Texas Lignite, and blended coal.

It is important to note that the preceding data relate to blending dissimilar coals, with particular emphasis on highly reactive subbituminous coals. The

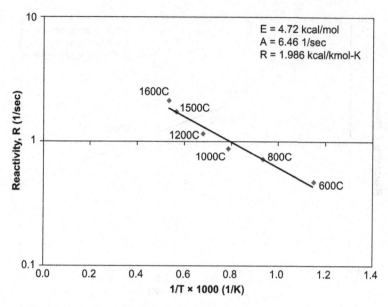

FIGURE 3.8 Devolatilization kinetics of Antelope (PRB) coal. *Source: [31].*

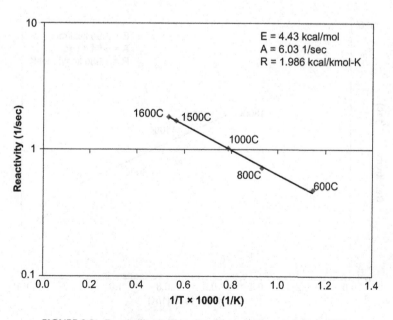

FIGURE 3.9 Devolatilization kinetics of Long Fork coal. *Source: [31].*

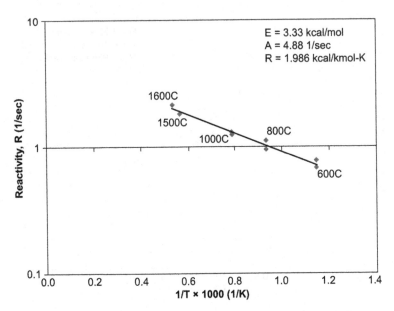

FIGURE 3.10 Devolatilization kinetics of 65% blended coal. *Source: [31].*

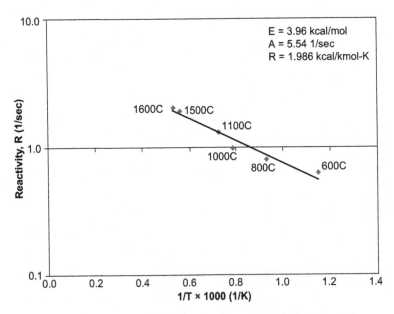

FIGURE 3.11 Devolatilization kinetics of 54% blended coal. *Source: [31].*

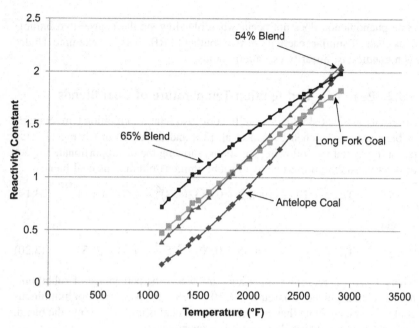

FIGURE 3.12 Comparative reactivity constants for devolatilization of Antelope, Long Fork, and blended coals. *Source: Adapted from [31].*

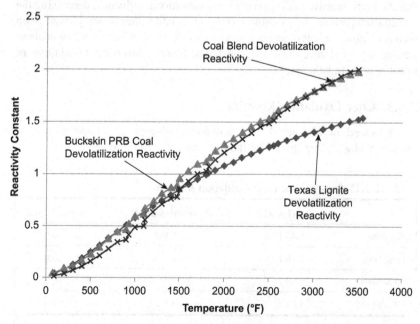

FIGURE 3.13 Devolatilization reactivities of Buckskin (PRB), Texas Lignite, and a 70/30 blend of the two coals. *Source: [32].*

same phenomenon does not occur when blending similar coals—two bituminous coals of similar reactivity or two southern PRB coals, for example. Under such conditions, linearity essentially holds.

3.5.2. Reactivity and Ignition Temperature of Coal Blends

Reactivity associated with devolatilization also relates to ignition temperatures of blends, as researched by Chi et al. [37] and discussed in Chapter 2. This research yielded the following equations, showing the disproportionate impact of reactive subbituminous coals in subbituminous-bituminous coal blends:

$$T_{ignition}(°F) = 1048.2 + 0.0132(\%S)^2 - 2.5976(\%S) \quad (3.19)$$

and

$$T_{ignition}(°C) = 564.35 + 0.0073(\%S)^2 - 1.4431(\%S) \quad (3.20)$$

where $T_{ignition}$ is ignition temperature and %S is the percentage of subbituminous coal in the blend. Consequently, 100 – %S is the percentage of bituminous coal in the blend. Note that as subbituminous coal is introduced into the blend, the ignition temperature decreases disproportionately.

Faundez et al. [38] found similar results to those of Chi et al. [37]. Faundez et al. measured ignition temperatures from blends of dissimilar coals and found that the more reactive coals exerted a disproportional influence, decreasing the ignition temperature of subbituminous–low-volatile bituminous blends. Such blends exhibited a heterogeneous ignition mechanism. When blending coals of similar rank and reactivity, ignition temperatures conformed to additive or linear averaging behavior.

3.5.3. Char Oxidation Kinetics

As discussed in Chapter 2, the activation energies associated with char oxidation kinetics are driven by the char from the more reactive coal. Again

TABLE 3.15 Summary of Char Oxidation Kinetics

	A (1/sec)	A (g/cm²-s-atm)	E (kcal/mol)
Antelope	1.45 E+3	1.37 E-2	24.9
Long Fork	1.36 E+5	1.37	30.8
65% Blend	2.08 E+5	1.69	26.0
54% Blend	3.65 E+3	2.85 E-2	25.0

Source: [31].

using the experiments conducted by The Energy Institute of Pennsylvania State University for DTE Energy, this can be seen. Further, the overall kinetic parameters are shown in Table 3.15. Note that in Table 3.14 the preexponential constant is also reported on a surface area basis (g/cm^2-s), using the surface area of the starting char sample (i.e., 1700°C char except for the Long Fork sample, where the 1600°C char was used) [31]. Char oxidation kinetics, as a whole, are driven by the char from the more reactive coal.

3.6. THE BEHAVIOR OF INORGANIC CONSTITUENTS

Inorganic constituents in coal are the source of slagging and fouling and much of the corrosion that occurs in coal-fired utility boilers. Harding and O'Connor [39] performed a detailed study of fuel problems of electric utilities cited by Barnes [40] and found that slagging and fouling were the dominant fuel problems among a list of 16 potential problems. Slagging and fouling caused more lost hours of generation than any other fuel-related problem.

Ash deposition costs U.S. utilities $943 million annually, according to Harding and O'Connor [39]; the causes and costs have not been reduced by fuel blending, yet over half of the coal-fired plants in the United States use some form of blending to achieve regulatory compliance and cost management. This blending occurs despite its implications for slagging and fouling. Consequently, the behavior of inorganic constituents in coal blends becomes a critical concern.

The combustion and gasification behavior of inorganic constituents in coal blends is almost universally found to be nonadditive or nonlinear. Gupta [41] provides several inferences concerning this phenomenon. Research by Su, Pohl, and Holcombe [42] provides similar insights in considering blends of Wyoming PRB and Oklahoma coals, and blends of PRB and New Mexico coals, and tests on nine component coals and five associated coal blends. Other researchers have found similar nonadditive or nonlinear behavior (see, for example, [11, 16, 43–49]). Much of this nonlinearity stems from the following factors:

- The exact compound in the coal containing specific minerals, as discussed in Chapter 2 (see also [10])
- The presence of both inherent and adventitious mineral matter in the coal (e.g., minerals contained in overburden, partings, and other similar materials that are supplied to the combustion or gasification unit with the organic matter of coal)
- The interaction between inorganic constituents, either in the liquid phase as slag or fouling material or in the gas phase resulting from high in-furnace temperatures (see particularly [44]), with such inorganic constituents coming from different coals in the blends
- The interaction between inorganic constituents and organic constituents in either the homogeneous gas-phase reactions or the heterogeneous gas–solids

reactions, with the organic constituents and inorganic constituents coming from different coals in the blend (e.g., reactions of sulfur displacing chlorine in alkali metal-chloride deposits in the heat recovery area, as discussed by Baxter [22, 23] and Tillman, Duong, and Miller [21]

These reactions may be favorable (e.g., mercury capture facilitated by oxidation with chlorine) or unfavorable (e.g., Fe_2O_3-CaO fluxing interactions to depress ash fusion and subsequent slagging temperatures).

Many of the mechanism principles for blends of coal ash formation and transformation have been discussed in Chapter 2, since these mechanisms are somewhat independent of the fuel source (coal, lignite, peat, biomass, waste). The mechanisms of deposition—slag and fouling formation—are most commonly associated with coal and coal blends due to the size of much of the combustion equipment.

3.6.1. Slagging and Blended Coals

Slagging is associated with deposition in the waterwall section of the boiler and those elements of the boiler that "see" the flames and associated radiant energy. Given that the inorganic matter may be inherent to the fuel or may exist as adventitious matter, multiple pathways exist for these components. Inorganic matter inherent to the fuel and bound to the organic components of the fuel will release during the combustion process typically as vapor. This is particularly the case because flame temperatures in large pulverized boilers have been measured at >2900°F (>1600°C). The fine particles (typically <200 mesh) undergo very rapid heating, commonly estimated at 10^4 to $10^{5\circ}$C/sec. The adventitious mineral matter may fracture into very small particles due to thermal shock and may also enter either a melt or a vapor phase. As these materials enter the vicinity of colder environments (e.g., near waterwall or pendent tubes), they may condense and deposit on such surfaces [40].

Barnes [40] puts the general radiative heat transfer equation for the furnace section of the boiler as follows:

$$Q = \sigma\epsilon\, A(T_f^4 - T_w^4) \tag{3.21}$$

where σ is a constant (Stefan-Boltzman constant), ϵ is the wall emissivity, T_f is the local flame temperature, and T_w is the wall or tube temperature. For the most part, the emissivity factor for clean waterwall tubes is approximately 0.7 or 0.8; if the tubes are covered with a light amount of nonreflective ash, the emissivity factor does not degrade significantly.

However, when coated with reflective ash, as is typical of PRB coals with high concentrations of calcium and magnesium (refer to Tables 3.6 and 3.7), the emissivity factor can deteriorate to about 0.45 to 0.55, even when the slag coating on the waterwall or pendent tubes is quite thin. Blending PRB coals, or some Western bituminous coals, with Eastern coals for fuel cost savings can result in this decrease in emissivity and result in both lower heat transfer in the waterwall section of the boiler and higher FEGT values.

The Barnes monograph [40] provides the key demonstration of the problems associated with blending a coal with a high iron ash (e.g., 15–25% Fe_2O_3) with a high calcium ash (e.g. 15–30% CaO). And these are the typical Eastern bituminous/PRB blends or the interior province (e.g., Illinois basin)/PRB blends. These are among the most common blends in utility power plants; the bituminous coal brings the Btu, while the PRB brings the reactivity, the low sulfur and nitrogen, and the low cost in \$/10^6 Btu. However, the iron oxide/calcium oxide blends are among the most prominent eutectics formed in the blending of coals and, specifically, the blending of coal inorganics.

Figures 3.14 and 3.15, from the American Ceramic Society/National Bureau of Standards work (see [50]), illustrate this eutectic through the ternary diagrams with silica. Practical research was conducted at the Monroe Power

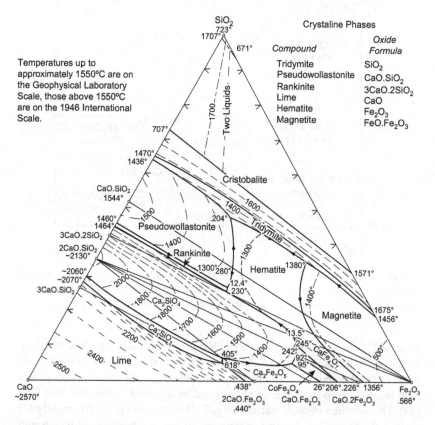

FIGURE 3.14 The CaO-Fe_2O_3-SiO_2 system. Note that one melt temperature is 1192°C or 2180°F. Such temperatures are common at or near the furnace exit of a boiler. *Source: [50]. Reprinted with permission of the American Ceramic Society, www.ceramics.org. All rights reserved.*

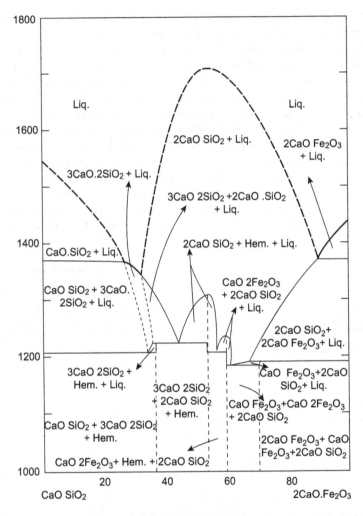

FIGURE 3.15 A second look at the CaO-Fe₂O₃-SiO₂ system. Note that although slagging temperatures may be high, they can be <2200°F (1200°C). *Source: [50]. Reprinted with permission of the American Ceramic Society, www.ceramics.org. All rights reserved.*

Plant of DTE Energy, where southern PRB coal—typically Antelope or Black Thunder—is blended with Central Appalachian coal—typically Long Fork or a like high-Btu/medium-sulfur product. The research at Monroe was consistent with results shown by Singer et al. [19] and by Su, Pohl, Holcombe, and Hart [49]. This research showed that an Fe_2O_3/CaO molar ratio of approximately 1.0 yielded the lowest ash fusion temperatures of any ash products of blends and that the curves rapidly rose as the ratio diverged from unity on either the <1.0 or >1.0 sides of a plotted curve.

Many coal-fired utility boilers blend high-iron bituminous coals with PRB coals for economic reasons. Slags are formed, as shown in Figures 3.16 and 3.17. Both are from plants firing a blend of bituminous and PRB coal, weighted heavily toward the PRB coal (and the calcium). However, as can be deduced from Tables 3.6 through 3.9 shown previously, the bituminous coals tend to have higher concentrations of ash than the southern PRB coals. Consequently blends of 60% to 70% PRB can generate Fe_2O_3/CaO ratios approaching 1.0.

Research published by Rushdi, Sharma, and Gupta [46] also addresses the nonlinearity of slag deposits as a function of fuel blending. This research shows that the composition and behavior of the slags are a consequence of elemental composition and temperature; elemental composition is a function of blending. Further, this research shows that the interaction among elements and compounds occurs both in the gas stream, as previously noted, and in the slag deposit on the tube itself.

This research is consistent with the findings of Baxter et al. [22, 23] concerning slag deposits—and the ability of sulfur to displace chlorine in alkali metal deposits on tubes. If the initial deposit contains potassium chloride (KCl) or sodium chloride (NaCl), in the presence of sufficient sulfur in the gas phase, the sulfur will address the deposit and displace the chlorine to convert the deposit to potassium or sodium sulfate (K_2SO_4, Na_2SO_4) and release the chlorine as hydrogen chloride (HCl). Rushdi, Sharma, and Gupta [46] found

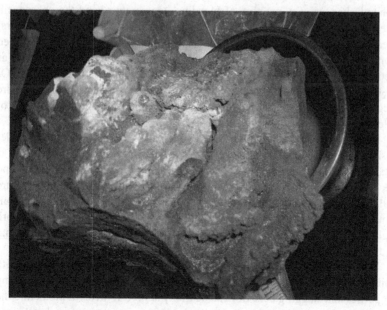

FIGURE 3.16 Slag produced by an iron oxide/calcium oxide eutectic at the Monroe Power Plant. *(Photograph by A. Widenman)*

FIGURE 3.17   Slag from a blend of bituminous coal and PRB coal frozen on the floor of a cyclone boiler. *(Photograph by K. Letheby)*

this interaction to go beyond the chlorine-sulfur phenomenon to include the minerals in the inorganic constituents of the coals as a whole.

Qiu et al. [45] studied blends of Chinese bituminous coals, with emphasis on the silica-alumina-calcium oxide system. Their research showed significant mineral changes as a function of blending, temperature, and the consequent interactions of minerals from each coal source. Significant changes in mineral species occurred with changing blend ratios. The ternary diagram published from their research shows a significant low-temperature eutectic region as a consequence of blending in the region of equal distribution of the SiO_2-CaO-Al_2O_3, and that eutectic region is favored because more calcium and silica are incorporated into the blend. The distinct nonlinearity is a function not only of calcium-iron eutectics but also of other mineral systems as well.

Kondratiev and Jak [47] provide an analysis of slagging behavior under gasification conditions for the Al_2O_3-CaO-FeO-SiO_2 system; iron oxide is treated as FeO because the analysis is for a gasification system. Kondratiev and Jak [47] developed mathematical models showing the impact of temperature and ash/slag composition, including totally liquid phase slags and slags containing both liquids and partially crystallized solid phase materials. Their work highlights the compositional issues, including the consequences of blending, and temperature considerations.

What becomes clear from the research just cited and from experience with boilers firing blended fuels is that predicting the effects of blends on slagging properties and slagging behavior can be tricky at best. It involves predicting the interactions between individual particles in the flue gas stream—and collisions between those particles. Sakurovs [44] has shown that ash fusion temperatures of blends depend, in reality, on the interactions between coal particles in the blends. Ash fusion is only additive or linear when coals of similar ranks are blended. Typical blends of bituminous and subbituminous coals do not exhibit such properties. The interactions as suggested by the research on iron oxide/ calcium oxide eutectics are sufficient to significantly impact ash fusion temperatures and properties. Even blends of two bituminous coals can show more fusibility than has been predicted by additivity. Further, this property is not particularly influenced by temperature [44].

As noted earlier, the mechanisms of deposition—and subsequent interactions between solid particles attached to heat transfer surfaces—are a function of particle size, source of the inorganic matter (bound into the coal, adventitious material), and temperature in the combustion environment. Mechanisms of deposition include inertial impaction, eddy impaction, thermophoresis, condensation, and chemical reaction [10, 40]. Baxter [10] provides an excellent, detailed review of these mechanisms. Of these, impaction mechanisms are most prevalent. All of these, however, depend on particle properties, including particle size, particle viscosity, the viscosity of any material already deposited on the heat transfer surface, and numerous other issues. It is sufficient here to note that impaction processes and mechanisms dominate for particles >10 μm. These particles dominate the ash particles in coal combustion [10].

3.6.2. Fouling and Blended Coals

Fouling, which involves deposition in the convective pass or heat recovery area (HRA) of a boiler, involves mechanisms similar to, yet distinct from, slagging. Fouling deposits are formed in regions of the boiler where there is no direct exposure to flames and radiant heat transfer. Unlike deposits of slag, where there is more than ample evidence of liquid materials, fouling deposits more characteristically may contain some liquid materials but are dominated by sintered or partially sintered inorganic matter.

Su, Pohl, and Holcombe [42] have demonstrated that fouling deposits, such as those shown in Figure 3.18, cannot be predicted in terms of formation, thickness, or other parameters on a linear basis. Their research with several pairs of coals, and two blends for each coal, shows the distinct nonlinearity associated with slags. Coals chosen for this research, from Australia, were designated A through H; blends were designated by the coupling of the two coal designations, with the first coal identified being blended at 67% and the second coal identified being blended at 33%. Figures 3.19 and 3.20 show two examples of the fouling deposit growth rate (mm/h) as a function of parent coals (A and E and F and G) and blends. Note the nonlinearity and, in one case, the fact that the growth rate for the blends exceeds the fouling deposit growth rate of either parent coal.

Fouling deposits, like slagging deposits, cannot be predicted from the weighted average of the coals in the blend. Rather, they are a function of the interaction of the coals in the blend. Particular attention is commonly paid to

FIGURE 3.18   Fouling deposits in a coal-fired boiler. *(Photograph by K. Letheby)*

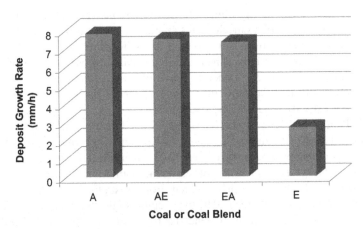

FIGURE 3.19   Example of fouling deposit growth rates for parent coals and blends. AE and EA are 67% and 33% blends, based on E. *Source: Adapted from [42].*

sodium oxide and potassium oxide concentrations. Bryers [51] identifies volatile sodium, calcium, chlorine, sulfur, and potassium all as major contributors to fouling deposits. Given this, it is not unexpected that fouling is a nonlinear phenomenon. Baxter [52] provides additional evidence for this

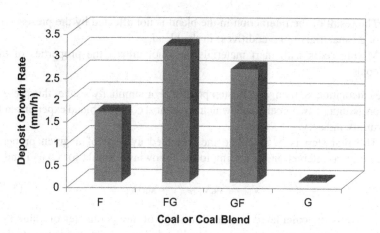

FIGURE 3.20   Example of fouling deposit growth rates for parent coals and blends. FG and GF are 67% and 33% blends. *Source: Adapted from [42].*

nonlinearity. One can conclude, then, that blending of coals—particularly dissimilar coals (by rank)—can yield significantly varying fouling deposits and deposit behavior. The consequences of blending must be evaluated on a case-by-case basis.

3.6.3. Quantifying the Inorganic Interaction

It is very clear that blending coal A with coal B does not result in an average of the two coals; the result is coal C. Similarly blending any given coal A with a dissimilar coal B, or multiple coals A, B, and C does not result in an average of all of the coals in the blend. Despite this, many coal tipples blend coals of varying types to meet utility specifications of calorific value, sulfur content (% or $lb/10^6$ Btu as SO_2), ash content, and so on. Few utilities have hard specifications on ash constituents, save for those burning high-sodium (northern PRB) coals or very high-iron or high-calcium coals. More specification constraints lead to higher fuel costs. Despite the lack of predictability for many of the characteristics of the blended product (e.g., coal C), it is useful to quantify the extent of interaction between the parent coals in combustion or gasification systems.

Sakurovs [53] has proposed a method for calculating the extent of interaction, assuming the following:

- The characteristic under consideration would be additive if there were no interaction between the coals.
- The consequence of the interaction—the effect on the combustion process—among coals in the blend is proportional to the product of the proportion of the (two or more) coals in the blend.

- The result of the interaction in the blend is not affected by the presence of other materials (e.g., additives) in the blend.
- Mixing coals with inert materials does not impact the properties of the coals.

The calculation is, then, a multistep process. For simplicity's sake, this presentation assumes a two-coal blend, which is the most common blend experienced in the utility industry.

The first step is to calculate the weighted average of a given property assuming no interaction, according to the following formula for coals i and j:

$$V_{ij} = x_{ib}V_i + (1 - x_{ib})V_j \tag{3.22}$$

where V_{ij} is the calculated weighted average of any parameter or value (V), and x_{ib} is the proportion of coal i in the blend. The second step is to calculate the difference (d_{ij}) between the measured value (V_m) and the calculated weighted average for any given parameter (V_{ij}), according to the following equation:

$$d_{ij} = (V_m - V_{ij})/[x_{ib}(1 - x_{ib})] \tag{3.23}$$

The term d_{ij} is the measure of the magnitude or consequence of the interaction between two coals in the blend. The measure of the interaction can then be expanded to a multifuel blend by summing the d values for the blend as a whole using the following equation [53]:

$$d_{blend} = \sum_{i=i+1}^{n} \sum_{i=1}^{n} x_i x_j d_{ij} \tag{3.24}$$

where d_{blend} is the measure of interaction for the blend as a whole.

Quantifying the interaction between two coals in a binary blend, or multiple coals in a more complex blend, is of more importance than intellectual interest. Measuring the d_{ij} or the d_{blend} for selected coals, using drop tube reactor kinetics data or other advanced characterizations of the combustibles and/or inorganic elements and compounds (e.g., fouling deposit growth rates), provides significant information for the selection of coals to be either combusted or gasified in a utility or industrial energy system. In seeking maximum combustion effectiveness per se, with emphasis on kinetics, maximizing d_{ij} is of benefit.

However, in the management of slagging and fouling, minimizing the d_{ij} value is of equal importance. This parameter, combined with the chemical composition of the coal, provides key insights into blend selection. For example, when blending PRB coal with bituminous coal in a furnace designed for Eastern bituminous coal, maximizing the d_{ij} value for fuel reactivity is of considerable importance. Otherwise the furnace may have insufficient volume to contain the combustion process. At the same time, attention must be given to the Fe_2O_3/CaO ratio and the implications for ash fusion temperatures. Blend selection based on advanced fuel characterization—with emphasis on behavioral characteristics—is of substantial importance. If additives are incorporated into the blend, then they must be accommodated by incorporation into one of the base coals.

Equations (3.23) and (3.24) work for conventional firing systems but are inappropriate for fluidized bed combustion and gasification. In fluidized bed systems, fresh fuel comprises only 3% to 5% of the total solid mass in the fluidized bed. The remainder is bed material—a composite of coal ash and limestone. The coal ash is inherently reactive. The limestone is added as a source of calcium for sulfur capture. Consequently, the fluidized bed combustion and gasification systems violate the third premise of the equations. Fluidized beds, as a whole, provide a special case for the behavior of inorganics when firing fuel blends. The behavior of calcium from both the coal and the limestone, alkali metals in the coals, and other inorganic matter creates unique relationships.

The low temperatures of combustion—1550°F to 1700°F (840–930°C)— also contribute to unique behaviors of many inorganic species. Finally, the presence of significant quantities of bed media contribute to physical scrubbing of the entire system and make the behavior of inorganic components distinct from those in conventional systems such as pulverized coal or cyclone boilers.

3.7. MANAGING THE COAL-ON-COAL BLENDING PROCESS

Managing the blending process means putting the chemistry just described into practice. This requires attention to the selection of a blending process and then monitoring the process. This has been the focus of DTE Energy [15–17, 32–35, 54] and numerous other utilities. It is interesting to consider the progression. PRB coal was originally considered an opportunity fuel. With the advent of the Clean Air Act, utilities learned to manage the slagging and fouling properties of PRB coal. Today it is used largely as the dominant coal in blends and supports 20% of the electricity generated in the United States. Figures 3.21 and 3.22 illustrate this progression, which is critical to the blending of coals.

3.7.1. Where Blending Can Occur

Blending can occur at any point in the supply of coal from mine to boiler or gasifier. Blending routinely occurs at coal tipples, particularly in the East where mines are small. There are numerous transfer facilities in the coal supply system—typically located on major waterways. These facilities, served by boat or barge, rail, and truck, can blend coal to virtually any specification or can blend based on coal sources and on client preferences (and contracts). The most

FIGURE 3.21 Mining Powder River Basin coal in Wyoming. Note the depth of the seam. *(Photograph by Richard Monts)*

FIGURE 3.22 Hauling PRB coal across Nebraska. Note that the line is triple tracked. All PRB coal is transported by rail. About 75 to 80 unit trains leave the Powder River Basin daily carrying this blend of coal. *(Photograph by Richard Monts)*

common blend location is in the coal yard of a power plant. This provides the maximum flexibility to the power plant.

Coal can be blended downstream of the coal yard [55]. One strategy that has been employed in recent years is to have the boiler as the place where blending occurs. Different coals are loaded into separate silos, run through separate pulverizers, and then fired in separate burners. This strategy requires silos rather than open bunkers for coal storage in the power plant. It is also more effective in tangentially fired boilers and less effective in wall-fired boilers. T-fired boilers have a single fireball where the fuels mix together; these furnaces are often considered a single burner with multiple fuel injection points.

Wall-fired units typically experience significant laning in the furnace and even through the convective pass. There is much less interaction between the flames in a wall-fired boiler than in a corner-fired boiler. This strategy has the advantage that individual pulverizers or mills can be set for the coal being fed. The disadvantages include the need for precision in filling silos and the lack of interaction between the fuels. The issue of precision in filling silos is not trivial; the demands of filling silos at power plants are such that this provides a basis for many mistakes.

3.7.2. Influence of Blending on Materials Handling Issues

The management of coal blending requires, from a physical systems perspective, the handling capabilities of the blends. This assumes that blending either occurs off-site or in the fuel yard. Vuthaluru et al. [55] studied the impact of blending on Australian coals. Citing the work of Conroy and Bennette [57], they indicate that the Hardgrove grindability index (HGI) values for blends of subbituminous coals are lower than the HGI values for parent subbituminous coals. The blends are more difficult to grind than either parent fuel.

For the Australian bituminous coals, however, Vuthaluru et al. [56] showed that the HGI for the blend is very close to the weighted average of the HGI for the parent coals. Linearity dominates the grindability of bituminous coals in these experiments. These researchers further found a very close correlation between vitrinite content in the fuel and HGI in coals and coal blends containing 40% to 70% vitrinite.

Experiments conducted by Kyle Taylor and Jordon Dosch reported in Tillman et al. [32] showed that blends of subbituminous and bituminous coals exhibited somewhat different results. These experiments used a subbituminous coal with a higher HGI than the bituminous coal, with the variation being approximately 6 HGI points (HGI of 51 for the subbituminous coal and 45 for the Central Appalachian coal). The results of these experiments showed that the mills preferentially ground the higher-HGI subbituminous coal. Measuring the calorific value of the screen fractions showed that the subbituminous coal typically reported to the finer fractions (+200 mesh, +400 mesh, and −400 mesh) and the bituminous coal typically reported to the coarser fractions

(+50 mesh and +100 mesh). This influenced combustion occurring beyond the nose of the boilers and across the dance floor.

Beyond grinding characteristics, Zhong, Ooi, and Rotter [58] evaluated general handling characteristics of coal blends. Their research focused on flow characteristics and flow blockages, finding that the best measurement for this purpose is the cohesive strength of the blend, which is best measured by determining unconfined compressive strength following consolidation to a particular stress level: the consolidation pressure.

The handling ability or flow characteristics are quite sensitive to surface moisture content—hence the selection of certain coals. Handling ability is also influenced by particle size and the incorporation of some coarse particles into the blend. Improving or deteriorating flow characteristics of a blend are sensitive to such physical parameters rather than directly influenced by rank. However, certain ranks of coal (e.g., some lignites) can contain high surface moisture percentages. Other coals may readily generate significant percentages of fines—also contributing to flow blockages. This has been a constant issue for the management of low-rank coals and bituminous coals during the wet winter season.

3.7.3. How Coal Blends Can Be Managed

Blending in the coal yard is probably the most common approach to blending; it provides for the maximum opportunity for control and blend manipulation. For many utilities, the basis of the approach is to assume that the coal quality parameters for each coal are essentially static or slow moving or that in-mine variation is tightly controlled. The blending strategy is based on initial characterization of each coal and then mechanically controlling the actual blend to a constant ratio of coals.

The use of on-line coal analyzers has emerged as an alternative approach to blending and to the control of blending. There is an extensive body of literature concerning this approach, of which several citations are used here [17, 32–35, 55, 59–65]. This approach has several demonstrated advantages, including the following:

- The ability of the plant operations staff to know what fuel is being fired and what fuel will be reaching the burners in the near future (e.g., 1 to 4 hours away)
- The ability of plant operations staff to adjust the blend—carefully and without "overdriving"—to accommodate variability in the fossil fuel supplied by each coal pile and thus the ability to optimize the blend to economic conditions

The approach taken at the Monroe Power Plant was described in Chapter 2 and is only summarized here. At MPP, coal is blended onto the main belt leaving the coal yard and being transported to silos, using belt scales and sophisticated controls that permit adjusting the mass percentage flow of each type of coal. The blended coal is then passed under an x-ray fluorescence (XRF) analyzer supplied by Quality Control, Inc. The XRF analyzer quantifies the coal characteristics every 90 seconds. The analyzer measures and reports the full proximate analysis (as-received), calorific value, and ash elemental analysis. The data are incorporated into the AccuTrack® system and are reported to the PI data system.

AccuTrack® is used to convert the proximate analysis to an ultimate analysis and also to calculate the base/acid ratio and both slagging and opacity measures. These data are reported to the shift supervisor, the supervising operators, and other plant operations personnel. They are also reported to plant management and any plant engineering personnel required to react to them. All plant operations personnel have been trained in the use of these data. Plant shift supervisors are empowered to adjust the blends to react to changes in the load required for each of the four 800-MW$_e$ wall-fired boilers.

Blends are also adjusted to react to fuel quality shifts in calorific value, moisture content (%), ash loading (lb/10^6 Btu), and base/acid ratio. Engineering personnel can make recommendations to the shift supervisors based on other data, including the iron/calcium ratio and opacity measures. This program permits the plant to maximize its use of lower-pollution-potential/lower-cost PRB coals without compromising its ability to meet the load requirements of the grid [54–55]. DTE has installed additional XRF capability at its Belle River Generating Station in St. Clair, Michigan.

Connemaugh Station in Pennsylvania also manages its coal quality using the AccuTrack® system, as does the Keystone Station. Both are Reliant Energy plants. In these stations, operators can evaluate fuel quality both in terms of what has been fired and what is coming, as shown in Figure 3.23.

The B.L. England Station of Connectiv also uses on-line analysis in the blending of PRB and Eastern bituminous coal. It blends at a rate of 30% PRB to meet the regulatory requirements of the New Jersey Department of Environmental Protection. It uses a nuclear analyzer (PGNA) to precisely measure the sulfur and ash contents of the coal blends. The analyzer also provides quantification of calorific value, moisture content, and ash composition in the form of mineral oxides. Sulfur accuracy is 0.08% for sulfur concentrations in the range of 0.2% to 1% as-received basis and 0.07% when sulfur concentrations are in the 1% to 3% range. Accuracy guarantees for the ash concentration are 0.825% for ash in the 5% to 10% range and 1.485% when ash is in the range of 10% to 18%. The analyzer accuracy has held up well.

PacificCorp experienced the closing of a local mine that served its Hunter Plant near Castle Dale, Utah. This closing, along with inherent variability in the coals available to the plant, caused numerous outages and derates due to slagging. PacificCorp installed a Prompt Gamma Neutron Activation Analysis (PGNAA) analyzer to measure major inorganic species in its coal ash and to manage the ash-softening temperature and reduce forced outages and derates caused by excessive slagging. Through coal blending as managed by the

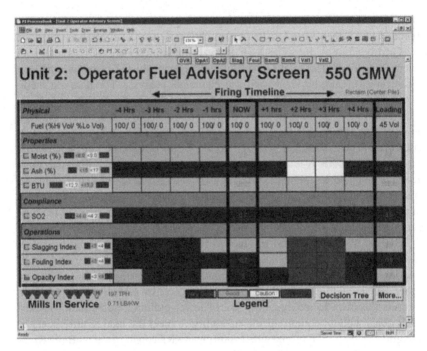

FIGURE 3.23 Operator screen at Connemaugh Station, providing fuel quality information for plant operations. *(Graph by ECG, Inc. Used with permission.)*

PGNAA, this objective was well achieved, with numerous additional benefits obtained as well [58, 60]. The analyzer supplied was the Thermo Electron (now Thermo Scientific) Coal Quality Manager (CQM), which can feed data to the Coal Blending Optimization System (COBOS), also supplied by Thermo Scientific [63]. Titus Station in Birdsboro, Pennsylvania, owned by Reliant Energy, also uses the CQM analyzer and COBOS system for fuel quality management.

One can cite numerous other utilities such as Dairyland Power, Southern Company, and Ameren that routinely blend for coal management. One can also cite numerous cases of the use of on-line coal analyzers for blend management. They have been used for single-coal quality management as well. In doing this, the utilities are following the example of the coal mines that have long used on-line coal analysis for fuel quality management.

3.7.4. Other Considerations

Other technical considerations are dominated by environmental issues. (Environmental issues are the focus of Chapter 6 and are also discussed extensively.) Blending can be used to reduce SO_2 emissions as a function of the blend; this exhibits linearity. NO_x emissions also can be managed to some extent with blending, both by control of fuel nitrogen content and by maximizing fuel volatility. Particulate emissions, particularly PM10 emissions, also can be addressed by coal blending [66, 67]. Finally, mercury emissions can be managed by blending, again by reducing the mercury content of the fuel and by providing sufficient halogens to oxidize the mercury and make it more readily captured in a baghouse or electrostatic precipitator.

Economics also dominates the issues of fuel blending and the ability to manage coal costs. These economics are best evaluated on a plant-by-plant basis to include the costs of transportation and of managing the blending system with the coal prices per se. There is a significant difference between the price of coal and the cost of coal.

3.8. CONCLUSIONS

Coal is the dominant fuel used by electric utilities, and electric utilities are the overwhelming users of this fuel. Coal accounts for more annual expenses to utilities than any other cost component. The total cost of fuel quality concerns—derates due to ash deposition, derates due to moisture and calorific value degradation, emissions management, and other concerns—is over $1.2 billion annually [39]. More than half of U.S. power plants and most of those in Europe and Asia blend coals to achieve environmental compliance, performance enhancement, and fuel cost control. For these reasons, coal-on-coal blending is of critical importance to electric utilities. Coal blending was initially instituted for both environmental and economic reasons; coal costs and technical performance issues are now dominant.

Some properties associated with blending are additive; they are the weighted average of the components of each element of the blend. These include proximate and ultimate analysis and calorific value. Ash elemental analysis and trace metals compositions are not linear with the weighted average of the fuels, but must be adjusted for the ash percentage in the fuels. When measured on a loading basis—on a lb component/10^6 Btu basis—these properties are not additive or linear.

Behavioral characteristics of blends are not linear or additive with respect to the parent coals. When one looks particularly at the kinetic parameters, it is apparent that blending can and does increase the reactivity of many, but not all, blends. Similarly, ash characteristics, particularly those associated with slagging and fouling, are not additive or linear. In both combustion kinetics and ash behavior, blending can increase the reactivity of the components through interactions between the component coals. This interaction can be quantified, however, and that quantification can help in the selection of coals and coal blends.

Given the importance of coal and the blending process, and given the sensitivity of unit performance to coal (blend) quality, management tools have been developed for operations personnel. One management tool is to blend outside the coal yard—to purchase the blend. Other approaches also exist, including blending in the coal yard. During the past 20 years, this process has been enhanced by the use of on-line coal analysis and associated data management. Numerous plants have incorporated such a strategy into their operations with considerable success. Management of the blend for such plants is a strategy for both environmental and economic optimization at the plant level.

REFERENCES

[1] US Energy Information Administration. Annual energy review 2009. Washington, DC: US Department of Energy; 2010.

[2] Thrush PW, editor. Dictionary of mining, mineral, and related terms. Washington, DC: US Bureau of Mines; 1968.

[3] Bryers R. Investigation of the reactivity of macerals using thermal analysis. Fuel Processing Technology 1995;44:25–54.

[4] Zhang J, Yuan JW, Sheng CD, Xu YQ. Characteristization of coals utilized in power stations in China. Fuel 2000;79:95–102.

[5] Su S, Pohl JH, Holcombe D, Hart JA. A proposed maceral index to predict combustion behavior of coal. Fuel 2001;80:699–706.

[6] Tillman DA. The combustion of solid fuels and wastes. San Diego: Academic Press; 1991.

[7] Tillman DA. Chapter I: introduction. In: Miller BG, Tillman DA, editors. Combustion engineering issues for solid fuel systems. Boston: Academic Press; 2008. p. 1–32.

[8] Bartok W, Sarofim AF, editors. Fossil fuel combustion: a source book. New York: John Wiley & Sons; 1991.

[9] Palmer HB, Beer JM, editors. Combustion technology: some modern developments. New York: Academic Press; 1974.

[10] Baxter LL. Ash deposit formation and deposit properties: a comprehensive summary of research conducted at Sandia's combustion research facility. Livermore, CA: Sandia National Laboratories; 2000. SAND2000–8253.

[11] Raask E. Mineral impurities in coal combustion: behavior, problems, and remedial measures. London: Hemisphere Publishing; 1985.

[12] Kitto JB, Stultz SC, editors. Steam: its generation and use, 41st ed. Barberton, OH: Babcock and Wilcox; 2005.

[13] Marx P, Morin J. Chapter 7: Conventional firing systems. In: Miller BG, Tillman DA, editors. Combustion engineering issues for solid fuel systems. Boston: Academic Press; 2008. p. 241–74.

[14] Miller BG, Miller SF. Chapter 8: fluidized bed firing systems. In: Miller BG, Tillman DA, editors. Combustion engineering issues for solid fuel systems. Boston: Academic Press; 2008. p. 275–340.

[15] Widenman A. Fuel characterization: a tutorial. Proceedings 35th international technical conference on coal systems and fuel utilization. Clearwater, FL; 2010, June 6–10.

[16] Widenman T. Ash fusion characteristics for binary blends of selected PRB and Pittsburgh seam coals. Proceedings 28th international technical conference on coal systems and fuel utilization. Clearwater, FL; 2003, March 8–13.

[17] Duong D, Tillman D, Widenman A. Fuel blending for combustion management. In: Miller B, Tillman D, editors. Combustion engineering issues for solid fuel systems. Boston: Academic Press; 2008. p. 171–98.

[18] Pierce BS, Dennen KO, editors. The National Coal Resource Assessment overview. Washington, DC: US Geological Survey Professional Paper; 2009.

[19] Singer JG. Combustion: fossil power systems, 3rd ed. Windsor, CT: Combustion Engineering; 1981.

[20] Miller BG. Coal energy systems. Boston: Academic Press; 2005.

[21] Tillman DA, Duong Dao, Miller Bruce. Chlorine in solid fuels fired in pulverized fuel boilers—sources, forms, reactions and consequences: a literature review. Energy & Fuels 2009;23:3379–91.

[22] Baxter LL, Miles TR, Miles Jr TR, Jenkins BM, Milne TA, Dayton DC, et al. The behavior of inorganic material in biomass-fired power boilers: field and laboratory experiences. Proceedings Engineering Foundation conference on industrial use of biomass. Snowbird, UT; 1996, April 28–May 3.

[23] Baxter LL, Miles TR, Miles Jr TR, Jenkins BM, Dayton DC, Milne TA, et al. The behavior of inorganic material in biomass-fired power boilers—field and laboratory experiences: vol. 2 of alkali deposits found in biomass power plants. Livermore, CA: Sandia National Laboratories; 1996. USDOE Report NREL/TP-433–8142.

[24] US Geological Survey. US coal resources databases (USCOAL); 2006.

[25] Osorio E, Gomes MLI, Vilela ACF, Kalkreuth W, Almeida DE, Antonio MA, et al. Evaluation of petrology and reactivity of coal blends for use in pulverized coal injection (PCI). Coal Geology 2006;68:14–29.

[26] Helle S, Gordon A, Alfaro G, Garcia X, Ulloa C. Coal blend combustion: link between unburnt carbon in fly ashes and maceral composition. Fuel Processing Technology 2001; 80:209–23.

[27] Su S, Pohl JH, Holcombe D, Hart JA. A proposed maceral index to predict combustion behavior of coal. Fuel 2001;80:699–706.

[28] Goodarzi F. Petrology of subbituminous feed coal as a guide to the capture of mercury by fly ash—influence of depositional environment. Coal Geology 2005;61:1–12.

[29] Haas J, Tamura M, Weber R. Characterization of coal blends for pulverized fuel combustion. Fuel 2001;80:1317–23.

[30] Miller BG. Clean Coal Engineering Technology. Boston: Academic Press; 2011.

[31] Johnson DK, Miller BG, Wasco RS. Pyrolysis and char oxidation kinetics of selected coals: final report. University Park: Pennsylvania State University Energy Institute; 2005.

[32] Tillman DA, Dobrzanski A, Duong D, Dosch J, Taylor K, Kinnick R, et al. Optimizing blends of Powder River Basin subbituminous coal and bituminous coal. Proceedings PRB users group annual meeting. Baltimore; 2006.

[33] Tillman DA, Dobrzanski A, Duong D, Dezsi P. Fuel blending with PRB coals for combustion optimization: a tutorial. Proceedings 31st international technical conference on coal utilization and fuel systems. Clearwater, FL; 2006, May 21.

[34] Tillman D, Duong D. Managing slagging at monroe power plant using on-line coal analysis and fuel blending. Fuel Processing Technology 2007;88(11–12):1094–8.

[35] Duong D, Tillman D, Miller B. Characterizing blends of PRB and Central Appalachian coals for fuel optimization purposes. Proceedings 31st international technical conference on coal utilization and fuel systems. Clearwater, FL; 2006, May.

[36] Bradley LC, Miller SF, Miller BG, Tillman DA. A study on the relationship between fuel composition and pyrolysis kinetics. Energy & Fuels 2011;25:1989–95.

[37] Chi T, Zhang H, Yan Y, Zhou H, Zheng H. Investigations into the ignition behaviors of pulverized coals and coal blends in a drop tube furnace using flame monitoring techniques. Fuel 2010;89:743–51.

[38] Faundez J, Arias B, Rubiera F, Arenillas A, Garcia X, Gordon AL, et al. Ignition characteristics of coal blends in an entrained flow furnace. Fuel 2007;86:2076–80.

[39] Harding NS, O'Connor DC. Ash deposition impacts in the power industry. Proceedings ash deposition conference. Snowbird, UT; 2006.

[40] Barnes I. Slagging and fouling in coal-fired boilers. London: IEA Clean Coal Centre; 2008.

[41] Gupta Rajender. Advanced coal characterization: a review. Energy & Fuels 2007; 21:451–60.

[42] Su S, Pohl JH, Holcombe D. Fouling propensities of blended coals in pulverized coal-fired power station boilers. Fuel 2003;82:1653–67.

[43] Sakurovs R, Lynch LJ, Maher TP. The prediction of the fusibility of coal blends. Fuel Processing Technology 1994;37:255–69.

[44] Sakurovs R. Direct evidence that the thermoplastic properties of blends are modified by interactions between the component coals. Fuel 1997;76:615–21.

[45] Qiu JR, Li F, Zheng Y, Zheng CG, Zhou HC. The influences of mineral behavior on blended coal ash fusion characteristics. Fuel 1999;78:963–9.

[46] Rushdi A, Sharma A, Gupta R. An experimental study of the effect of coal blending on ash deposition. Fuel 2004;83:495–506.

[47] Kondratiev A, Jak E. Predicting coal ash slag flow characteristics (viscosity model for the Al_2O_3-CaO-'FeO'-SiO_2 system. Fuel 2001;80:1989–2000.

[48] Barroso J, Ballester J, Ferrer LM, Jimenez S. Study of coal ash deposition in an entrained flow reactor: influence of coal type, blend composition, and operating conditions. Fuel Processing Technology 2006;87:737–52.

[49] Su S, Pohl JH, Holcombe D, Hart JA. Slagging propensities of blended coals. Fuel 2001; 80:1351–60.

[50] Levin EM, Robbins CR, McMurdle HF. Phase equilibrium diagrams. vol 1. Columbus, OH: American Ceramic Society; 1964.

[51] Bryers RW. Fireside slagging, fouling, and high-temperature corrosion of heat transfer surface due to impurities in steam-raising fuels. Progress in Energy and Combustion Science 1996;22:29–120.

[52] Baxter LL. Experimental and theoretical comparisons of the combustion and ash deposition behavior of blended coals and that of blend components. In: Bryers RW, Harding NS, editors. Coal-blending and switching of low-sulfur Western coals. New York: American Society of Mechanical Engineers; 1994. p. 255–64.

[53] Sakurovs R. A method for identifying interactions between coals in blends. Fuel 1997;76:623–4.

[54] Trupiano L. Development of furnace/boiler cleanliness/slagging predictor from pi and fuels data. Proceedings electric power conference. Chicago; 2007, April.

[55] Peltier R, Wicker K. PRB coal makes the grade. Power 2003;147(8):28–36.

[56] Vuthaluru HB, Brooke RJ, Zhang DK, Yan HM. Effects of moisture and coal blending on hardgrove grindability index of Western Australian coal. Fuel Processing Technology 2003;81:67–76.

[57] Conroy A, Bennette C. The combustion behavior of Australian export and overseas low rank coals. ACARP; 1997. Final Report C3097.

[58] Zhong Z, Ooi JY, Rotter JM. Predicting the handlability of a coal blend from measurements on the source coals. Fuel 2005;84:2267–74.

[59] Snider K, Richard W, Evans M. Using an on-line elemental coal analyzer to reduce lost generation due to slagging. Proceedings international on-line coal analyzer technical conference. St. Louis; 2004, Nov. 8–10.

[60] Foster S. What is involved in effective calibration of a pgna analyzer in the factory and the field. Proceedings international on-line coal analyzer technical conference. St. Louis; 2004, Nov. 8–10.

[61] Belbot M, Vourvopoulos G, Paschal J. A commercial on-line coal analyzer using pulsed neutrons. Proceedings application of accelerators in research and industry—American Institute of Physics 16th international conference; 2001.

[62] Benson S. Application of on-line coal analyzer to plant performance. Proceedings Minnesota energy ingenuity conference. Great River, MN; 2008, Nov. 6–7.

[63] Stallard S, Smolenack A, Anderson A. More from your on-line coal analyzer. Aurora, CO: CyberTech, Inc; 2011.

[64] Woodward RC, Evans MP, Empey ER. A major step forward for on-line coal analysis. Proceedings international on-line coal analyzer technical conference. St. Louis; 2004, Nov. 8–10.

[65] Bhamidipati VN, Rose CD, Russell JM. Compliance blending of PRB coal at B.L. England Station using cross-belt analyzer. Proceedings international on-line coal analyzer technical conference. St. Louis; 2004, Nov. 8–10.

[66] Wang Q, Zhang L, Sato A, Ninomiya Y, Yamashita T. Effects of coal blending on the reduction of PM_{10} during high-temperature combustion. Part 1: mineral transformations. Fuel 2008;87:2997–3005.

[67] Wang Q, Zhang L, Sato A, Ninomiya Y, Yamashita T. Effects of coal blending on the reduction of PM_{10} during high-temperature combustion. Part 2: a coalescence-fragmentation model. Fuel 2009;88:150–7.

Chapter 4

Blending Coal with Biomass

Cofiring Biomass with Coal

4.1. INTRODUCTION

Lignocellulosic materials have historically provided an abundance of natural fuel for mankind. One of the most important events was the discovery and then the control of fire generation through the ignition of vegetation or lignocellulosic materials [1, 2]. Thus, the first fuel of importance in the preindustrial economies was wood. Until 1850, wood was the dominant energy source in the United States—used in fueling steam engines, riverboats, and railroad boilers. The advent of pulping technology provided more sophisticated material use of wood.

Mechanical pulping was invented in Germany in 1844 and was brought to the United States in 1867. Subsequently chemical pulping technologies (e.g., kraft pulping) were deployed. From 1850 through 1900, energy consumption in the United States and other industrialized parts of the world advanced and depended increasingly on coal. By 1950, wood and biomass energy declined to <0.5 quadrillion (10^{15}) Btu/yr. By the latter half of the twentieth century, woody biomass became increasingly important again, with its contribution increasing to $>3 \times 10^{15}$ Btu/yr [1]. The increase was largely based on process energy needs in the pulp and paper industry and elsewhere. Through cofiring, biomass utilization may become increasingly important, with utilities making a more significant contribution to biomass energy utilization. (See also the following papers and reports by various authors: [3–12].)

4.2. BIOMASS AND COAL BLENDING

The cofiring of biomass with coal is essentially the blending of two dissimilar fuels, with one being the dominant fuel and the other its supplement. Typically coal is the dominant fuel, and the principles of blending apply. Both biomass fuels and coal vary significantly in properties, thereby making some combinations of biomass and coal advantageous, while others can be disasters—particularly to the operators. The method of weighted averages is important and necessary in evaluating blends, but it does not provide a complete picture of the combustion process.

Tillman and others have shown that the burning of dissimilar fuels can produce significantly different burning behaviors when compared to the averaged fuel blend [13]. Attention must be paid to basic chemical properties such as proximate, ultimate, and ash elemental analyses, along with trace elements. In addition, the reactivity of the fuel mass—kinetics, volatile evolution, and chemical fractionation—can help to understand how a particular blend will combust and how it behaves. Furthermore, boiler configuration—wall-fired versus tangentially fired, mill capacity, fan capacity, steam condition, electrostatic precipitators (ESP) versus fabric filters, and others—must be considered. When cofiring biomass with coal, the dominant fuel is still coal, so both fuels must be well understood. Coal varies significantly from one coal rank to the next. Biomass types and even the species within the general categories vary drastically [14].

Coal can be classified into four essential groups: lignite, subbituminous, bituminous, and anthracite. As the rank increases, the fixed carbon concentration increases, while the moisture and oxygen contents decrease. Even within the bituminous coal rank, the Central Appalachian and Illinois Basin, the fuels can have significantly different characteristics, both from the organic and inorganic perspective. An Illinois Basin coal will tend to have higher slagging and fouling tendency than a Central Appalachian coal due to its high iron content. Northern Appalachian coals also exhibit slagging and fouling tendencies not found in Central Appalachian coals.

Beyond differences between coals of different coal-producing regions, there can be significant variability between deposits in given coal-producing regions, and there can also be significant in-seam variability of coals [14]. The first is demonstrated by the differences between Jacobs Ranch and Black Thunder Powder River Basin (PRB) coals despite the fact that the two mines are adjacent to each other. The second has been demonstrated repeatedly by research from CONSOL and other coal-producing companies.

Biomass fuels tend to have modest heat contents when compared to coal, with a high volatility. Like coal, biomass fuels can vary significantly from one type to the next. Silica content in biomass can range from 10% to >50%. Silica concentration depends on harvesting practice and a host of other variables. For sawdust, silica concentration tends to be low. Nitrogen concentrations can be lower or higher than coal, whereas sulfur concentrations are typically lower in the biomass fuels. Chlorine concentration can also vary significantly among different biomass fuels. Therefore, when blending dissimilar fuels, the effects of blending merit detailed analysis and understanding in order to adequately assess the benefits and/or issues associated with each blend.

4.2.1. Properties of Biomass and Coal

Variations in biomass and coal properties can be significant. These properties include both physical and chemical composition. The differences affect system design and consequently the performance of the combustion system. The bulk density of a fuel significantly influences the materials-handling design. For biomass fuels, the bulk density is typically in the range of 80 to 320 Kg/m^3 (5–20 lb/ft^3) when compared to coal at a typical range of 800 to 960 Kg/m^3 (50–60 lb/ft^3) not compacted. This disparity in bulk density and increased moisture concentration in biomass fuels presents one of the most challenging issues when cofiring biomass with coal [15, 16].

For switchgrass, the bulk density is very low and tends to be considerably less dense than coal. A mixture of 10% switchgrass with 90% coal (by mass) equates to approximately 50% mixture (by volume) of the two fuels [17, 18]. Studies of blending switchgrass with coal indicate that due to the fibrous nature of grass and the low bulk density of the grass-coal mixture, the mixture does not flow easily through existing coal bunkers.

Chemical composition of fuels significantly impacts the combustion process and consequently the products of combustion. By understanding the chemical composition of fuels and the influences of these components on the combustion process, system optimization can be more easily achieved.

The organic constituents of coal and biomass vary significant from one to another and within the same fuel category. Tables 4.1 and 4.2 are typical proximate and ultimate analyses for several coals and biomass fuels. Fixed carbon contents for the coals are significantly higher than the biomass fuels. Conversely, volatile matter concentrations for the biomass fuels are much greater. Nitrogen and sulfur concentrations for the biomass fuels also tend to be much lower. Thus, by cofiring biomass with coal, SO$_x$ emissions are inherently decreased.

Lignin and holocellulose (cellulose and the hemicelluloses) are typically considered to be the major precursors to coal. Since coal is an organic rock, the study of petrography is important in understanding the organic behavior of this fuel. Coal petrography is the science of describing coal as an organic mineral. It is intended as a description of the origin, history, occurrence, chemical composition, and classification of the coal [19, 20].

Petrography employs a nomenclature to define coal that distinguishes the rock types (lithotypes) and their microscopic constituents, known as macerals. In general, there are three maceral groups: vitrinite, liptinite, and inertinite. Table 4.3 depicts the ultimate analysis for each of the coal macerals with similar carbon concentrations. The vitrinite macerals tend to be fairly rich in oxygen, while the liptinite macerals are rich in hydrogen. Inertinite macerals tend to have higher concentrations of carbon. They also tend to have lower hydrogen contents compared to the other macerals. It has been observed that the different maceral groups have different behaviors.

In general, wood residues are lignocellulosic in nature. Chemically they include cellulose, the hemicelluloses, lignins, various extractives, and some mineral compounds. The types of hemicelluloses and lignins vary as a function of wood species, as do the types and percentages of extractives [21–23]. Extracted hardwoods contain approximately 43% cellulose, 35% hemicelluloses,

TABLE 4.1 Typical Proximate and Ultimate Analyses for Several Coals

	Gulf Coast Lignite	Black Thunder Subbituminous	Central Appalachian (Long Fork)	Illinois #6
Proximate Analysis (wt%)				
Fixed Carbon	29.16	34.94	50.09	44.98
Volatile Matter	31.58	30.72	31.23	35.32
Ash	12.52	5.19	11.52	7.43
Moisture	26.74	29.15	7.16	12.27
Ultimate Analysis (wt%)				
Carbon	30.14	51.30	66.93	66.04
Hydrogen	4.28	2.87	4.43	4.41
Oxygen	24.96	10.46	7.55	5.66
Nitrogen	0.56	0.68	1.34	1.40
Sulfur	0.80	0.35	1.07	2.79
Ash	12.52	5.19	11.52	7.43
Moisture	26.74	29.15	7.16	12.27
Higher Heating Value—as Received (Btu/lb)	7,613	8,888	12,114	11,731
Volatile/Fixed Carbon Ratio	1.08	0.88	0.62	0.79

Source: [1, 16].

and 22% lignin. For softwoods, their composition is approximately 43% cellulose, 28% hemicelluloses, and 29% lignin. When compared to the woody material, bark contains significantly more extractives and lignin, contributing to higher heats of combustion.

Bark also contains lower amounts of holocellulose, the total combined fractions of cellulose and hemicelluloses, which causes a lower heat of combustion [1, 24–26]. For the woody biomass material, the moisture concentration is a function of the living and growing process and the manufacturing process imposed on the wood. Much of the moisture in sawdust can be attributed to processing activity, since many debarkers and saws are typically water-cooled [1].

For switchgrass and related herbaceous fuels, moisture concentration is typically lower than most biomass. For residue materials such as rice hulls,

TABLE 4.2 Typical Proximate and Ultimate Analyses for Several Biomass Fuels

	Sawdust	Urban Wood Waste	Switchgrass	Corn Stover
Proximate Analysis (wt%)				
Fixed Carbon	9.35	12.58	12.18	15.64
Volatile Matter	55.03	52.56	65.19	71.08
Ash	0.69	4.08	7.63	7.05
Moisture	34.93	30.78	15.00	6.23
Ultimate Analysis (wt%)				
Carbon	32.06	33.22	39.68	43.39
Hydrogen	3.86	3.84	4.95	5.15
Oxygen	28.19	27.04	31.93	37.25
Nitrogen	0.26	1.00	0.65	0.83
Sulfur	0.01	0.07	0.16	0.10
Ash	0.69	3.99	7.63	7.05
Moisture	34.93	30.84	15.00	6.23
Higher Heating Value— as Received (Btu/lb)	5431	5788	6601	7088
Volatile/Fixed Carbon Ratio	5.89	4.20	5.35	4.54

Source: [1, 16. 26].

cotton gin trash, vineyard prunings, and others, the higher heating values are substantially lower than those of woody biomass, switchgrass, or coal [1]. Straw, for instance, has high concentrations of volatiles and high chlorine and high alkaline contents [27]. Variations in straw ash composition depend heavily on where it is grown (weather conditions, soil, etc.). Nitrogen concentration in straw varies depending on fertilizer concentration.

The terms *mineral matter* and *ash* are typically used interchangeably, but this usage is not completely correct. Ash is the residue that remains after complete combustion of the organic portion of the coal occurs. The constituents of ash are formed as a result of chemical changes that occur in the mineral matter during combustion. Mineral matter in coal is typically classified as either inherent mineral matter or extraneous mineral matter. Inherent mineral matter is the inorganic material that originated from plant material and formed the organic debris in the source bed [20]. Extraneous mineral

TABLE 4.3 Ultimate Analysis of Coal Macerals with Similar Carbon Content

Composition	Liptinite	Vitrinite	Inertinite Others	Fusinite
Carbon	82–83	83.0	83–85.0	94.0
Hydrogen	8.7–9.0	5.5	2.7–4.0	2.8
Oxygen	6.0–7.3	9.0–10.1	9–12	2.3
Nitrogen	0.5–1.4	1.3–2.0	1.3–1.9	0.9
Sulfur	0.5–0.6	0.5	0.5	—
Volatile Matter	80	33–40	10–15	5

Source: [20].

matter is the inorganic portion that was brought into the coal deposit by various external means and makes up the larger portion of the total inorganic constituent.

Like coal, the ash constituents of biomass fuels can vary significantly from one type to the next. In addition, the total ash concentration can vary significantly. For the clean woody materials, ash concentrations typically are low when compared to the herbaceous materials. Certain constituents such as iron, calcium, and others can cause excessive slagging, fouling, corrosion, and erosion problems. Therefore, it becomes essential to understand the nature and behavior of the mineral contents in order to evaluate the inorganic material [20].

Coal ash constituents can be classified as either basic or acidic. Types of ash that are highly acidic or highly basic have the tendency to exhibit high ash fusion and melting temperatures. Basic constituents within the ash will flux or reduce the melting temperature and viscosity. Because the percentage of basic and acidic components are nearly equal, the fusion temperatures and ash viscosity are at a minimum. Kitto and Stultz [28] discuss this in more detail.

Tables 4.4 and 4.5 show representative ash elemental values for several coals and biomass fuels; for the Illinois Basin coal, this fuel tends to have a significant concentration of iron. If this coal is blended with a high-calcium and high-ash biomass fuel, the effects can be detrimental. Calcium and iron are both fluxing agents, so the combination, if within the eutectic zone, depresses the melting point of the fuels.

The varying coal ranks exhibit different structures but are considerably condensed when compared to the biomass fuels. Clusters of aromatic rings vary in size from 1 to 4 or, at times, 5, increasing with rank [26]. Aromaticity is a primary determinant of reactivity. Table 4.6 depicts the differences between aromaticity for various solid fuels.

TABLE 4.4 Typical Ash Elemental Analyses for Several Coals

	Gulf Coast Lignite	Black Thunder Subbituminous	Central Appalachian (Long Fork)	Illinois #6
Ash Elemental Analysis (wt%)				
SiO_2	38.17	35.32	51.99	44.32
Al_2O_3	14.01	17.29	26.25	19.33
TiO_2	1.15	1.16	1.07	0.93
Fe_2O_3	7.39	5.42	8.38	22.17
CaO	13.69	21.10	2.31	3.19
MgO	2.51	3.32	1.42	0.66
Na_2O	0.60	1.07	0.71	0.41
K_2O	0.51	0.56	3.26	2.09
P_2O_5	0.39	0.99	0.56	0.13
SO_3	14.41	13.10	2.20	2.48
Base/Acid Ratio	0.46	0.59	0.20	0.44

Source: [1, 16].

Woody material is principally constructed of C-C bonds and C=C bonds that are linked together by ether-type bonds (R-O-R). In addition, they consist of 60% to 75% polycyclic aliphatic hydrocarbons and 15% to 24% aromatic hydrocarbons (lignin). Ether bonds generally are much weaker than polycyclic aromatic hydrocarbons that typically form the coal structure [29]. Thus, woody materials will release higher-volatile yields when compared to coal. In addition, the oxygen within the functional groups changes to less reactive forms as rank increases. For example, the –OH functional group present in lower-rank coals are more reactive than the less reactive form of –O present in higher-rank coals. Highly reactive functional groups are typically present in biomass fuels.

Most chemical reactions are sensitive to temperature changes. The dependence of reaction rates was first quantified by Arrhenius and is still used to this day. This relationship is depicted by the following equation:

$$k = Ae^{-\frac{E_a}{RT}} \tag{4.1}$$

where A is the preexponential constant, E_a is the activation energy, and R is the gas constant [30]. The Arrhenius equation is commonly used to look at the reactivity of the fuel and blends on a comparative basis. It provides information on the rate of reaction and how quickly the fuel will complete combustion.

TABLE 4.5 Typical Ash Elemental Analyses for Several Biomass Fuels

	Sawdust	Urban Wood Waste	Switchgrass	Corn Stover
Ash Elemental Analysis (wt%)				
SiO_2	23.70	55.52	65.18	52.13
Al_2O_3	4.10	6.51	4.51	3.77
TiO_2	0.36	0.63	0.24	0.13
Fe_2O_3	1.65	2.13	2.03	1.79
CaO	39.95	17.50	5.60	8.75
MgO	4.84	2.47	3.00	3.36
Na_2O	2.25	0.69	0.58	1.52
K_2O	9.81	6.88	11.60	17.30
P_2O_5	2.06	0.33	4.50	2.68
SO_3	1.86	0.89	0.44	3.71
Base/Acid Ratio	2.08	0.47	0.33	0.58

Source: [1, 14, 16, 26].

TABLE 4.6 Aromaticity Values for Various Solid Fuels

Fuel	Aromatic Rings per Cluster in the Fuel				Percent of Carbon Atoms in Aromatic Rings
	1	2	3	4	
Woody Biomass	100%	0%	0%	0%	20–25%
Texas Lignite	42%	26%	21%	11%	46%
Wyoming Subbituminous	39%	35%	—	—	40%
Illinois Bituminous	8%	50%	36%	8%	60–70%

Source: [2, 26, 31].

The reactivity of individual fuels can significantly influence the way the fuel mass behaves. Thus, measuring the reactivity of fuels is important to further understand fuels and their behavior within the combustor. One of the key areas in understanding fuel pyrolysis and reactivity is in the fuel structure.

Lignocellulosic materials, which are common in woody material, are not directly combustible but will decompose to form volatile pyrolysis under a sufficient source of energy. For the cellulosic component, it is mostly converted to the combustible and noncombustible volatiles. Lignin components, on the other hand, contribute to the char fraction [25–26, 31]. Lignin is the lesser reactive material in biomass fuels, consisting of a basic skeleton of four or more substituted phenylpropane units.

Heteroatoms such as N and Cl are a function of the living process and thus exist as a consequence of protein structure. These protein structures can be found in the inner bark of wood and some lignin precursors, and are thus almost always in the amine structure ($-NH_{1-3}$) [31]. Nitrogen in coal may be less volatile and less accessible when compared to biomass fuels [26, 31]. This is due to the nitrogen that is present; Figures 4.1 and 4.2 are representations of the structure for softwood lignin and subbituminous coal. The softwood structure is open, allowing for easy access when compared to the coal structure.

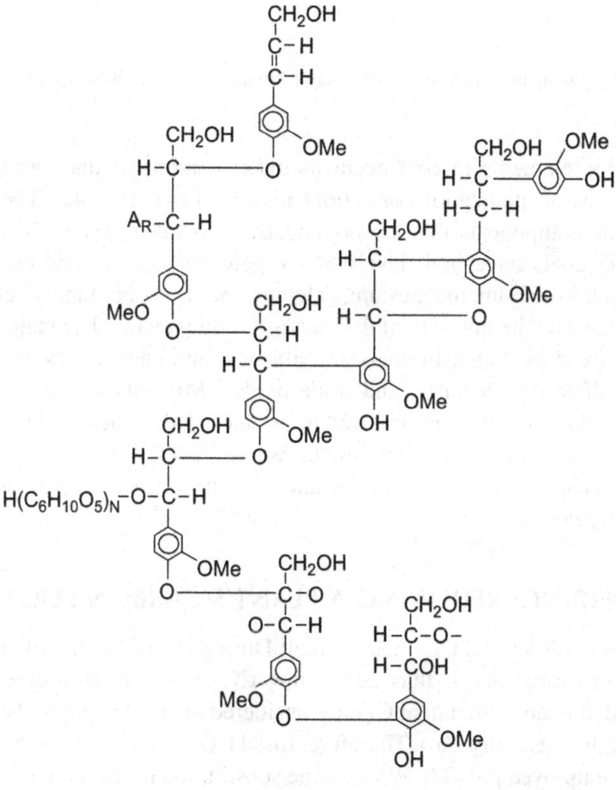

FIGURE 4.1 Adler structure of softwood lignin. *Source: [32].*

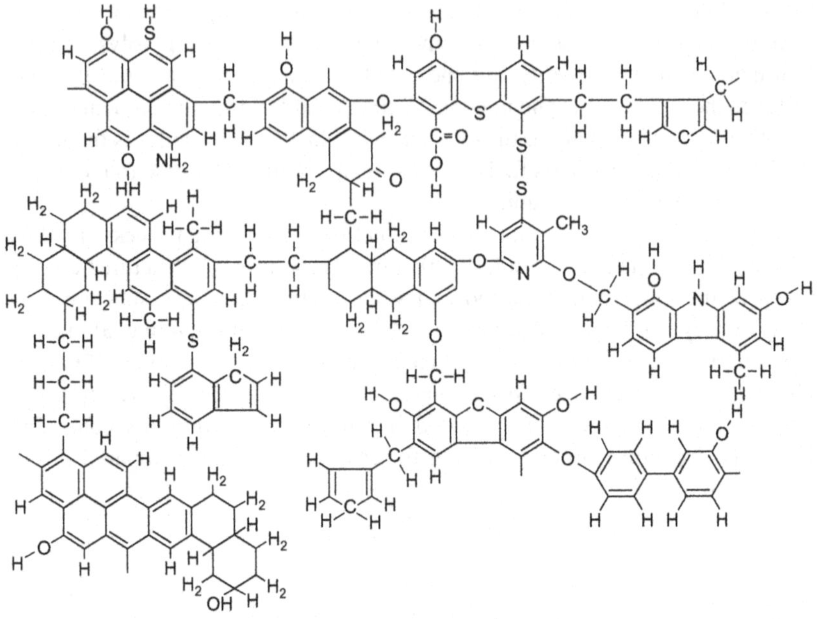

FIGURE 4.2 Speculative structure of high-volatile bituminous coal by Wiser. *Source: [33], cited in [32].*

Inorganic elements in coal occur as discrete minerals that can be either organically associated cations or cations dissolved in pore water. The fraction of inorganic components that are organically associated vary with coal rank. Lower-rank coals have high levels of oxygen that act as bonding sites for cations, such as sodium, magnesium, calcium, and so on. For higher-rank coals, minerals make up the majority of the inorganic components. The major mineral groups that can be found in coals are silicates, aluminosilicates, carbonates, sulfides, sulfates, phosphates, and some oxides. Minerals that are associated with the organic fractions are referred to as "included," whereas those that are not associated with the organic fractions are "excluded." The organically associated elements will react and/or interact with other ash-forming constituents during combustion [15].

4.3. COFIRING: REDUCING A PLANT'S CARBON FOOTPRINT

Biomass is considered a CO_2 neutral fuel. During the life of the plant, photosynthetic processes occur, thus consuming CO_2 from the atmosphere. Once combusted, the same amount of CO_2 is considered to be released, so the net CO_2 production is essentially zero. Therefore, fossil CO_2 is decreased when biomass cofiring is employed [34, 35]. Wood waste substitution for coal directly reduces fossil CO_2 emissions by approximately 1.05 ton fossil CO_2/ton wood burned.

4.3.1. The Carbon Cycle

Carbon inventories for the atmosphere, biosphere, soils and rocks, and oceans are linked by a set of natural and anthropogenic biogeochemical processes. This set of parameters make up the carbon cycle [36]. Atmospheric carbon inventory is made up of essentially carbon dioxide, with a current concentration of approximately 388 ppm or 0.04%. This inventory has increased by almost 40% since the preindustrial age, which is believed to be caused by fossil fuel combustion and changes in land use and land management practices. Terrestrial carbon inventory, estimated at 2200 Gt-C, is made up of living biomass and organic carbon in soils and sediments. The inventory has decreased by approximately 10% since the preindustrial times due to land use and land management practices such as deforestation, conversion of grassland to agricultural use, and other practices. Oceanic carbon inventories sum up to about 39,000 Gt-C.

Three key processes distribute the carbon between organic and inorganic fractions, transport, and deposit of the carbon in the sediments. These are the biological and solubility pumps and the thermohaline circulation. The earth's crust, representing the upper portion of the lithosphere, is estimated to hold $\sim 5 \times 10^7$ Gt-C in sedimentary rocks—organic carbon and limestone. Carbon inventories are subject to constant flux due to the interlinking of natural processes [36]. Figure 4.3 illustrates these parameters, referred to as inventories (shown in bold) and fluxes; units are in Gt-C per year.

4.3.2. The Role of Biomass in Coal-Fired Plants

For the U.S. market, the cofiring of a modest amount of biomass (5–20%, heat basis) will be important to the energy industry. Cofiring biomass with coal can help reduce emissions and thus help to preserve the coal-firing fleet in the U.S. market, particularly in a carbon constrained world. Biomass firing, like coal, can be used as a base-loaded technology, so its impacts on the plant can be minimal. At times, its use can be beneficial to the plant. During wet coal events, the use of biomass can potentially help boost the plant's capacity and increase flame stability [37, 38, 39].

4.4. OTHER REASONS FOR COFIRING

Biomass cofiring is increasingly becoming an essential aspect of the Renewable Portfolio Standards (RPS). An RPS creates a market demand for renewable and clean energy generation. States that have RPS requirements mandate between 4% and 30% of electric generation from renewable sources by a specified time [40]. The percentages and specified dates are state dependent and vary significantly between states. The cofiring of biomass with coal can help to achieve these goals in a cost-effective manner while enabling the use of the existing coal fleet.

The reduction of greenhouse gases has been observed by both demonstration and experimental tests. These reductions have included SO_x and, if properly

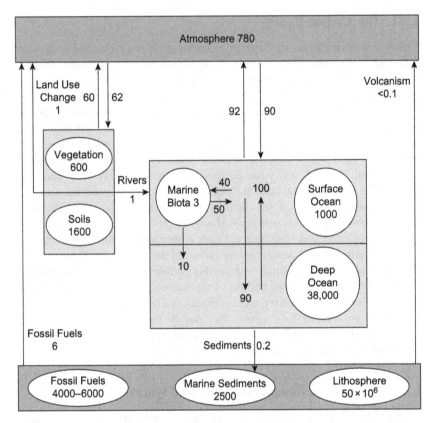

FIGURE 4.3   Inventories and fluxes in the carbon cycle: 2008 estimates. *Source: Adapted from [36].*

designed, NO_x, mercury, and CO_2. Biomass fuels are inherently low in sulfur and mercury. As the biomass share increases in thermal output, SO_2 emissions decrease proportionally [41]. And since biomass is considered a carbon neutral fuel, biomass cofiring will be very important in a carbon constrained world in order to preserve the coal industry. When designed properly with the combustion effects well understood, NO_x reductions have been documented by many demonstration tests. Dependent on the biomass combusted, these benefits may vary. The mechanism of transport and injection into the combustor also determines the amount of benefits obtained [42, 43]. Therefore, biomass cofiring becomes crucial in the effort to reduce greenhouse gas emissions.

4.4.1. SO_2 Management

Almost all of the sulfur in a fuel will convert to SO_2 given enough oxygen availability and residence time. The benefits of sulfur reductions are easily obtained through cofiring biomass with coal. Since most biomass fuels have inherently low sulfur concentrations, SO_x reductions have been observed by many cofiring applications.

4.4.2. NO_x Management

Volatile nitrogen can form reactive species such as amine species (e.g., NH_2, NH_3), which can react in the presence of oxygen, forming NO, some N_2O, and a small portion of N_2 [44]. Additionally, biomass and coal both evolve nitrogen as HCN and as tar nitrogen [45, 46]. The principle of NO_x reduction is to design a system and control the combustion process in order to essentially form more N_2 than other forms of nitrogen species. In a pulverized coal (PC) fired application, one of these techniques involves initiating the combustion process in a fuel-rich region.

Laboratory studies and full-scale demonstration tests have shown that cofiring can achieve rapid fuel nitrogen volatilization in a fuel-rich environment, thus promoting fuel nitrogen reduction. If designed properly, accounting for fuel chemistry and unit configuration and operations, the benefits of NO_x reduction can be achieved [14, 47, 48].

Cofiring may be one of the lowest-cost approaches to reducing fossil CO_2. Local jobs in the forest product and transportation industries can be created, and landfill use of fuels that typically do not have many uses can be reduced. Wood waste is commonly landfilled, so by burning the material, the total reduction can potentially be 3 ton fossil CO_2/ton wood waste burned [31, 37].

4.5. COFIRING IN THE UNITED STATES AND EUROPE

Biomass firing is heavily practiced in the European community and has been demonstrated successfully in many installations. One of the most well-documented and well-studied demonstrations is the one conducted at Denmark's Studstrup Power Station. The demonstration test cofired coal and straw from January 1996 to February 1998 [27]. The Studstrup Power Station boiler used in this demonstration has since been decommissioned. The European market promotes and incentivizes heavy use of biomass combustion in order to meet regulatory demands.

As an approach to increase biomass usage in the U.S. electric utility sector, cofiring development was accelerated in the 1990s. The dominant cofiring experience has been wood waste or switchgrass as the biomass fuel, with Eastern bituminous coal. Other base fuels, such as Black Thunder (PRB) and Shoshone (Western bituminous) coals, have also been tested [16, 42, 49]. The cofiring of wood and coal has been evaluated and demonstrated successfully in numerous electric utilities. By 2003, more than 30 individual cofiring tests had been performed in more than 19 states.

These tests have covered wall-fired and tangentially fired (T-fired) PC boilers as well as cyclone firing [38, 50–52]. Additionally, the recently designed 600-MW VCHEC power plant of Dominion Energy, now under construction, has the capability to cofire up to 20% wood waste with coal (heat input basis). Switchgrass, the second most significant biomass fuel used in cofiring applications, is a native grass that was adapted by the eastern United States. Several cofiring tests and facilities have used switchgrass successfully.

The experience over the years has been significant and successful, but for biomass cofiring to be heavily implemented in the United States, there must exist positive economics through regulation and incentives.

4.6. CHARACTERISTICS OF BIOMASS

There are many definitions for "biomass." Politically, these definitions can depend on the country and state that is being considered. From a theoretical perspective, based on agricultural and silvicultural sciences, biomass fuels can be generated directly and subsequently from the growth and harvest of plant materials [50]. Biomass fuel can include anything from wood, alfalfa, fast-growing grasses such as switchgrass, fruit pulps, sawmill residue, and a host of other fuels. They can also include fecal matter from animals that eat plant materials. Much of the material in urban waste (e.g., paper waste, yard waste, food waste) also originates as biomass.

4.6.1. Types of Biomass

Wood fuels are primarily made up of residues from the wood processing industry. They can also include in-forest residues and urban wood wastes. Herbaceous biomass fuels such as agricultural wastes (e.g., corn stover) and crops grown as fuel such as switchgrass are fundamentally different from woody biomass fuels. This category of biomass can be subdivided into field materials and industrial wastes; it includes a host of materials, including bagasse, switchgrass, miscanthus, giant reed, corn stover, rice straw, rice hulls, oat hulls, corn cobs, vineyard prunings, orchard prunings, stone fruit pits, and many others [1].

Woody Biomass

Woody biomass, particularly mill or forest residue material, is considered a premium biomass fuel. It amounts to approximately one-quarter of the total wood production in the United States [50]. Residues from the wood processing industry typically come from sawmills, veneer and plywood mills, particle board plants, medium density fiberboard (MDF) mills, and different types of pulp mills. These residues include bark, sawdust, planer shavings, sander dust, chips, and any mixture of them, referred to as "hog fuel." In-forest residues have been recovered in the past and used for fuel applications, but these events are rare due to the high cost of these fuels.

Another class of woody biofuels is urban wood wastes, which include manufacturing residues, construction demolition, land-clearing materials, wood from pallet processors, pallets and other dunnage from the commercial transportation industry, and "clean" or untreated woody materials. Treated wood wastes are a part of this category and include railroad ties, used utility poles, and other related products. The treated wood wastes are distinct and separate from the "clean," untreated wood waste [1, 15].

Pellets versus Unprocessed Woody Biomass

Pelletizing biomass fuels has several benefits. A significant amount of moisture can be driven off during the pelletizing process, thereby increasing the calorific value of the fuel. Furthermore, by increasing the bulk density, the materials handling aspect is easier and less expensive. However, the cost of producing the pellets must be considered when evaluating the potential of utilizing pellets versus regular woody biomass fuels [53, 54].

There is no specification for the heat content of pellets, but much of the literature indicates a range between 7450 Btu/lb to above 8000 Btu/lb, depending on the wood species and moisture content. Table 4.7 lists several other properties for pelletized biomass fuels that were obtained from the Pellet Fuels Institute (PFI).

TABLE 4.7 Residential/Commercial Fuel Standards from the Pellet Fuels Institute

Fuel Property	Super Premium	Premium	Standard	Utility
Bulk Density (lb/ft³)	40−46	40−46	38−46	38−46
Diameter (in.)	0.25−0.285	0.25−0.285	0.25−0.285	0.25−0.285
Diameter (mm)	6.35−7.25	6.35−7.25	6.35−7.25	6.35−7.25
Pellet Durability Index	$\geq$97.5	$\geq$97.5	$\geq$95.0	$\geq$95.0
Fines (% at pellet mill gate)	$\leq$0.5	$\leq$0.5	$\leq$0.5	$\leq$0.5
Inorganic Ash (%)	$\leq$0.5	$\leq$1.0	$\leq$2.0	$\leq$6.0
Length (% > 1.5 in.)	$\leq$1.0	$\leq$1.0	$\leq$1.0	$\leq$1.0
Moisture (%)	$\leq$6.0	$\leq$8.0	$\leq$8.0	$\leq$10.0
Chloride (ppm)	$\leq$300	$\leq$300	$\leq$300	$\leq$300

Source: [53].

Agricultural Biomass

The classification of agricultural biomass fuel covers a host of field wastes and materials, including corn stover, switchgrass, rice straw, wheat straw, hay, and others. For switchgrass and related herbaceous fuels, moisture concentration is typically lower than most biomass. Typically, straws from different sources have high concentrations of volatiles and high chlorine and high alkaline contents [23, 27]. Variations in straw ash composition depend heavily on where it is grown (weather conditions, soil, etc.). Nitrogen concentration in straw varies, depending on fertilizer concentration. These biofuels are of importance to countries such as Denmark and China. Table 4.8 depicts the proximate and ultimate analysis along with calorific value for mulch hay and reed canary grass.

Agribusiness Biomass-Processing Wastes

This classification of biomass fuel includes rice hulls, corncobs, corn seeds, cotton gin trash, and other material. Typically, the higher heating values are substantially lower than those of woody biomass, switchgrass, or coal [1].

TABLE 4.8 Typical Proximate Analysis, Ultimate Analysis, and Calorific Values for Selected Agricultural Biomass

Parameter	Mulch Hay	Corn Stover
Moisture (%)	19.5	65.2
Proximate Analysis (Dry Basis)		
Volatile Matter (%)	77.6	76.1
Fixed Carbon (%)	17.1	19.8
Ash (%)	5.3	4.1
Ultimate Analysis (Dry Basis)		
Carbon (%)	46.5	45.8
Hydrogen (%)	5.7	6.1
Oxygen (%)	40.6	42.9
Nitrogen (%)	1.7	1.0
Sulfur (%)	0.2	0.1
Ash (%)	5.3	4.1
Higher Heating Value (Btu/lb, Dry Basis)	8058	7239

Source: [15].

TABLE 4.9 Typical Proximate Analysis, Ultimate Analysis, and Calorific Values for Selected Agribusiness Biomass

Parameter	Rice Hulls	Cotton Gin Trash
Moisture (%)	7–10	7–12
Proximate Analysis (Dry Basis)		
Volatile Matter (%)	63.6	75.4
Fixed Carbon (%)	15.8	15.4
Ash (%)	20.6	9.2
Ultimate Analysis (Dry Basis)		
Carbon (%)	38.30	42.77
Hydrogen (%)	4.36	5.08
Oxygen (%)	35.45	35.38
Nitrogen (%)	0.83	1.53
Sulfur (%)	0.06	0.55
Ash (%)	21.00	14.69
Higher Heating Value (Btu/lb, Dry Basis)	6400	6700

Source: [1].

Feedstocks such as rice hulls contain less than 10% moisture [32]. Table 4.9 shows typical proximate analysis, ultimate analysis, and heating value for rice hulls and cotton gin trash.

Orchard and Vineyard Materials

Orchard and vineyard materials are another subcategory of agribusiness residues. These materials tend to have heating values that are much closer to those of the woody material. The ash concentrations are lower when compared to those of the agribusiness-processing wastes. Table 4.10 shows some typical proximate analysis, ultimate analysis, and heating values of both orchard and vineyard prunings.

Fecal Materials

Fecal matter is of interest to the biomass community due to regulatory pressures, forcing utilities to consider their use in boilers and gasifiers. Fecal matter tends to be one of the more difficult fuels to handle; it is considered to be among the most difficult opportunity fuels in the biomass community [1, 55].

TABLE 4.10 Typical Proximate Analysis, Ultimate Analysis, and Calorific Values for Selected Orchard and Vineyard Materials

Parameter	Orchard Prunings	Vineyard Prunings
Moisture (%)	20–40	20–40
Proximate Analysis (Dry Basis)		
Volatile Matter (%)	80.75	83.91
Fixed Carbon (%)	17.84	14.71
Ash (%)	1.41	1.38
Ultimate Analysis (Dry Basis)		
Carbon (%)	50.18	49.20
Hydrogen (%)	5.96	6.00
Oxygen (%)	41.40	43.13
Nitrogen (%)	0.90	0.25
Sulfur (%)	0.09	0.04
Ash (%)	1.47	1.38
Higher Heating Value (Btu/lb, Dry Basis)	7230	8190

Source: [32].

Typically, this fuel type has high nitrogen and ash concentration. In addition, the ash compositions tend to be highly reactive. Table 4.11 depicts some of the fuel compositions of several fecal materials. Table 4.12 shows chemical fractionation analyses for sheep manure.

4.6.2. Standard Characteristics of Biofuels

Generally, woody material is sulfur free, low in fuel nitrogen, and low in ash. On the other hand, clean urban waste can be high in fuel nitrogen and ash compared to mill or forest residue [1]. The woody biomasses are also typically highly reactive with significant volatility. Despite the generalizations, woody biomass fuels are complex and variable. Fuel characteristics depend on source (including the species), type of processing facility, type of processing, and other factors [1, 56]. Herbaceous biofuels are significantly higher in ash than woody biomass. The ash can have significant slagging and fouling potential, which can be of concern. Certain herbaceous biofuels can have high concentrations of nitrogen and chlorine. Materials such as switchgrass have very low-bulk densities, thus presenting both handling and transport issues.

TABLE 4.11 Typical Proximate Analysis, Ultimate Analysis, and Calorific Value for Selected Fecal Material

	Chicken Manure	Sheep Manure	Dairy Free– Stall Manure
Moisture (%)	25.30	47.8	69.8
Proximate Analysis (Dry Basis)			
Volatile Matter	66.05	65.10	30.10
Fixed Carbon	13.23	14.00	7.40
Ash	20.72	20.90	62.50
Ultimate Analysis (Dry Basis)			
Carbon	40.39	40.60	22.60
Hydrogen	5.05	5.10	2.90
Oxygen	29.05	30.70	10.80
Nitrogen	3.99	2.10	1.10
Sulfur	0.80	0.60	0.10
Ash	20.72	20.90	62.50
Higher Heating Value (Btu/lb)	6617	6895	3644

Source: [1].

TABLE 4.12 Chemical Fractionation Analysis for Sheep Manure

Average CHF Analysis	Removed by H_2O and NH_4OAc (%)	Removed by HCl (%)	Remaining (%)
Silicon	0	0	100
Aluminum	32	0	68
Calcium	58	41	1
Magnesium	77	21	2
Sodium	96	2	2
Potassium	97	1	2

Source: [1].

TABLE 4.13 Typical Bulk Density for Several Biomass Fuels

Biomass Fuel	Typical Bulk Density (kg/m³)	Typical Bulk Density (lb/ft³)
Sawdust	288	18
Urban Wood Waste	288	18
Switchgrass	80–130	5–8
Corn Stover	112	7
Peach Pits	480.5	30
Walnut Shells	533	33.3
Cotton Gin Trash	98	6.1
Rice Hulls	288	18
Vineyard Prunings	213	13.3
Hybrid Corn Seed	400.5	25

Source: [1, 16, 26].

When cofiring, the differences among the biomass fuels must be considered, and these differences can be as significant as the differences among those of coal. The interaction between the coal and biomass must be considered. More often than not, coal constitutes the greater fraction within the fuel blend and therefore cannot be ignored.

The bulk density of woody biomass is typically in the range of 255 to 320 Kg/m³ (16–20 lb/ft³). For switchgrass and related herbaceous biomass fuels, the bulk densities are typically much lower than the woody material. Field crops such as switchgrass will have densities in the range of 80 Kg/m³ (5 lb/ft³) loose and 130 Kg/m³ (8 lb/ft³) baled. For vineyard prunings, stone fruit pits, and related products, the densities are higher and approach those of the woody biomass. Table 4.13 shows the typical bulk density for several biomass fuels. In comparison, the bulk density of coal is typically in the range of 50 to 70 lb/ft³. This is significantly greater than many biomass fuels; the lower-bulk densities must be considered when designing a system to handle biomass cofiring [57, 58].

4.6.3. Fuel Porosity and Its Implications

Total void volume or porosity can be calculated from the moisture concentration of the fuel. For coals, moisture concentrations range from 25% to 30% (and sometimes higher) in the lignite and subbituminous ranks to as little as 1%

to 2% among the higher ranks [59]. For the biomass fuels, moisture concentration varies significantly and can be much higher when compared to coal.

Woody biomass is a porous material, and its porosity is caused by hollow fibers that make up the woody material. The porosity or void volume exists as macropores such as resin canals and related structures that are used in the transportation of moisture and nutrients within the tree. Herbaceous biofuels are also porous solids with significant void volumes. As a result of the high concentration of inorganic matter typically associated with these biofuels, the calculations of porosity are less precise. Porosity influences the moisture concentration and other constituents that influence the bulk density and chemical properties of the fuel [1]. Bulk density is important when considering the design of the handling and transport system. The chemical constituents influence the combustion behavior of the fuel.

4.6.4. Proximate and Ultimate Analysis and Higher Heating Value

The proximate and ultimate analysis and higher heating value of a fuel provide a broad understanding of its combustion characteristics. General parameters such as fixed carbon, volatile matter, ash concentration, moisture, and others provide a crude estimate of how the fuel will behave when it is burned and can provide certain traditional measures such as ash loading and estimates for emission constituents such as sulfur. Proximate and ultimate analysis along with higher heating values for several of the biomass fuels was given in the preceding tables. These parameters are necessary but are not sufficient to define and detail the behavior of a particular fuel and/or fuel blend. Other measures such as reactivity for both the organic and inorganic fractions will be discussed later.

4.6.5. Ash Elemental Analysis

Ash elemental analysis is another important parameter that must be understood. This measure quantifies the constituents that make up the inorganic fraction of the fuel. Understanding these parameters is essential to understanding the fouling and slagging behavior of the fuel. Table 4.14 show the ash elemental analysis for several biomass fuels. Agricultural biomass fuels in significant quantities, for example, can produce significant slagging and fouling problems in a boiler. Alkali metal concentration (most notably potassium) in agricultural biomass fuels can make up a large portion of the inorganic fraction. Potassium is an essential part of the nutrient cycle. Consequently, the combustion of certain biomass fuels needs to be limited in different combustion mechanisms and firing configurations. The combustion process must be monitored operationally, and the designs of the systems must mitigate these issues.

TABLE 4.14 Ash Elemental Analysis for Several Biomass Fuels

Elemental Composition (wt%)	Hardwood Sawdust	Fresh Switchgrass	Wheat Straw	Hazelnut Shell
SiO_2	23.70	65.18	55.70	33.70
Al_2O_3	4.10	4.51	1.80	3.10
TiO_2	0.36	0.24	0.00	0.10
Fe_2O_3	1.65	2.03	0.70	3.80
CaO	39.90	5.60	2.60	15.40
MgO	4.84	3.00	2.40	7.90
Na_2O	2.25	0.58	0.90	1.30
K_2O	9.81	11.60	22.80	30.40
P_2O_5	2.06	4.50	1.20	—
SO_3	1.86	0.44	1.70	1.10

Source: [1, 60].

4.6.6. Trace Elements

Typically, trace metal concentrations in woody biomass are lower than those present in coal. For agricultural materials, trace metals are typically a consequence of fertilizer practices and any economic activities that occur within the area. Some fertilizers contain as much as 4570 mg/kg Cr, 430 mg/kg Pb, and 3000 mg/kg Zn. Arsenic concentrations in agricultural material are typically lower than those associated with Western low-rank coals and lignites. Table 4.15 shows typical trace metal concentrations for woody biomass and agricultural materials. For comparison, Table 4.16 shows typical trace metal concentrations for several coals.

4.7. REACTIVITY MEASURES FOR BIOMASS

Reactivity for both the combustible and inorganic fractions of the fuel plays a significant role in how the fuel will behave in the boiler. Understanding both the organic and inorganic portions—particularly when blending—will help determine the optimal range of parameters by which a boiler can operate.

4.7.1. Reactivity of Combustibles

Understanding the kinetics of fuel pyrolysis and combustion is essential in technology development and efficient utilization of solid fuels. Reactivity

TABLE 4.15 Trace Metal Concentrations in the Ash from Woody Biomass and Agricultural Materials

Metal	Agricultural Material (mg/kg)	Woody Biomass (mg/kg)
Antimony	10	—
Arsenic	3.4—12	0.5—6.3
Barium	41—220	51.5—130
Beryllium	0.01—0.06	≤0.1
Cadmium	0.36—1.1	1.5—6.7
Chromium	11—20	16.8—128.4
Cobalt	2.9—14	—
Copper	14—31	5.6—76.9
Lead	21—55	2.7—70
Mercury	BDL	<0.05
Molybdenum	—	3—14
Nickel	4.4—5.8	11—137
Selenium	BDL	5.0
Vanadium	11—20	27
Zinc	40—190	99—560

BDL = Below detection limit
Source: [1].

influences the process of combustion, NO_x emissions, carbon burn-out, unburned carbon in flyash, and other related phenomena [61–64].

From a practical standpoint, volatility is influenced by two factors: fuel particle size and combustion temperature. Particles that are smaller release more volatiles and at a faster rate. In addition, higher temperatures cause more volatiles to be released, whereas lower-temperature environments promote char formation [26, 31, 65].

The combustion effects of cofiring biomass involve changing the process of combustion within any firing mechanism [26]. Biomass introduction into coal-fired PC boilers adds a fuel where the gas-phase combustion and volatilization are dominant. This is in contrast to char formation and gas-solids oxidation as the dominant reaction in coal combustion. The volatilization stage can be further enhanced through smaller particle size. For instance, in sawdust

TABLE 4.16 Trace Metal Concentrations in the Ash from Several Coals

Metal	Illinois #6 (mg/kg)	Western Lignite (mg/kg)
Antimony	0.09	—
Arsenic	0.33	32.6
Barium	40	1306
Beryllium	0.77	20.3
Cadmium	0.70	1.4
Chromium	23	57.7
Manganese	49	—
Copper	16	—
Lead	5.8	74.2
Mercury	0.28	0.3
Nickel	17	1093
Selenium	1.5	3.9
Vanadium	—	147
Strontium	17	—
Zinc	31	237.3

BDL = Below detection limit
Source: [66].

combustion, the drying and volatile evolution stages are overlapped. Volatile evolution occurs significantly during the drying stage, releasing combustible gaseous compounds. Volatile matter/fixed carbon ratios of 4 to 5 further indicate the dominance of gas-phase oxidation of the volatile species [26].

Reactivity measures for the combustible portion of the fuel include, but are not exclusive to, the structure of the fuel, drop tube reactor kinetics, the evolution of specific elements, and ratios from other measures. These measures can help to determine how certain fuels will behave within the boiler.

4.7.2. Structure and Reactivity

One of the key areas in understanding fuel pyrolysis and reactivity is in the fuel structure. The structure of a fuel will determine how easily a fuel combusts, the flame intensity, and where and how the elements or compounds exist. Consequently, it has a significant impact on the behavior of the fuel [67–69].

Lignocellulosic materials, which are common in woody material, are not directly combustible but will decompose to form volatile pyrolysis under a sufficient source of energy. For the cellulosic component, it is mostly converted to the combustible and noncombustible volatiles. Lignin components, on the other hand, contribute to the char fraction [25, 26, 31]. Lignin is the lesser reactive material in biomass fuels, consisting of a basic skeleton of four or more substituted phenylpropane units.

4.7.3. Drop Tube Kinetics

Drop tube reactor studies are used to determine fuel reactivity based on weight loss measurements at various temperatures. Studies of various solid fuels and blends have been conducted over the years [70]. Figures 4.4 through 4.6 are devolatization kinetics for sawdust and urban wood waste; kinetic parameters such as preexponential constant and activation energy show that combustion will occur rapidly and completely in the boiler.

Evolution of Specific Elements

The evolution of particular species such as carbon and nitrogen can be measured and analyzed by use of a drop tube reactor analysis [71]. Figures 4.7 through 4.9 show the nitrogen and carbon evolution pathways for Long Fork bituminous coal, sawdust, and fresh switchgrass. Rapid release of the nitrogen—particularly for the sawdust fuel—can promote nitrogen evolution

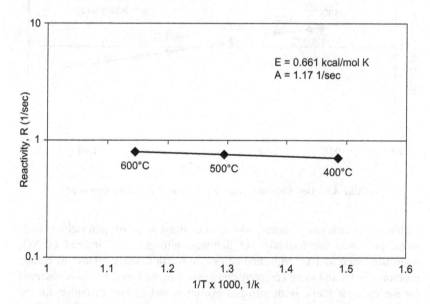

FIGURE 4.4 Low-temperature devolatilization reactivity of sawdust. *Source: [1].*

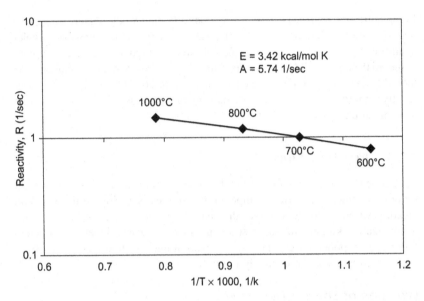

FIGURE 4.5   Higher-temperature devolatilization reactivity of sawdust. *Source: [1].*

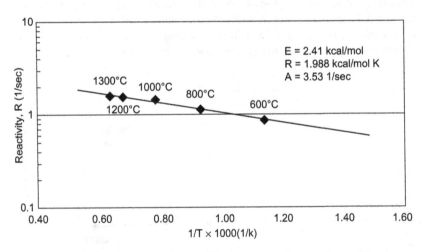

FIGURE 4.6   Devolatilization reactivity of urban wood waste. *Source: [1].*

within a fuel-rich environment, which, in contrast to an oxygen-rich environment, promotes the formation of diatomic nitrogen (N_2) instead of NO_x molecules such as NO, NO_2, and so on. In addition, early release of carbon promotes faster and more complete combustion in the furnace; this is observed for the biomass fuels. Both nitrogen evolution and carbon evolution for the Long Fork bituminous coal are delayed when compared to the biomass fuels.

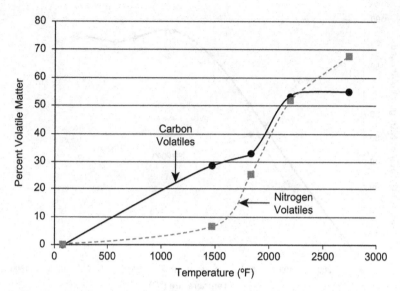

FIGURE 4.7 Nitrogen and carbon volatile evolution from Long Fork as a function of temperature. *Source: [1].*

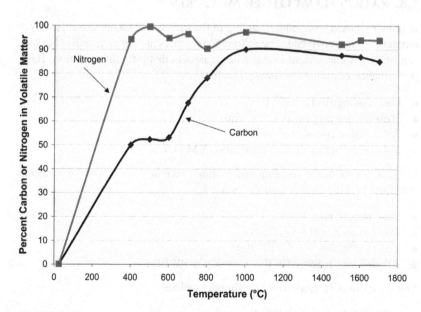

FIGURE 4.8 Nitrogen and carbon volatile evolution from sawdust as a function of temperature. *Source: [1].*

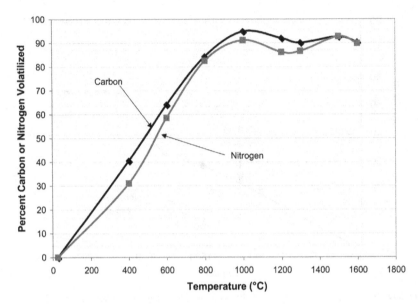

FIGURE 4.9 Nitrogen and carbon volatile evolution from fresh switchgrass as a function of temperature. *Source: [1].*

4.8. RATIOS FROM OTHER MEASURES

In order to have a basic understanding of the fuel or fuel blend, several ratios or measures can be analyzed. The following measures are not exhaustive, but they can provide some insight into the basic aspects that affect combustion. Basic performance or system parameters include

- Fuel loading (lb fuel/10^6 Btu)
- Moisture loading (lb H_2O/10^6 Btu)
- Ash loading (lb ash/10^6 Btu)
- Volatile matter to fixed carbon ratio (VM/FC)

In order to look at the slagging and fouling potentials, the following measures can provide some insight into the process:

- Base to acid ratio (B/A)
- Iron to calcium (Fe_2O_3/CaO)
- Chlorine loading (lb Cl/10^6 Btu)
- Alkali metal loading (lb [$Na_2O + K_2O$]/10^6 Btu)

Some measures of emissions constituents include

- Sulfur loading (lb SO_2/10^6 Btu)
- Nitrogen loading (lb N/10^6 Btu)

Understanding the inorganic behavior of the fuel is just as important as understanding the organic behavior of the fuel. The behavior and constituents can dictate the phenomena of slagging, fouling, and deposition in a boiler.

The reactivity of a fuel ash can be measured and analyzed through a method known as chemical fractionation (CHF). CHF is used to selectively extract different elements based on solubility. The coal is initially extracted with water to remove water-soluble elements such as Na and other elements that are most probably associated with the groundwater in the coal. Next, extraction is conducted with 1 M of ammonium acetate, removing elements such as Na, Ca, and Mg, which are typically bounded as salts. Residues from the extraction with 1 M of ammonium are then extracted with 1 M HCl in order to remove acid-soluble species such as Fe and Ca.

Upon completion of all three extractions, insoluble minerals such as clays, quartz, and pyrite are known as residues. Species that are water soluble or are easily extracted during the first two stages are highly reactive. Tables 4.17 through 4.19 depict CHF data for several U.S. coals [15]. Tables 4.20 and 4.21 depict CHF data for several biomass fuels [1].

Scanning electron microscopy (SEM) coupled with energy dispersive spectroscopy (EDS) is a technique that measures the local chemical composition along with shape and size of the individual mineral grains in the fuel particles. This provides both morphological and chemical information about

TABLE 4.17 Chemical Fractionation Analysis for Representative Lignite

Average CHF Analysis	Removed by H_2O (%)	Removed by NH_4OAc (%)	Removed by HCl (%)	Remaining (%)
Silicon	0	0	0	100
Aluminum	1	0	15	84
Iron	5	4	4	87
Titanium	3	5	2	91
Phosphorus	13	24	56	7
Calcium	2	49	48	1
Magnesium	4	60	32	4
Sodium	20	66	8	6
Potassium	8	31	6	55

Source: [72].

TABLE 4.18 Chemical Fractionation Analysis for Representative Wyoming Subbituminous Coal

Average CHF Analysis	Removed by H$_2$O (%)	Removed by NH$_4$OAc (%)	Removed by HCl (%)	Remaining (%)
Silicon	0	0	0	99
Aluminum	1	4	30	65
Iron	4	9	59	28
Titanium	4	8	7	80
Phosphorus	6	15	62	17
Calcium	2	62	33	4
Magnesium	3	71	20	6
Sodium	30	60	4	3
Potassium	15	21	6	51

Source: [72].

TABLE 4.19 Chemical Fractionation Analysis for Representative Central Appalachian Bituminous Coal

Average CHF Analysis	Removed by H$_2$O (%)	Removed by NH$_4$OAc (%)	Removed by HCl (%)	Remaining (%)
Silicon	0	0	0	100
Aluminum	0	0	2	98
Iron	11	0	14	74
Titanium	0	0	6	94
Phosphorus	4	4	12	80
Calcium	30	55	1	14
Magnesium	8	8	4	80
Sodium	37	20	3	41
Potassium	3	1	2	94

Source: [72].

TABLE 4.20 Chemical Fractionation Analysis for Representative Switchgrass Fuel

Average CHF Analysis	Removed by H_2O and NH_4OAc (%)	Removed by HCl (%)	Remaining (%)
Silicon	0	0	100
Aluminum	0	0	100
Calcium	82	10	8
Magnesium	80	7	13
Sodium	52	0	48
Potassium	78	15	7

Source: [1].

TABLE 4.21 Chemical Fractionation Analysis for Representative Pine Shavings Wood Fuel

Average CHF Analysis	Removed by H_2O and NH_4OAc (%)	Removed by HCl (%)	Remaining (%)
Silicon	29	0	71
Aluminum	30	0	70
Calcium	92	6	2
Magnesium	72	10	18
Sodium	77	0	23
Potassium	65	0	35

Source: [1].

the material. SEM/EDS analysis is one of several techniques that can be used to identify mechanisms of ash agglomeration behavior. Through the use of a computer, SEM/EDX can be computer controlled, otherwise known as CCSEM [46]. The electron beam scans over the field of view, locating bright inclusions that correspond to mineral or ash species. The mineral grain is then sized with EDS, providing the chemical composition of the grain [15].

4.9. COMPARISONS OF BIOMASS TO COAL

The comparison of biomass and coal relies on the coal data presented in Chapter 3. Data are presented here for completeness, recognizing—as with chemical fractionation—unique problems when blending biomass and coal. Such problems may be more pronounced when blending biomass and coal relative to blending two coals.

When compared to those for coals, heating values for biomass fuels can be significantly less. Moisture concentrations of the biomass fuels are typically higher than those of most coals, although lignite can be the exception. The disparity between biomass fuels and coals decreases as rank decreases. Thus, the effects of blending decrease when low-rank coals are cofired with biomass fuels.

4.9.1. Central Appalachian Bituminous Coal

Tables 4.1, 4.2, 4.4, and 4.5 earlier in this chapter gave some of the basic fuel analyses for a typical Central Appalachian bituminous coal and several of the biomass fuels. More data are available in Chapter 3. Fixed carbon concentration for the Central Appalachian bituminous coal is significantly higher than that for any of the biomass fuels. Conversely, volatile matter concentration of the biomass fuels is significantly higher than that of coal. These differences are depicted by the ratio of volatile matter to fixed carbon. Sulfur, and in many cases nitrogen, concentration is much less for the biomass fuels. The base/acid ratio of the biomass fuels typically is greater than that of Central Appalachian coal. Potassium concentration inherent to the biomass fuels merits attention, particularly when the ash concentration is of significant value.

4.9.2. Illinois Basin Coal

Illinois Basin coals are bituminous coals. The disparity between various forms of biomass and Illinois Basin coals is significant. Inherent to the Illinois Basin coals are the higher sulfur and iron concentrations; this has presented slagging problems to utilities where the boilers were not designed to handle these fuel properties. When cofired with high-calcium and high-ash biomass fuels, the effects can be detrimental, since both iron and calcium are fluxing agents. It is essential to look at the interaction between calcium and iron when the base fuel utilized is an Illinois Basin coal.

4.9.3. Powder River Basin Coal

The Powder River Basin coals are a lower-rank subbituminous coal that is of increasing importance to the electricity sector. Driven by two main factors— lower pricing and lower sulfur concentration—the use of PRB coal is almost equivalent and will probably exceed that of Eastern bituminous coal. PRB coals tend to exhibit higher moisture concentrations with a lower ash content when compared to the higher-rank coals. Calcium concentration also tends to be higher; this is not true of all PRB coals or subbituminous coals. When blending with other biomass fuels, it is important to note that the calcium concentration does not increase significantly. High calcium concentrations will contribute to the effects of reflective ash, thereby decreasing heat transfer in the furnace. This will push the flame upward in the boiler, increasing the slagging and fouling issues and subjecting the tubes to temperatures they were not designed to withstand.

4.9.4. Lignite

Ash concentration inherent to the lignite coals can be significantly higher than that inherent to other coals, and the moisture concentration is comparable to those of PRB coals. Furthermore, the higher heating values for the lignite coals are significantly lower than those of other higher-rank coals. Calcium concentration tends to also be slightly higher than those of the higher-rank coals but is less than those of the PRB coals. Certain biomass fuels will have heating values and other properties comparable to those of the lignite coal. Consequently, the effects of blending diminish as the rank decreases; this is most apparent when blending biomass fuels with lignite.

4.10. THE CHEMISTRY OF COFIRING

Certain aspects of fuel blending can be analyzed by performing averaging calculations, although this is not possible with other aspects [16]. The impact of cofiring biomass with coal is shown for a woody biomass with a bituminous Central Appalachian coal in Table 4.22. There is a significantly higher mass flow when more biomass is added to the blend; a 28% increase in fuel flow can be seen with a 20% biomass blend with coal. Moisture concentration increases with an increase in the percentage of biomass in the fuel mass. The higher heating value of the fuel mass is decreased. In addition, slagging and fouling tendencies are increased. Several indicators such as base/acid ratio, alkali metal loading, and Fe_2O_3/CaO ratio provide such indications [16].

Cofiring in wall-fired PC boilers through separate injection can concentrate the biomass in a desired location within the flame of a burner. Under certain circumstances, the volatiles from biomass are concentrated and can be released on injection into the furnace. The release of volatiles from biomass can promote early ignition of the total fuel mass. This early ignition in turn can consume additional oxygen, increasing the fuel staging effects of low-NO_x burners [48].

4.10.1. Reactivity and Cofiring

The reactivity of a fuel is highly dependent on the structure of the fuel. The manner in which fuels behave can be better understood by studying the kinetics

TABLE 4.22 Impact of Cofiring Woody Material with Central Appalachian Bituminous Coal

Parameter	0% Wood Cofiring	5% Wood Cofiring (Heat Input Basis)	10% Wood Cofiring (Heat Input Basis	20% Wood Cofiring (Heat Input Basis)
Mass Percentage of Wood	0%	11.25%	21.12%	37.59%
Blend Heating Value (Btu/lb)	12,114	11,317	10,618	9450
Base/Acid	0.20	0.22	0.24	0.29
Fe_2O_3/CaO	3.63	2.43	1.78	1.10
lb Cl/10^6 Btu	0.10	0.10	0.09	0.08
lb (Na_2O + K_2O)/10^6 Btu	0.38	0.39	0.39	0.41
lb SO_2/10^6 Btu	0.88	0.85	0.81	0.73
lb Fuel/10^6 Btu	83	88	94	106
lb H_2O/10^6 Btu	5.91	9.79	13.67	21.43
lb Ash/10^6 Btu	9.51	9.26	9.02	8.53
lb N/10^6 Btu	1.11	1.11	1.11	1.12
VM/FC	0.62	0.73	0.84	1.09

Source: [16].

and evolution of the specific elements such as carbon, nitrogen, sulfur, and others. These measures provide more detailed understanding of the combustion process.

Understanding the interaction between coal and biomass is important in order to successfully implement cofiring applications. For that reason, Foster Wheeler and The Energy Institute of Pennsylvania State University conducted drop tube reactor analysis in order to study the interaction between biomass and coal. Reactivity measures were conducted for the parent fuels and blends of mixed hardwood/softwood sawdust and a Central Appalachian coal. The two blends tested were 20% sawdust/80% coal and 35% sawdust/65% coal, and were conducted based on a heat input basis. Table 4.23 and Figures 4.10 and 4.11 compare the parent fuel and blend reactivities based on the Arrhenius equation.

These plots show that the 35% sawdust blend is more reactive than either of the parent fuels. For the 20% sawdust blend, activation energy was lowest among the parent fuels and blends, thus indicating that pyrolysis will initiate

TABLE 4.23 Reactivity Constants of Parent Fuels and Blends

Material/Fuel	Preexponential Constant (A)	Activation Energy (E)
Sawdust	2.75 sec^{-1}	1.81 kcal/mol
Long Fork Coal	6.96 sec^{-1}	4.65 kcal/mol
20% Sawdust Blend	2.04 sec^{-1}	1.23 kcal/mol
35% Sawdust Blend	3.76 sec^{-1}	2.26 kcal/mol

Source: [27].

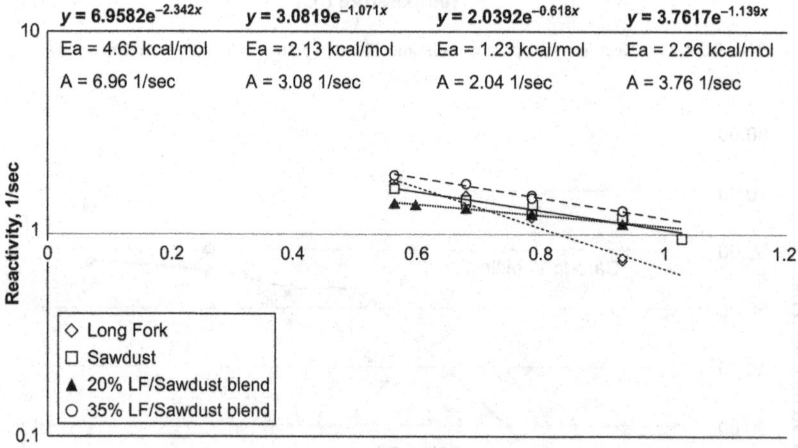

FIGURE 4.10   Reactivity plots of parent fuels and blends. *Source: [73].*

earlier and more rapidly with blends rather than with either parent fuel. This increased reactivity was consistent with previous studies of blending a reactive PRB subbituminous coal with a less reactive Eastern bituminous coal [15]. These data imply that the blends relative to the parent fuels will experience easier and more rapid ignition and thus more rapid and complete burn-out of the total fuel mass.

Despite the increased fuel and air mass, combustion can still be managed in the furnace. Since there is more complete burn-out, unburned carbon concentration can be managed. In addition, more intense flames may be observed despite the increase in moisture concentration [64]. These results are consistent with full-scale experience that shows that combustion with certain biomass fuels increases the reactivity of the fuel mass.

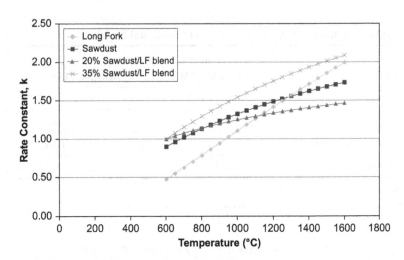

FIGURE 4.11 Reactivity of parent fuels and blends versus temperature. *Source: [64].*

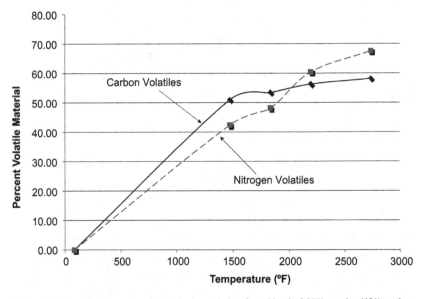

FIGURE 4.12 Nitrogen and carbon volatile evolution from blend of 35% sawdust/65% coal as a function of temperature. *Source: [64].*

4.10.2. Evolution of Specific Elements and Compounds

Laboratory studies have shown that cofiring can achieve rapid fuel nitrogen volatilization in a fuel-rich environment, thus promoting fuel nitrogen reduction. Figure 4.12 shows the nitrogen and carbon evolution pathways of the Long

Fork bituminous coal and the blend of 35% sawdust with 65% coal; the plot looks at the early stages of volatile evolution. These curves support the NO_x reductions observed in numerous cofiring demonstrations. As biomass is added to the blend, the rate of nitrogen volatile evolution is increased when compared to the rate with coal alone. These results indicate that biomass cofiring can enhance volatility of the fuel mass.

NO_x reductions have been observed in several demonstration tests, but the system must be designed properly, accounting for fuel chemistry and injection location. These tests demonstrated the benefits of separate injection of the biomass at the center of the fireball. The presence of the biomass promoted early ignition of the total fuel mass. In a wall-fired application, the introduction of the biomass down the center pipe created a flame within a flame. The early ignition of the biomass at the center promoted the combustion of the coal flame. Thus, more nitrogen could be released within a fuel-rich region [48].

4.11. BURNING PROFILES OF BIOMASS–COAL BLENDS

Thermal analytical techniques such as thermogravimetric analysis (TGA) have been used to characterize the thermal behaviors of coals and, more recently, biofuels. This technique was adapted by Babcock and Wilcox as one method in evaluating fuels for boiler design purposes. Four characteristic temperatures are typically interpreted. The first initiation temperature occurs where the weight first begins to fall; the second initiation temperature occurs where the weight loss accelerates due to the onset of char combustion. The third is the peak temperature where the weight loss is maximum; and the fourth is the burn-out temperature where the weight is constant, indicating the completion of combustion [24, 25].

The commonly derived combustion parameters are intrinsic reactivity, volatiles release profiles, and burning profiles [74]. TGA plots are not a direct substitute for kinetics, but they can illustrate the combustion process [26]. Burning profiles are useful in measuring a fuel's chemical reactivity. They provide information on the oxidation rates from ignition to completion in a comparative manner. Fuels that have similar profiles will tend to have similar performance [75].

Derivative thermogravimetric analysis (DTG) can be used to characterize and predict oxidation behaviors of solid fuels. A plot of the rate at which a solid fuel changes weight as a function of temperature, heated at a constant rate, is termed a burning profile [75]. Burning profiles provide information on the combustion rate: reaction intensity, heat release rate, and residence time requirements relative to a known fuel combustion characteristic.

Four characteristic temperatures are typically interpreted. The first initiation temperature occurs where the weight first begins to fall; the second initiation temperature occurs where the weight loss accelerates due to the onset of char combustion. The third is the peak temperature where the weight loss is maximum, and the fourth is the burn-out temperature where the weight is constant, indicating the completion of combustion. The first peak represents the moisture loss; in some cases, completion temperature can be as high as 204°C (400°F), but this temperature is typically achieved by about 96°C (205°F). Excluding the initial moisture loss, the entire burning profile represents oxidation of the fuel. Upon completion of ignition, the rate of weight loss varies widely, and the height of the oxidation peak is proportional to the reaction intensity.

Experimental analysis for blends of selected biomass fuels—wood waste, corn stover, and switchgrass—with an Eastern bituminous coal and a PRB coal were conducted by Foster Wheeler NAC (see [76]). TGA and DTG analysis were conducted for the parent fuels and then the blends of 10%, 20%, and 30% biomass, by mass input. The following paragraphs discuss some of the results obtained from the study [76].

Figures 4.13 and 4.14 show the TGA and DTG results for 100% wood waste, 100% Eastern bituminous coal, and their blends at a ramping rate of 20°C/min. It was apparent from the TGA curves of the blends that the initiation temperature for char oxidation occurred much sooner in the experimental results compared to calculated weighted average values. This indicated that the blends were more reactive than predictions based on weighted average calculations. When comparing the DTG curves obtained from experimental results with calculated weighted average values, the effects of blending exceeded the weighted average. This was even more pronounced—for both pyrolysis and char oxidation—as the percentage of wood waste in the blend increased.

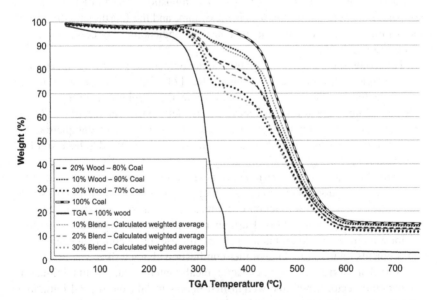

FIGURE 4.13 TGA curve for 100% wood waste, 100% Eastern bituminous coal, and their blends at 20°C/min ramp rate. *Source: [76].*

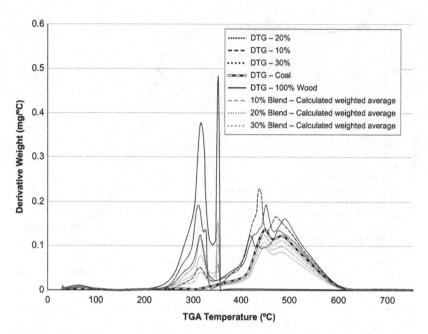

FIGURE 4.14 DTG curve for 100% wood waste, 100% Eastern bituminous coal, and their blends at 20°C/min ramp rate. *Source: [76].*

Figures 4.15 and 4.16 depict the TGA and DTG profiles for 100% switchgrass, 100% Eastern bituminous coal, and their blends. The effects of blending appeared to be suppressing the reactivity of the fuel mass, particularly in the pyrolysis stage of the switchgrass blends. Based on traditional analyses such as oxygen to carbon atomic ratio, hydrogen to carbon atomic ratio, volatile matter to fixed carbon ratio, and volatile matter concentration, switchgrass was expected to be the most reactive fuel of the three biomass fuels studied. However, the results indicated the opposite effect. It is believed that the effects of weathering on the switchgrass fuel itself caused the results observed when blending switchgrass with the Eastern bituminous coal.

Figures 4.17 and 4.18 depict the TGA and DTG profiles for 100% corn stover, 100% Eastern bituminous coal, and their blends. Similar to those for wood waste, experimental results for the blends significantly exceeded the weighted average calculation for both pyrolysis and char oxidation. This effect was more pronounced as the percentage of biomass in the blend increased.

When blending a more reactive fuel such as PRB coal with the biomass fuels, the interaction between the fuels appeared to be enhanced. TGA and DTG analysis for 100% wood waste, 100% PRB coal, and their blends are depicted in Figures 4.19 and 4.20. At the peaks of the DTG curves, the measured values for the burning profiles exceeded those of the calculated

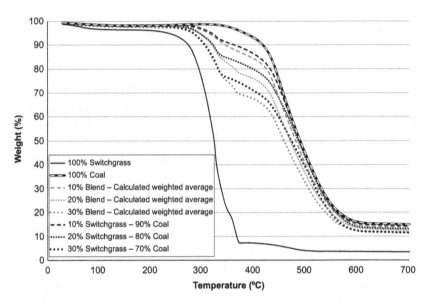

FIGURE 4.15 TGA curve for 100% switchgrass, 100% Eastern bituminous coal, and their blends at 20°C/min ramp rate. *Source: [76].*

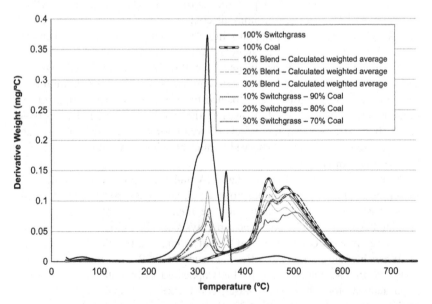

FIGURE 4.16 DTG curve for 100% switchgrass, 100% Eastern bituminous coal, and their blends at 20°C/min ramp rate. *Source: [76].*

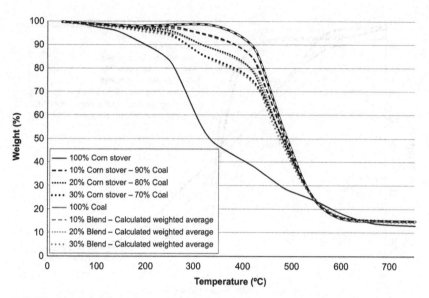

FIGURE 4.17 TGA curve for 100% corn stover, 100% Eastern bituminous coal, and their blends at 20°C/min ramp rate. *Source: [76].*

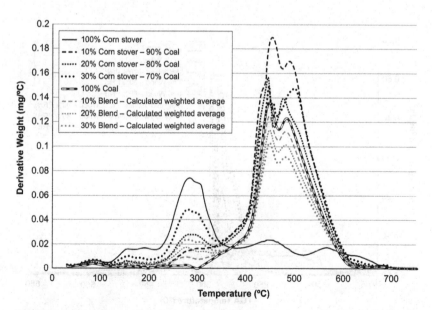

FIGURE 4.18 DTG Curve for 100% corn stover, 100% Eastern bituminous coal, and their blends at 20°C/min ramp rate. *Source: [76].*

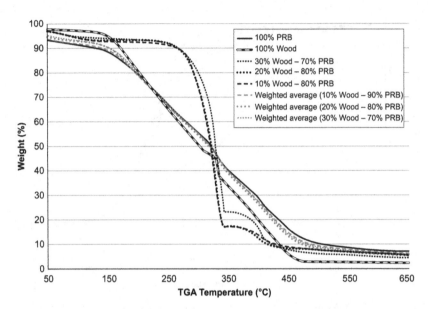

FIGURE 4.19   TGA curve for 100% wood waste, 100% PRB coal, and their blends at 20°C/min ramp rate. *Source: [76]*.

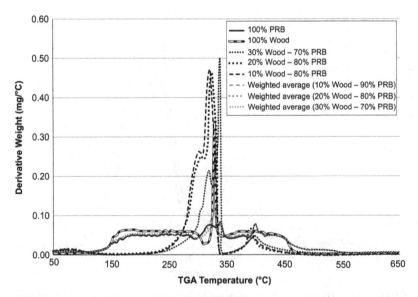

FIGURE 4.20   DTG curve for 100% wood waste, 100% PRB coal, and their blends at 20°C/min ramp rate. *Source: [76]*.

weighted average. They exceeded even those of the parent fuels. This indicates that the blends are very reactive.

Figures 4.21 and 4.22 show the TGA and DTG analysis of 100% corn stover, 100% PRB coal, and their blends. Similar to the wood waste blending

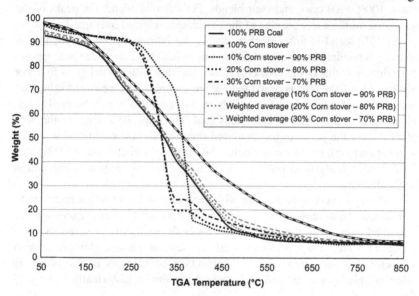

FIGURE 4.21 TGA curve for 100% corn stover, 100% PRB coal, and their blends at 20°C/min ramp rate. *Source: [76].*

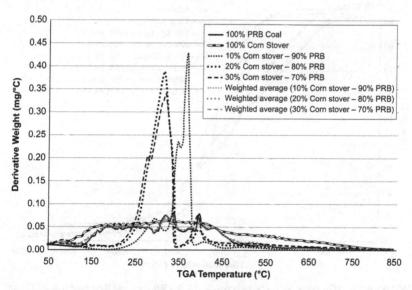

FIGURE 4.22 DTG curve for 100% corn stover, 100% PRB coal, and their blends at 20°C/min ramp rate. *Source: [76].*

cases, the measured value exceeded those of the calculated weighted average. The blends of 20% and 30% show a temperature of initiation that is lower than that of the 10% blend, thus indicating that they are more reactive.

Figures 4.23 and 4.24 show the TGA and DTG analysis of 100% switchgrass, 100% PRB coal, and their blends. For the 10% blend, the peaks of the measured profile exceeded that of the calculated weighted average. The profile of the 10% blend is different than that of the 20% and 30% blends; further analysis is required to refine and clarify this result. However, the profile appears to be driven by that of the PRB coal rather than that of the switchgrass fuel. For the 20% blend, the effects of weathering are apparent. The measured values are below that of the calculated weighted average that was observed when blending with the Eastern bituminous coal. The opposite is true for the 30% blend, where the peaks of the measured curve exceed the calculated weighted average and even of the parent fuels. The blends of switchgrass and PRB coal need further analysis in order to get a more thorough understanding and to refine the data.

Cofiring effects were evident with each one of the biomass fuel blends. For wood waste and corn stover, the effects of blending exceeded the weighted average of the parent fuels for both the pyrolysis stage and char oxidation. This may be influenced by several factors. Volatile matter concentration increased with the percentage of biomass in the blend. In addition, highly oxygenated fuels have been shown historically to be very

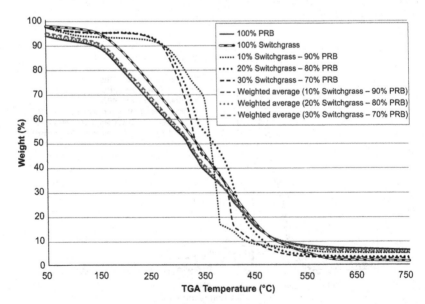

FIGURE 4.23   TGA curve for 100% switchgrass, 100% PRB coal, and their blends at 20°C/min ramp rate. *Source: [76]*.

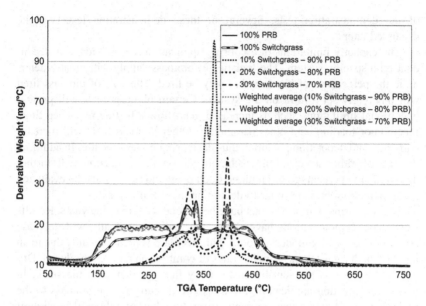

FIGURE 4.24 DTG curve for 100% switchgrass, 100% PRB coal, and their blends at 20°C/min ramp rate. *Source: [76].*

reactive. The structures of the biomass fuels are open carbon chains, and the degree of aromaticity has been shown to be low when compared to coals and other solid fossil fuels.

However, for switchgrass, the effects of blending appeared to be suppressing the reactivity of the fuel mass, particularly during the pyrolysis stage. This most likely was caused by weathering effects. When blending the biomass fuels with PRB coal, the reactions appeared to be driven by the PRB coal more than that of the biomass fuels. This was evident with the measured profiles of the blends exceeding that of the parent fuels and when blending the weathered switchgrass with the PRB coal.

4.12. IMPLICATIONS FOR BIOMASS–COAL COFIRING SYSTEMS

The study of kinetics and evolution of elements have indicated that very reactive biomass fuels such as fresh switchgrass and sawdust can be beneficial to the combustion process. They can promote the early release of volatile matter, thereby initiating the combustion process much sooner than anticipated by measures such as calculated weighted averages. Therefore, if a system is designed with these principles in mind, the reduction of certain emission constituents such as NO_x can be achieved. Other factors, such as flame stability, along with better and more complete combustion, have been observed in many of the demonstration tests. Several of these demonstration tests will be discussed later.

The capacity limitations of a cofiring application are individual or dependent on a specific system. The logistics of biomass supply will greatly determine the percentage of biomass that may be fired. The type of biomass fired will inherently set a limit on the system. Certain biomass fuels will present operational limitations such as increased mass flow affecting residence time, precipitator loading, and other parameters. Other biomass fuels will increase corrosion and deposition potential, thus decreasing boiler tube life. In addition, the use of either a co-milling technique or separate injection will further determine the percentage of biomass. Each plant and boiler must be evaluated on an individual basis in order to determine these limitations.

Many cofiring demonstrations have been conducted over the years. Results from these demonstrations have shown that cofiring typically will decrease boiler efficiency by a modest amount. This decrease is predominantly due to an increase in dry gas losses as higher rates of combustion air are introduced. Dry gas loss represents the sensible heat of dry flue gas that exits the stack. In a separate injection application, the transport air carrying the biomass to the boiler is typically at or near ambient temperature, heated only by the action of the blowers. Therefore, less heat is recovered from the flue gas, which normally preheats the combustion air. Hydrogen and moisture losses also contribute to boiler efficiency reductions [77, 78].

Overall, emission reductions have been observed when cofiring biomass with coal. Depending on the biomass combusted, these benefits may vary. The mechanism of transport and injection into the combustor also determines the benefits obtained.

Biomass is considered a CO_2 neutral fuel. During the life of the plant, photosynthetic processes occur, consuming CO_2 from the atmosphere. Once combusted, the same amount of CO_2 is considered to be released, so the net CO_2 production is essentially zero. Therefore, fossil CO_2 is decreased when biomass cofiring is employed. Wood waste substitution for coal directly reduces fossil CO_2 emissions by 1.05 ton fossil CO_2/ton wood burned. Typically, wood waste is landfilled, so by burning the material, the total reduction can potentially be 3 ton fossil CO_2/ton wood waste burned [26, 37].

Almost all of the sulfur in a fuel will convert to SO_2 given enough oxygen availability and residence time. The benefits of sulfur reduction are easily obtained through cofiring biomass with coal. Since most biomass fuels have inherently low sulfur concentrations, SO_x reductions have been observed by many cofiring demonstrations.

Volatile nitrogen can form reactive species such as amine species (e.g., NH_3, HCN, and tar nitrogen) that can react in the presence of oxygen, forming NO, some N_2O, and a small portion of N_2 [44]. The principle of NO_x reduction is to design a system and control the combustion process in order to essentially form more N_2 than other forms of nitrogen species. In a PC fired application, one of these techniques involves initiating the combustion process in a fuel-rich region.

Laboratory studies have shown that cofiring can achieve rapid fuel nitrogen volatilization in a fuel-rich environment, thus promoting fuel nitrogen reduction [1]. NO_x reductions have been observed in several demonstration tests, but the system must be designed properly, accounting for fuel chemistry and injection location. These tests demonstrated the benefits of separate injection of the biomass at the center of the fireball. The presence of the biomass promoted early ignition of the total fuel mass [48].

Figure 4.25 shows NO_x reductions at the Allen Fossil Plant (cyclone boiler), with similar curves derived for the Seward Generating Station (wall-fired PC) demonstration. The measured NO_x reduction exceeded the theoretically calculated values. In addition, the mass percentage of biomass had a significant influence on the reduction—more than the calorific percentage of biomass [26].

Trace metal concentration for biomass fuels is typically lower when compared to the coals. Therefore, when cofiring, the emission of trace metals will tend to decrease as the fraction of biomass fuels present in the fuel mass increases. There are significant differences between woody materials and agricultural materials. This is dependent on the intensity with which foresters and farmer apply fertilizer and other material to the soil, frequency of harvesting, and other factors [66]. Trace metals vary significantly in the composition, depending on the location of the forest under consideration. If the forest is near significant sources of trace metals in the atmosphere or the forest is fertilized with material that has significant concentrations of trace metals, these woody biomass fuels will produce more trace metal emissions relative to others. Trace metals for agricultural crops can be much higher than those for woody material. This is a result of higher fertilization rates. Metals contained in the fertilizer can be taken up by the crops, thus resulting in organometallic compounds in the biomass [66].

4.12.1. Biomass–Coal Blend Issues

Reactivity of alkalinity is important to the combustion process and, consequently, deposition and corrosion concerns. Studies by Baxter et al. [45] and others have demonstrated that the alkali metals—particularly sodium and potassium—are far more reactive in biomass than they are in the coals. This increased reactivity, as measured by chemical fractionation, can cause combustion difficulties that can result in increased deposition and corrosion [23, 26]. Deposition and corrosion potentials are very fuel dependent, so a thorough understanding of fuel and ash behavior is important. Each new blend of fuel should be analyzed and assessed, looking at the characteristics of the blend.

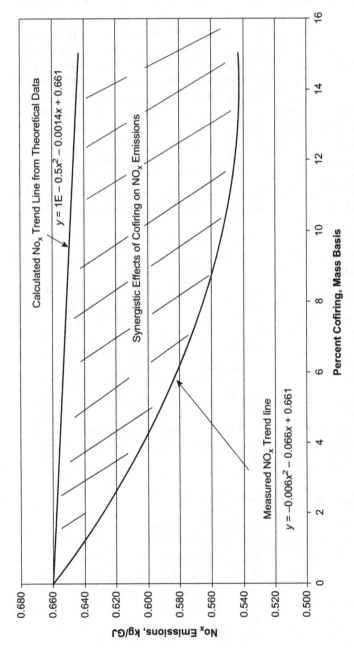

FIGURE 4.25 NO$_x$ reductions at the Allen Fossil Plant demonstration. *Source: [26].*

Many of the demonstration tests with wood and switchgrass have indicated that there is no significant corrosion issue when firing these fuels. However, when firing other biomass fuels—particularly when the chlorine concentration can be higher—the potential for corrosion does exist and therefore should be monitored. Testing at the Studstrup Power Station in Denmark, cofiring 10% straw with coal, showed that corrosion was generally low and was only moderately increased when compared to coal firing. In addition, at 20% straw cofiring with coal, approximately 92% of the total chlorine exited the system as HCl. This indicated that the chlorine did not form significant quantities of KCl [27]. Significant quantities of alkali chloride compound formation would have resulted in increased deposition and corrosion
issues.

Capacity limitations of a cofiring application are system-dependent. The logistics of the biomass supply will greatly determine the percentage of biomass that may be fired. The type of biomass fired will inherently set a system's limit. Certain biomass fuels will present operational limitations, such as increased mass flow, affecting residence time, precipitator loading, and other parameters. Other biomass fuels will increase corrosion and deposition potential, thus decreasing tube life of the boiler. In addition, the use of either a co-milling technique or separate injection will further determine the percentage of biomass. Each plant and boiler must be evaluated on an individual basis to determine these limitations.

4.12.2. Biomass–Coal Blend Systems

Cofiring has been demonstrated by many through several different firing methods. These include pulverized coal—wall-fired and tangentially fired (T-fired)—cylones, and fluidized bed boilers. Firing methods were discussed previously in Chapters 1 and 2 and are reviewed in detail here from a cofiring perspective.

Pulverized Coal-Fired Boilers

The conventional pulverized fuel systems include both wall-fired and tangentially fired arrangements. Cofiring is achieved in these systems through different methods, but the principles of combustion are similar. In wall firing, the fuel is mixed with combustion air in individual burners. Coal is pulverized in the coal mill to a very fine particle size. The coal and primary air are then introduced to the coal nozzle. Swirled air, coupled with the burner throat's flow-shaping contour, enables recirculation patterns that extend into the furnace. Burners are typically located on the front wall or both the front and rear walls, which is typically referred to as opposed firing [79]. Figure 4.26 depicts a side view of a front wall-fired boiler.

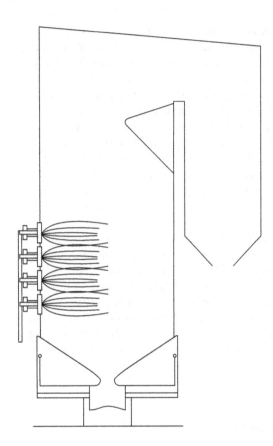

FIGURE 4.26 Front wall-fired boiler, side view. *Source: Adapted from [28].*

In tangential firing, the furnace is essentially used as the burner or firing mechanism. The heat absorption patterns are more uniform and definable [79]. Fuel and combustion air are introduced in the corners of the furnace, along a line tangent to a small circle. Intensive mixing occurs where the streams meet, and a rotational motion is imparted on the flame body. The flame body then spreads out and fills the furnace area. Figure 4.27 shows the firing configuration—from the top view—of a tangentially fired boiler.

Cyclone-Fired Boilers

Cyclone furnaces were originally designed to take advantage of several aspects. Some of these advantages include lower fuel preparation capital and operating costs, smaller furnaces, and less flyash and convection pass fouling, since less fuel ash leaves the furnace. Cyclone furnaces fire relatively large crushed coal particles—typically 95% passing through a 4-mesh screen; however, this is dependent on the particular fuel that is combusted. A molten, sticky slag

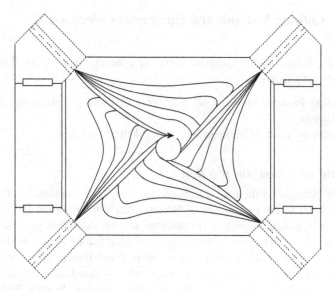

FIGURE 4.27　Top view of a tangentially fired boiler. *Source: Adapted from [79].*

layer is formed in the cyclone whereby large fuel particles are captured and held, providing longer residence times and thus enabling the larger particles to burn to completion. The fines from the fuel burn in suspension and provide the radiant heat required to help form the slag layer [28].

Fluidized Bed Boilers

Fluidized bed technology has been in existence since the early 1930s; it was the basis for the Winkler gasification technology in Germany. Fluidized bed combustion was developed subsequently and was first applied in the 1970s to large-scale utility boilers in order to explore different options of burning solid fuels such as high-sulfur coal or waste coal. Fluidized beds combust fuel in an air-suspended mass or bed of particles; by controlling bed temperatures and using reagents such as limestone, emissions of nitrogen oxides and sulfur dioxide can be minimized. The fluidizing process induces an upward flow of gas through a stacked height of solid particles. With enough gas velocity, the gas/solids mass exhibits liquid-like properties, called a fluidized bed.

When the bed becomes less uniform and bubbles of air form while the bed becomes more violent with increasing air flow, this is known as a bubbling fluidized bed (BFB) [28]. As the flow of air increases further, particles blow out of the bed and the container; the velocity then reaches the region that is experienced in circulating fluidized beds (CFBs). In a CFB, the solids are captured, separated from the air, and returned to the bed; they are circulated around a loop. Unlike a BFB, in a CFB there is no distinct transition between the dense bed at the bottom and the dilute zone above.

4.12.3. Cofiring Methods and Equipment—Mechanical Systems

Different transport options are available in cofiring applications. The two dominant methods are

- Blending biomass on the coal pile (or co-milling if pulverized coal is employed).
- Separate injection of biomass into the coal-fired boiler.

Blending with Coal on the Belt

Blending biomass on the coal pile is the least expensive option, but there is an upper limit to the percentage of biomass that can be burned [80]. In the co-milling application, early experimental results showed degradations in sieve analysis when cofiring percentages exceeded 5% on a mass basis. This was documented by TVA, where bowl mills could handle up to 5% sawdust—on a mass basis—while achieving industry standard sieve analysis. The product particle size could achieve >70% passing through 200 mesh screens [26]. The addition of wood waste to coal increases pulverizer amps and impacts feeder speeds in ball and race mills. Thus, auxiliary power consumption is increased [81]. Demonstrations have also shown that mill outlet temperatures for bowl mills can decrease. These impacts can cause mill derates [26, 82].

The use of comingling or the introduction of biomass with coal on the coal belt is a necessity when cyclone firing is employed. The combustion of biomass with coal has been documented to increase fuel reactivity, thereby promoting more rapid and more complete combustion in the cyclone barrels [83–86]. The use of comingling is ideal and well suited for cyclone firing [26]. Particle size is one of several key aspects to monitor and control; if the fuel particles are too large, it can be harmful to the combustion process.

Tests at the Allen Fossil Plant showed that even though the combustion process is essentially completed in the cyclone barrel when significant quantities of 1-in. wood chips are combusted, the cyclone appeared to "throw out" large unburned particles. The larger particles, on drying and devolatilization, became entrained in the gas stream and traveled to the furnace exit while still burning [38].

Blending on the belt can be achieved and many have demonstrated it, but there are capacity limitations. This limit has been documented by many to be in the range of 5% to 10% by mass input [26, 35]; the limitation is typically set by problems experienced in the coal mills. The Naantali-3 T-fired boiler that is located in Naantali, Finland, experienced mill fineness deterioration when biomass was fed in with the coal. Biomass accumulation in the mill was detected, and increased smoke formation was also observed during the cofiring tests. Negative effects were seen for boiler efficiency and unburned carbon concentration [35]. Separate injection is one of the best options for higher percentages of biomass in the fuel mass.

When comingling biomass with coal on the coal belt in PC firing, many demonstration tests have shown that the capacity limitation is typically in the range of 5% to 10% by mass input [38]. This limitation is set due to the detrimental effects observed with pulverizing certain biomass fuels in the mills.

For cyclone firing configurations, there is a capacity limit to the amount of biomass that can be combusted. This is typically driven by the fuel's higher heating value, since many of the cyclone boilers were designed for bituminous coals; a total heating value of approximately 9000 Btu/lb is a limit documented by several [38]. In addition, due to the inherently low ash concentration of biomass fuels—particularly the woody material—the ash concentration of the fuel mass must be monitored. Experience by others has limited the ash concentration of the fuel blend to be at or above approximately 5%.

Separate Injection of Biomass into Coal-Fired Boilers

In separate injection, the biomass is separately prepared and then fired into the boiler with the coal. The biomass fuel bypasses the pulverizers and can be introduced into the boiler by several mechanisms. It can be introduced within the burner enclosure (i.e., through the inner barrel of the burners) or through dedicated burners for a wall-fired application. In a tangentially fired boiler, the boiler essentially functions as a single burner with multiple injection points. Therefore, the biomass can be introduced as another injection point. The approach of separate injection requires more equipment, so it is more capital intensive. However, this approach has several benefits [26]:

- Higher cofiring percentages can be achieved.
- If designed properly, NO_x reduction has been demonstrated in many tests.
- During wet coal events or because of pulverizer capacity limitations, biomass introduction can recover some of the capacity.

Separate injection also demands more attention to the fuel preparation process. In addition, the interplay between fuel preparation and furnace residence time must be understood [26].

Separate injection is ideal for pulverized coal boilers. Through separate injection, coal mills can be bypassed, thus minimizing the concerns typically associated with comingling biomass with coal. These issues include but are not limited to mill fineness deterioration, increased mill amps, mill fires, and others. Separate injection enables the increased use of biomass in the fuel blend [42, 78, 85]. Capacity limitations through this method are set by the heat transfer of the boiler and, likewise, the higher heating value of the fuel blend along with other factors.

4.13. CASE STUDIES IN COFIRING

Over the last two decades, there have been many cofiring demonstrations, and the experience has been substantial. This section summarizes some cases that have taken place in Europe and the United States.

4.13.1. Cofiring Experiences

The significance of renewable energy—particularly what is produced through biomass combustion—is a reality for the European Union. The Renewable Energies White Paper set an indicative target of 12% of EU total energy consumption from renewable energy sources by 2010. In the directive on promoting electricity from renewable energy sources, the EU proposed a target of 22.1% of gross electricity consumption from renewable energy by 2010 in all EU countries. The directive classified the biodegradable fraction of industrial and municipal wastes as biomass, naming it *bio-energy* [87].

For the United States, there is significant interest in pursuing biomass cofiring. The knowledge and experience of biomass cofiring exists, but the key driver—through the implementation of policies and regulations—does not currently exist. If the incentives are present, biomass cofiring will be a reality and can be a significant part of electric generation.

Albright Generating Station

Albright Generating Station Boiler #3 is a 140-MW$_e$ (gross) T-fired four-cornered furnace located in Albright, West Virginia. In order to control NO$_x$ formation, three levels of separated overfire air (SOFA) were installed. Sawdust was cofired with coal from May 30, 2001, to July 27, 2001; the biomass percentage ranged from 0% to 10% on a mass basis [42, 88]. The demonstrations were designed to address certain issues associated with generation capacity, operability, and efficiency when cofiring biomass in boilers that were originally designed to burn Eastern bituminous coals.

The Albright demonstration cofired Pittsburgh seam coal with locally available biomass. Table 4.24 presents the ultimate analysis and higher heating value for the coal. Table 4.25 shows the ultimate analysis, higher heating value, and ash elemental analysis of the sawdust.

The Albright system was relocated from Seward Generating Station and constructed in West Virginia during the spring of 2001. As a result of the poor soil conditions encountered, construction required pouring an extensive sub-foundation made from flowable fill. This flowable fill was produced from flyash supplied by the plant. Figures 4.28 and 4.29 show the biomass processing facility and screening of biomass at the Albright Generating Station.

Albright separately injected biomass from the coal. Sawdust was received through a walking floor receiver that was capable of holding the entire load of a single truck. Screw conveyors transported sawdust to a 30-ton/hr disc screen,

TABLE 4.24 Albright Generating Station
Pittsburgh Seam Coal Ultimate

Ultimate Analysis	
Carbon (%)	68.15
Hydrogen (%)	4.35
Oxygen (%)	4.63
Nitrogen (%)	1.30
Sulfur (%)	1.48
Moisture (%)	7.40
Ash (%)	12.69
Higher Heating Value (Btu/lb)	13,000

Source: [2, 88].

where the sawdust was screened to size: <6.35 mm ($<\frac{1}{4}$-in.). The use of disc screens with covers was found to generate less dust in the work area compared to trommel screens. Sized material was conveyed to a silo, while oversized material was reground through a two-stage grinder. From the silo, sawdust was reclaimed and fed through a surge hopper onto a weigh-belt conveyor that fed a live bottom bin. The sawdust was then discharged into two rotary airlocks, each supported by blowers. Injection of the sawdust was in opposing corners of the T-fired boiler, right at the center of the fireball. The use of a separate rotary airlock and blower along with biomass being fed from a separate part of the live bottom bin ensured that there was an equal supply of biomass on each side of the boiler.

The practice of cofiring reduced furnace gas residence times that could have impacted emissions, but the volatility of sawdust promoted more rapid ignition. The rates for both devolatilization and char oxidation were much higher when compared to those for Eastern bituminous coal. Thus, the opacity and CO for the demonstration showed that up to 10% mass basis, minimal impacts were observed. SO_2 emission decreased, mostly because the sawdust had essentially no sulfur (Figure 4.30). Significant NO_x reductions were observed during the demonstration test. Figure 4.31 shows the NO_x concentrations versus the cofiring percentage. For this particular test and fuel blend, NO_x reductions were $\geq 3\%$ for every additional percent sawdust (heat input basis) added to the blend [42, 84]. Load did not appear to be a significant contributor to NO_x effects within the load range tested.

Furthermore, significant impacts on the induced draft (ID) fan amps were not observed. Unburned carbon (UBC) concentrations were measured for both the bottom ash and flyash. UBC concentrations did not vary as a function of cofiring percentage. Mercury emission was also reduced by the cofiring of

TABLE 4.25 Albright Generating Station
Sawdust Ultimate and Ash Elemental Analysis

Ultimate Analysis	
Carbon (%)	29.87
Hydrogen (%)	3.51
Oxygen (%)	26.66
Nitrogen (%)	0.12
Sulfur (%)	0.01
Moisture (%)	39.53
Ash (%)	0.30
Higher Heating Value (Btu/lb)	5087
Ash Elemental Analysis	
SiO_2 (%)	24.37
Al_2O_3 (%)	3.83
TiO_2 (%)	0.17
Fe_2O_3 (%)	2.34
CaO (%)	33.51
MgO (%)	2.56
Na_2O (%)	0.29
K_2O (%)	14.94
SO_3 (%)	1.53
P_2O_5 (%)	1.54
Base/Acid Ratio	2.62

Source: [2, 88].

sawdust with coal; the sawdust itself had mercury concentrations between 0.003 and 0.009 mg/kg (dry basis).

Biomass cofiring demonstrated at the Albright Generating Station caused no significant loss of capacity. There was a modest decrease in boiler efficiency, which was only shown through theoretical calculations. A boiler efficiency decrease translated into a modest decrease in net station heat rate. Both SO_2 and NO_x emissions were favorable, and the decrease in SO_2 and NO_x was observed without negatively impacting CO or opacity.

FIGURE 4.28   Albright Generating Station receiving a load of sawdust. *Source: [83].*

FIGURE 4.29   Screening of biomass at the Albright Generating Station. *Source: [2].*

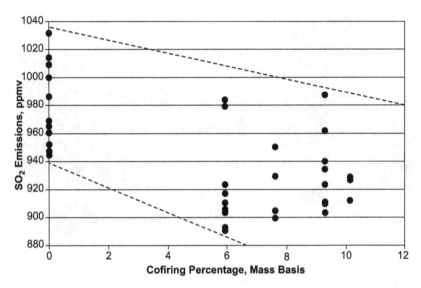

FIGURE 4.30 SO_2 emissions at the Albright Generating Station as a function of cofiring percentage. *Source: [83].*

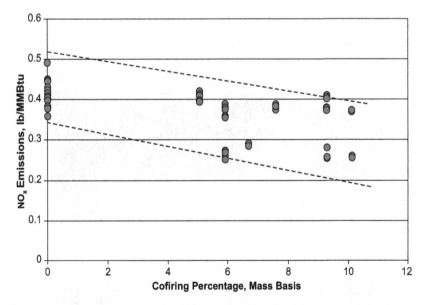

FIGURE 4.31 NO_x emissions versus cofiring percentage for the Albright Generating Station demonstration. *Source: [83].*

Bailly Generating Station

The Bailly Generating Station is located in Chesterton, Indiana. The facility has two cyclone boilers that normally fire a blend of 70% Illinois Basin high-sulfur coal with 30% low-sulfur Shoshone coal. Boiler #7 is a 160-MW$_e$ cyclone unit that generates about 1.2×10^6 lb/hr of 2400 psig/1000°F/1000°F steam. The objective of the program was to reduce all fossil emissions without negatively impacting unit performance [14].

The program cofired petroleum coke with coal, biomass with coal, and then a blend of petroleum coke, urban wood waste, and coal. The blends that were studied during the testing period included the following:

- 100% coal (70% Illinois coal/30% Shoshone coal)
- 10% petroleum coke/90% coal
- 15% petroleum coke/85% coal
- 20% petroleum coke/80% coal
- 25% petroleum coke/75% coal
- 5% wood waste/95% coal
- 7.5% wood waste/92.5% coal
- 10% wood waste/90% coal
- 5% wood waste/15% petroleum coke/80% coal
- 7.5% wood waste/22.5% petroleum coke/70% coal
- 10% wood waste/20% petroleum coke/70% coal

Table 4.26 presents the typical proximate analysis, ultimate analysis, and higher heating value for the fuels combusted during the test period.

The cofiring facility included a fuel receiving area with a pole barn and trommel screen to ensure that the urban wood waste was sized to <3/4 in. Once the wood waste was screened, it was then blended with petroleum coke for storage and transport. The facility also included an aboveground reclaim system for the opportunity fuel and a conveyor linking the reclaim system to the main coal belt. A Stamler aboveground reclaim system that is typically designed for coal mines was chosen. The machine has a stub conveyor that elevates the fuel and discharges it onto a belt conveyor that links the opportunity fuel reclaim to the main belt conveyor.

The system design was labor intensive, with the recognition that it was constructed for a demonstration test. Manual or bucket blending of the fuel at the pole barn and manual feeding of the Stamler reclaimer were utilized. The two facilities were about 2000 feet apart due to site conditions and constraints. Therefore, conveying of the material between the two locations required trucks; this was due to cost considerations [14].

Foster Wheeler constructed the facility during the fourth quarter of 1998 and the first quarter of 1999. Construction was completed in February 1999. Figures 4.32 and 4.33 show the blending facility at the Bailly Generating Station.

TABLE 4.26 Typical Proximate Analysis, Ultimate Analysis, and Higher Heating Value for the Fuels Combusted at Bailly Generating Station

	Wood Waste	High-Sulfur Coal	Low-Sulfur Coal	Petroleum Coke
Promimate Analysis				
Fixed Carbon (%)	12.58	41.94	42.15	78.28
Volatile Matter (%)	52.56	34.43	37.56	13.90
Ash (%)	4.08	9.66	5.63	1.34
Moisture (%)	30.78	13.97	14.66	6.48
Ultimate Analysis				
Carbon (%)	33.22	62.30	63.17	81.11
Hydrogen (%)	3.84	4.34	4.68	3.39
Oxygen (%)	27.04	5.06	9.68	1.34
Nitrogen (%)	1.00	1.22	1.44	1.23
Sulfur (%)	0.07	3.45	0.74	5.11
Ash (%)	3.99	9.66	5.63	1.34
Moisture (%)	30.84	13.97	14.66	6.48
Higher Heating Value (Btu/lb)	5788	11,113	10,900	14,308

Source: [14].

Results from the testing indicated that wood waste, petroleum coke, and the blended opportunity fuel did not reduce unit capacity. The triburn opportunity fuel increased boiler efficiency by approximately 0.5% and did not require additional excess O_2 for combustion; air heater exit temperature was not impacted. The combination of petroleum coke and wood waste helped to offset the carbon in flyash; petroleum coke increased unburned carbon in the flyash, while the wood waste decreased unburned carbon in the flyash. Slag formation and characteristics were not impacted by the wood waste, petroleum coke, or triburn opportunity fuel [14].

Figure 4.34 presents the trend lines for NO_x emissions at full load operating conditions, measured in $lb/10^6$ Btu. Based on the data, a dominant mechanism was presented. Wood waste introduces significant concentrations of volatiles, causing the fuel to complete combustion in the cyclone barrel, thus reducing combustion occurring in the primary furnace. For the petroleum coke, the fuel burns in the slag layer in a heterogeneous gas–solids reaction. Unless there is

FIGURE 4.32 Blending on the belt at Bailly Unit #7. *Source: [83].*

FIGURE 4.33 Biomass/petroleum coke processing and blending facility at Bailly Generating Station. This building housed a trommel screen. *Source: [83].*

a significant amount of fines, the presence of petroleum coke reduces the amount of combustion occurring in the primary furnace. Consequently, the mechanism reduced the temperature in the primary furnace, thus reducing NO_x formation [14].

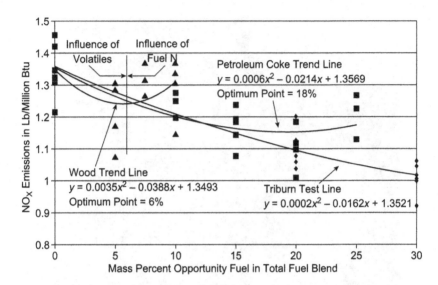

FIGURE 4.34   Bailly Generating Station NO_x emissions at full load operating conditions. *Source: [83].*

Figure 4.35 presents the impact of cofiring biofuel, petroleum coke, and tri-firing petroleum coke and biofuel with coal on CO, THC, and SO_3 emissions. Peculiarly, SO_3 increases when wood waste is burned, decreases slightly with petroleum coke, and increases with trifiring. Since unburned carbon absorbs SO_3, the unburned carbon concentration affects the results. When cofiring biomass, unburned carbon decreases significantly. Conversely, the concentration of unburned carbon increases when petroleum coke is cofired. When triburning,

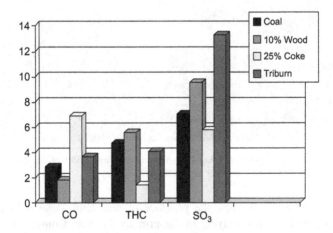

FIGURE 4.35   CO, total hydrocarbon, and SO_3 emissions when cofiring wood waste, cofiring petroleum coke, and trifiring all fuels. Emissions expressed in ppmvd at 3% O_2. *Source: [83].*

there is more sulfur in the system, so more SO_3 is formed, since unburned carbon decreases and thus reduces the amount of SO_3 absorbed into the flyash [14].

The Bailly Generating Station program met its objectives. It demonstrated a cost-effective method for burning biomass, reducing fossil CO_2 emissions. There was improved boiler efficiency, reduced NO_x emissions, reduced CO_2 emissions, and reduced metal emissions. These results were achieved while minimizing the effects on opacity, CO, THC, and SO_3 emissions. In addition, the triburn opportunity fuel combined the advantages of biomass with petroleum coke, while minimizing the disadvantages that are typically inherent to each of the individual opportunity fuels.

Virginia City Hybrid Energy Center

The Virginia City Hybrid Energy Center (VCHEC) plant—located in Virginia—will have a net generating capacity of approximately 585 MW_e (660 MW_e gross). The plant consists of two circulating fluidized bed boilers with one steam turbine. This plant, with the capacity to burn up to 20% biomass (heat input basis), will be one of the cleanest plants of its kind [73, 89].

VCHEC will burn waste coal (bituminous gob) with up to 20% biomass (wood waste) in the fuel blend. Waste coal has inherently high ash concentration and significantly lower heating values when compared to subbituminous and bituminous coals. Waste wood—from forest residues and other sources—will tend to have slightly higher concentrations of ash ($\sim$2%) when compared to clean wood ($\leq$1%) and lower moisture contents [73].

The biomass system consists of two wood feed systems per boiler. The system incorporates a wood feed bin, drag chain conveyor, robbing screws, and rotary airlocks. The drag chain screw conveyor delivers the wood chips to one, two, or three outlets. Robbing screws, located at each outlet of the drag chain, remove the wood chips at a controlled rate. The wood chips then go through the rotary airlock, located below each of the robbing screws. The use of rotary airlocks helps to prevent positive pressure in the furnace from causing backflow of furnace gas [73].

VCHEC will be one of the cleanest coal-burning units in the United States. The use of waste coal along with wood waste made the project environmentally attractive. Completion date for the plant is projected for the summer of 2012 [89]. (See also the following papers that describe fluidized bed combustion: [90–92].)

Gadsden Unit 2

Alabama Power's Gadsden Unit 2—located in Gadsden, Alabama—is a tangentially fired PC unit that normally fires an Eastern bituminous coal. Forty tests were conducted with switchgrass over a six-week period. Up to 10% of the energy was generated by switchgrass during the testing period; 6% to 8% was contributed by the switchgrass at full load operation. The system was able to produce anywhere

TABLE 4.27 Representative Chemical Analyses of Bituminous Coal and Switchgrass Used at Gadsden Plant

	Pratt Seam Bituminous Coal	Mechanically Harvested Switchgrass	Manually Harvested Switchgrass
Promimate Analysis			
Fixed Carbon (%)	62.64	15.49	5.39
Volatile Matter (%)	24.93	69.03	81.79
Ash (%)	11.28	5.95	3.93
Moisture (%)	1.15	9.53	8.89
Ultimate Analysis			
Carbon (%)	81.00	40.54	43.55
Hydrogen (%)	3.99	5.28	5.13
Nitrogen (%)	1.63	0.92	0.79
Sulfur (%)	1.57	0.20	0.10
Oxygen [by difference] (%)	−0.62	37.58	37.61
Ash (%)	11.28	5.95	3.93
Moisture (%)	1.15	9.53	8.89
Higher Heating Value (Btu/lb)	13,513	7333	7421
Ash Elemental Analysis			
SiO_2	46.51	62.71	48.56
Al_2O_3	29.78	8.39	1.08
TiO_2	1.67	0.53	0.07
Fe_2O_3	13.64	5.82	0.34
CaO	1.12	4.62	12.77
MgO	0.85	3.65	20.41
Na_2O	0.30	0.59	2.38
K_2O	2.28	1.95	2.92
P_2O_5	0.55	2.73	5.89
SO_3	0.05	2.28	3.83

Source: [17].

from 4 to 4.5 MW_e renewable energy [17, 77]. The objective of the demonstrations was to understand the effects of cofiring switchgrass in a PC unit. The switchgrass was grown on 120 hectares of farmland located in Winterboro and Lincoln, Alabama. The coal was a "synfuel," with the parent coal of the synfuel derived from a Southern Appalachian bituminous coal [17]. Switchgrass was successfully cofired with the synfuel. Table 4.27 shows representative chemical analyses of the bituminous coal and switchgrass fuels used at the Gadsden plant.

Initially, plans were to mix biomass with the coal on the pile and then transport the fuel mix into the plant on the existing coal conveying system. The mixture would then be sent through the pulverizing system. However, laboratory tests indicated the issues of flowability when switchgrass is introduced to the fuel mix. The tests indicated that the fuel mix would not properly flow through the coal storage bunkers, even at low mass concentrations of 5%. Consequently, a separate handling system was employed for these tests. Switchgrass was injected at two opposite corners of the four-cornered furnace [18, 93].

Bales of switchgrass were processed at the site with the use of tub grinders and two screens—one at 1/2-in. and the other at 1-in. perforations. A metering bin with four chain-driven augers controlled the rate of switchgrass flow to the furnace. Auger outputs were entrained into the intake of a fan, which transported the switchgrass to the boiler. Twelve-inch-diameter galvanized steel ducts carried the switchgrass into the plant. The lines were then divided into two 8-in.-diameter ducts—one to the burner on the left front corner and the other to the right rear corner. Figure 4.36 illustrates the direct injection system at Gadsden.

Sulfur and mercury emission was reduced when compared to 100% synfuel firing. Unburned carbon was lower for the cofiring case than for the synfuel case. There was no noticeable change in NO_x emission with switchgrass cofiring. However, opacity was observed to slightly increase during the test. Higher carbon concentration of the fine material is believed to be the contributing factor, not the higher particulate emissions. One thousand bales of switchgrass were cofired with minimal slagging and fouling consequences over an extended time period [77]. Boiler efficiency decreased by approximately 0.3% to 1.0% as economizer O_2 increased to 3.5% from 2.7% normal operation with the synfuel. Efficiency for switchgrass cofiring was lower due to higher dry gas losses associated with the cold air used to transport the biomass into the furnace. Additional losses were contributed by higher hydrogen and moisture losses.

Overall, switchgrass was successfully cofired with coal at the Gadsden plant. In some of the tests, up to 10% of the energy was produced from switchgrass. For full load tests, 6% to 8% of the energy input was from switchgrass. A decrease in certain emission constituents—sulfur dioxide and mercury—was experienced during the test. Effects on NO_x were not observed, but a slight increase in opacity was experienced.

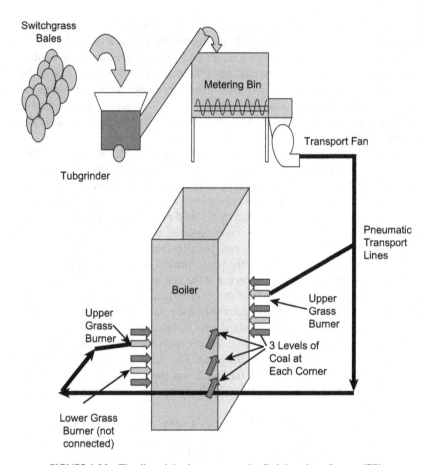

FIGURE 4.36 The direct injection system at the Gadsden plant. *Source: [77].*

Studstrup Power Station

Studstrup Power Station Unit #1 was a 150-MW$_e$ wall-fired boiler located in Denmark. A demonstration test cofiring coal and straw was conducted from January 1996 to February 1998. Studstrup Power Station has since been decommissioned. The objective of the program was to evaluate the influence of cofiring on boiler performance, combustion chemistry, surface deposits and corrosion, residue quality, emissions, and selective catalytic reduction (SCR) performance [27].

Straw is the largest biomass resource in Denmark, but supply varies in quantity each year. Straw is a low-grade and inhomogeneous fuel. This fuel tends to have high volatile matter, chlorine, and alkaline concentrations. Straw quality is heavily dependent on leaching of salt, which occurs during the harvesting process. The impact of rainfall can influence ash removal; potassium

TABLE 4.28 Typical Fuel Analyses for the Coals and Straw Burned during Studstrup Demonstration

	Coal Type #1	Coal Type #2	Straw
Ultimate Analysis			
Carbon (%)	61.31	56.88	40.38
Hydrogen (%)	3.76	3.90	5.26
Oxygen [by Difference] (%)	9.84	16.70	34.90
Nitrogen (%)	1.32	1.33	0.51
Sulfur (%)	0.90	1.95	0.11
Chlorine (%)	0.02	0.22	0.48
Ash (%)	12.35	7.62	5.96
Moisture (%)	10.50	11.40	12.40
Lower Heating Value (Btu/lb)	9196	9751	6440
Ash Elemental Analysis			
SiO_2	59.70	50.00	34.00
Al_2O_3	19.20	20.00	0.94
TiO_2	0.75	0.89	0.06
P_2O_5	0.18	0.23	2.83
SO_3	1.98	3.40	4.74
Fe_2O_3	8.10	14.70	0.65
CaO	2.05	3.30	7.30
MgO	1.76	0.90	2.00
Na_2O	0.63	1.05	0.85
K_2O	2.15	2.00	29.8

Source: [27].

and some chlorine can be washed out [27]. Table 4.28 shows the typical fuel analyses for the two coals and straw utilized by Studstrup during the demonstration test.

During the first half of the two-year demonstration, the straw handling system experienced availability problems. This was mainly because the suppliers did not have any previous experience handling straw. Wet inclusions in the bales caused some of the processing issues. Therefore, quality checks of the incoming straw became important [27].

The straw handling equipment consisted of a storage facility and a processing building. Bales were delivered to Studstrup by truck, and an overhead crane unloaded the trucks. The crane was able to unload between 10 and 12 bales at a time. In the unloading process, the bales were weighed and the moisture content was measured by a microwave technique. The processing plant consisted of four parallel lines, each with a capacity of handling 5 tonnes of straw per hour. Tier conveyors controlled the flow of straw to the boiler, which was situated before a shredder. These shredders were modified from heavy-duty garbage disposal machines. From the shredders, the straw was conveyed through a stone trap (used to remove heavy tramp material) and into the hammer mill, where the straw was ground into smaller fractions—less than 30 to 50 mm. Ground straw then went through rotary airlocks, where it was pneumatically transported in four parallel pipelines [27].

The boiler was equipped with 12 burners on three elevations located on the back wall. Straw was introduced at the middle elevation, down the inner barrel of the burners, with coal introduced in the outer annulus.

Overall, boiler performance was satisfactory. The need for increased soot blowing was evident, particularly when a 20% straw share (by heat input) was used. Unburned carbon also increased; up to 20% unburned carbon in the bottom ash was observed. Bottom ash concentration varied from approximately 5% at 100% coal firing, 12% at 10% straw share, and 15% to 20% at 20% straw share (by heat input). Nitrogen concentration in the straw averaged 0.35 g/MJ (0.81 lb/10^6 Btu) compared to 0.61 g/MJ (1.42 lb/10^6 Btu) for the coal. No significant change to NO_x emission was observed by cofiring coal with straw. Oxygen concentration and temperature were the key parameters. This may be driven by the structure of straw compared to sawdust; how and where the nitrogen is located will affect its release rate. Negative results were observed for the SCR catalysts. Fly ash concentration increased by two to three times at the inlet; further investigation is required [27].

High-temperature corrosion testing that was conducted by the plant showed that the co-combustion of 10% straw resulted in generally low corrosion rates. At a 20% straw share, the rate increased by a factor of 1.5 to 3 in the superheaters, with steam temperatures up to 540°C (1004°F). However, this rate is still considered a low to medium corrosion rate when compared to corrosive potentials of coals. At the higher straw share, corrosion is reduced simply by the removal of the deposits through sootblowing or similar mechanisms. In addition, at a 20% straw share, 92% of the total chlorine input was released as HCl. This contributed to the reduced corrosion rates. Formation of HCl in the flue gas is preferred compared to KCl formation, which can exacerbate deposition and corrosion problems [27].

The two-year demonstration was successful; straw handling and combustion are possible. The combustion of straw is possible without major issues;

fouling is not critical, but slagging does increase with increasing straw share. Coal quality is an important parameter to also monitor. Corrosion increases slightly, but these levels stay between a low to medium corrosive coal type.

Naantali-3 Generating Station

Naantali-3 is a T-fired PC boiler, located in Naantali, Finland, that produces 315 MW_{th} of electricity, district heat, and steam. Cofiring tests were carried out in April 1999 and March–April 2000. The objective of the tests was to evaluate the impact of cofiring of sawdust on boiler performance, flame stability, and emissions [35].

Coal and sawdust were the test fuels utilized at Naantali-3 for the tests. The fraction of sawdust cofired ranged from 2.5% to 8% (heat input basis). Pine sawdust came from a local sawmill that was located 50 kilometers from the power plant. Table 4.29 presents the fuel analysis for the sawdust and coals used during the test.

TABLE 4.29 Representative Fuel Analysis for the Sawdust and Coals at Naantali-3

	Sawdust Pine	Russian Coal	Polish Coal
Moisture (%)	51−63	9−11	9−13
Proximate Analysis (Dry Basis)			
Volatile Matter (%)	85−87	26−29	31−34
Ash (%)	0.25−0.30	9−11	12−15
Fixed Carbon (%)	12.14−14.50	54.60−63.80	52.7−61.2
Ultimate Analysis (Dry Basis)			
Carbon (%)	51.50	76.50	70.00
Hydrogen (%)	6.20	4.50	4.20
Oxygen (%)	41.80	4.90	9.20
Nitrogen (%)	0.20	1.90	1.20
Sulfur (%)	<0.10	0.30−0.40	0.80−1.00
Chlorine (%)	<0.01	0.25	0.21
Ash (%)	0.25−0.30	9−11	12−15
Lower Heating Value (Btu/lb)	2452−3012	10,755−11,186	10,325−10,755

(Continued)

TABLE 4.29 Representative Fuel Analysis for the Sawdust and Coals at Naantali-3—cont'd

	Sawdust Pine	Russian Coal	Polish Coal
Ash Elemental Analysis			
K_2O (%)	2–12	—	1.8–2.4
P_2O_5 (%)	4–8	—	0.6–0.7
CaO (%)	22–62	—	2.5–3.6
MgO (%)	6–16	—	1.3–2.6
Fe_2O_3 (%)	0.5–0.8	—	8.1
SO_3 (%)	2–5	—	2.5–3.2
SiO_2 (%)	1.5–5	—	50–55
Al_2O_3 (%)	1.1–2.5	—	22–25
TiO_2 (%)	<0.6	—	0.9–1.3
Na_2O (%)	0.4–0.8	—	0.3–1.2
Ash Fusion Temperature (Reducing Atmosphere)			
Initial Deformation (°C)	1495	—	1205
Softening (°C)	1510	—	1240
Hemispherical (°C)	1520	—	1275
Fluid (°C)	1520	—	1340

Source: [35].

Coal and unscreened biomass were blended in the coal yard; the mixture was fed into the boiler through the coal mills. Three mills (Loesche roller mills) fed three burner levels, with each level consisting of four burners.

Coal and wood were mixed by a bulldozer in the yard. The coal and sawdust were spread in layers on top of each other by bucket charger and bulldozer. The thicknesse of each fuel and layer was based on the blending ratio. This method of blending is only a rough estimate; the level of accuracy is marginal.

Coal mill performance was the limiting factor. Mill fineness deteriorated with the addition of biomass in the fuel mass and increased with increasing percentages of sawdust. Biomass accumulation in the mill was detected; increased smoke formation was observed during the cofiring tests. Coal mill grindability was affected; biomass does not pulverize well, creating a mat on the grinding table, thus inhibiting the grinding effects. Negative effects were also seen for boiler efficiency and unburned carbon concentration [35].

For the most part, sawdust cofiring was successful. However, some negative effects were observed—particularly in the coal mills. Therefore, it was concluded that in order to achieve high cofiring percentages, separate injection of the biomass was necessary [35].

4.14. CONCLUSIONS

Biomass cofiring with coal is the combustion of dissimilar fuels. Since the base fuel is still coal, the chemistry for both the biomass and the coal must be understood. This includes understanding both the organic and the inorganic behavior of the fuels and the interaction that exists between the fuels.

Cofiring has been demonstrated successfully through numerous different tests and installations. Many of the technical issues associated with materials handling and transport along with slagging and fouling problems have been well studied and documented. However, each installation and plant must be analyzed and understood on an individual basis when considering biomass cofiring.

The implementation and increased use of biomass cofiring is in existence in the European Union. However, for the United States, there must exist regulatory drivers and economic incentives for the use of biomass cofiring to increase. The use of biomass to augment the generation of electricity is technically feasible and may increase in the future, particularly in a carbon constrained world.

REFERENCES

[1] Tillman DA, Harding NS. Fuels of opportunity: characteristics and uses in combustion systems. Amsterdam: Elsevier; 2004.

[2] Tillman DA, Miller BG, Johnson DK, Clifford DJ. Structure, reactivity, and nitrogen evolution characteristics of a suite of solid fuels. Proceedings 29th international technical conference on coal utilization and fuel systems. Clearwater, FL; 2004, April 18–22.

[3] Bram S, De Ruyck J, Lavric D. Using biomass: a system perturbation analysis. Applied Energy 2007;86(2):194–201.

[4] Cobb JT, Elder WW, Freeman MC, James RA, McCreery LR, Biedenbach W, et al. Demonstration program for wood/coal cofiring in Western Pennsylvania. Proceedings 8th biennial conference (Bioenergy '98), p. 251–62. Madison, WI; 1998, October 4–8.

[5] De S, Assadi M. Impact of cofiring biomass with coal in power plants—a techno-economic assessment. Biomass and Bioenergy 2009;33(2):283–93.

[6] Freeman MC, Goldberg PM, Plasynski SI. Biomass cofiring R&D and utility experiences: what's happened, what's next? Proceedings 8th biennial conference (Bioenergy '98), p. 263–73. Madison, WI; 1998, October 4–8.

[7] Gani A, Morishita K, Nishikawa K, Naruse I. Characteristics of co-combustion of low-rank coal with biomass. Energy and Fuels 2005;19(4):1652–9.

[8] Hall DO, Overend RP, editors. Biomass: regenerable energy. London: John Wiley & Sons; 1987.

[9] Loo SV, Koppejan J. Handbook of biomass combustion and co-firing. The Netherlands: Twente University Press; 2002.

[10] Marshall L, Fralick C, Lyng R. A utility perspective on the use of biomass co-firing. Proceedings 33rd international technical conference on clean coal and fuel systems. Clearwater, FL; 2008, June 1–5.

[11] Savolainen K. Co-firing of biomass in coal-fired utility boilers. Applied Energy 2003;74(3-4): 369–81.

[12] Tillman DA, Hughes E, Gold BA. Cofiring of biofuels in coal fired boilers: results of case study analysis. Proceedings first biomass conference of the Americas: energy, environment, agriculture, and industry, p. 368–81. Burlington, VT; 1993, August 30–September 2.

[13] Tillman D, Dobrzanski A, Duong D. Management of PRB–bituminous coal blends. Proceedings electric power conference. Atlanta; 2007, May 24.

[14] Tillman D. Biomass cofiring: field test results. Palo Alto, CA: Electric Power Research Institute; 1999. Report TR-113903.

[15] Miller BG, Tillman DA, editors. Combustion engineering issues for solid fuel systems. Boston: Academic Press; 2008.

[16] Tillman DA, Duong D, Fuel selection for cofiring biomass in pulverized coal and cyclone fired boilers. Proceedings 34th international technical conference on clean coal and fuel systems. Clearwater, FL; 2009, May 31–June 4.

[17] Boylan D, Bush V, Bransby DI. Switchgrass cofiring: pilot scale and field evaluation. Biomass and Bioenergy 2000;19(6):411–17.

[18] Bush VP, Smith HA, Bransby DI, Boylan DM. Final technical report: EPRI-USDOE cooperative agreement: evaluation of switchgrass as a co-firing fuel in the Southeast 2001. Contract No. DE-FC36–98GO10349.

[19] Cai H-Y, Megaritis A, Messenbock R, Dix M, Dugwell DR, Kandiyoti R. Pyrolysis of coal maceral concentrates under PF-combustion conditions (I): changes in volatile release and char combustibility as a function of rank. Fuel 1998;77(12):1273–82.

[20] Speight JG. The chemistry and technology of coal. New York: Marcel Dekker; 1983. p. 3–33, 72–90.

[21] Broido A. Kinetics of solid-phase cellulose pyrolysis. In: Shafizadeh F, Sarkanen Kyosti V, Tillman DA, editors. Thermal uses and properties of carbohydrates and lignins. New York: Academic Press; 1976. p. 19–36.

[22] Haygreen JG, Bowyer JL. Forest products and wood science: an introduction. Ames: Iowa State University Press; 1982.

[23] Jenkins BM, Baxter LL, Miles Jr TR, Miles TR. Combustion properties of biomass. Proceedings biomass usage for utility and industrial power: an Engineering Foundation conference. Snowbird, UT; 1996, April 28–May 3.

[24] Shafizadeh F, DeGroot WF. Thermal analysis of forest fuels. In: Tillman DA, Kyosti Sarkanen V, Anderson LL, editors. Fuels and energy from renewable resources. New York: Academic Press; 1977. p. 95–114.

[25] Shafizadeh F, DeGroot WF. Combustion characteristics of cellulosic fuels. In: Shafizadeh F, Kyosti Sarkanen V, Tillman DA, editors. Thermal uses and properties of carbohydrates and lignins. New York: Academic Press; 1976. p. 1–18.

[26] Tillman DA. Biomass cofiring: the technology, the experience, the combustion consequences. Biomass and Bioenergy 2000;19(6):365–84.

[27] Wieck-Hansen K, Overgaard P, Larsen OH. Cofiring coal and straw in a 150 MW_e power boiler experiences. Biomass and Bioenergy 2000;19(6):395–409.

[28] Kitto JB, Stultz SC, editors. Steam: its generation and use, 42nd ed. Barberton, OH: Babcock & Wilcox; 2005.

[29] Meesri C, Moghtaderi B, Gupta RP, Rezaei HR, Wall TF. Co-firing of biomass with coal: combustion issues. Proceedings 18th international Pittsburgh Coal Conference. Newcastle, NSW; 2001, December 3–7.

[30] Hammes GG. Principles of chemical kinetics. London: Academic Press; 1978.

[31] Tillman DA. Cofiring biomass with coal: issues for technology commercialization. Proceedings 17th international Pittsburgh coal conference. Pittsburgh; 2000, September 11–14.

[32] Tillman DA. The combustion of solid fuels and wastes. San Diego: Academic Press; 1991.

[33] Sliepcevich CM. Assessment of technology for the liquefaction of coal. Washington, DC: National Academy of Sciences; 1997.

[34] Hoekman KS. Biofuels in the U.S.—challenges and opportunities. Renewable Energy 2009;34(1):14–22.

[35] Savolainen K, Sormunen R. Co-firing of biomass and coal: a means to reducing greenhouse gas emissions. Proceedings 18th international Pittsburgh coal conference. Newcastle, NSW; 2001, December 3–7.

[36] Rackley SA. Carbon capture and storage. Boston: Butterworth-Heinemann; 2010. p. 5–11.

[37] Hus PJ, Tillman DA. Cofiring multiple opportunity fuels with coal at Bailly Generating Station. Biomass and Bioenergy 2000;19(6):385–94.

[38] Tillman D, Hughes E. Biomass cofiring: update 2002. Palo Alto, CA: Electric Power Research Institute; 2003. Report TR-1004319.

[39] Tillman D, Hughes E. Biomass cofiring at NIPSCO Michigan City Station. Palo Alto, CA: Electric Power Research Institute; 1997. Report TC4603–001.

[40] EPA. Renewable portfolio standards fact sheet. www.epa.gov/chp/state-policy/renewable_fs.html; 2009.

[41] Spliethoff H, Hein KRG. Effect of co-combustion of biomass on emissions in pulverized fuel furnaces. Proceedings biomass usage for utility and industrial power: an Engineering Foundation conference. Snowbird, UT; 1996, April 28–May 3.

[42] Tillman D, Hughes E. Annual report on biomass cofiring program. Palo Alto, CA: Electric Power Research Institute; 2001. Report TR-1004601.

[43] Tillman DA, Payette K, Battista J. Designer opportunity fuels for the Willow Island Generating Station of Allegheny Energy Supply Company, LLC. Proceedings 17th international Pittsburgh Coal conference. Pittsburgh; 2000, September 11–14.

[44] Konttinen J, Hupa M, Kallio S, Winter F, Samuelsson J. NO formation tendency characterization for biomass fuels. Proceedings 18th international conference on fluidized bed combustion. Toronto; 2005, May 22–25.

[45] Baxter LL, Mitchell RE, Fletcher TH, Hurt RH. Nitrogen release during coal combustion. Energy & Fuels 1996;10(1):188–96.

[46] Baxter LL. Biomass combustion and cofiring issues overview: alkali deposits, flyash, NOx/SCR impacts. Proceedings international conference on co-utilization of domestic fuels. Gainesville, FL; 2003, February 5–6.

[47] Gold BA, Tillman DA. Wood cofiring evaluation at TVA power plants EPRI project RP 3704-1. Proceedings EPRI conference on strategic benefits of biomass and waste fuels, p. 4-33–4-45. Washington, DC; 1993, March 30–April 1.

[48] Laux S, Tillman D, Seltzer A. Design issues for co-firing biomass in wall-fired low NOx burners. Proceedings 27th international technical conference on coal utilization and fuel systems. Clearwater, FL; 2002, March 4–7.

[49] Aerts DJ, Ragland KW. Co-firing switchgrass in a 50 MW pulverized coal utility boiler. Proceedings 8th biennial conference (Bioenergy '98). Madison, WI; 1998, October 4–8.

[50] Battista JJ. Hughes EE. Survey of biomass cofiring experience in the U.S. Proceedings 17th international Pittsburgh Coal Conference. Pittsburgh; 2000, September 11–14.

[51] Battista J, Tillman D, Hughes E. Cofiring wood waste with coal in a wall-fired boiler: initiating a 3-year demonstration program. Proceedings 8th biennial conference (Bioenergy '98). Madison, WI; 1998, October 4–8.

[52] Tillman D, Battista J, Hughes E. Cofiring wood waste with coal at the seward generating station. Proceedings 23rd international technical conference on coal utilization and fuel systems. Clearwater, FL; 1998, March 9–13.

[53] Pellets Fuel Institute, www.pelletheat.org/; 2010, July 25.

[54] Rhen C, Ohman M, Gref R, Wasterlund I. Effect of raw material composition in woody biomass pellets on combustion characteristics. Biomass and Bioenergy 2007;31(1):66–72.

[55] Miller BG, Mille SF, Harlan DW. Utilizing animal-tissue biomass in coal-fired boilers: an option for providing biosecurity. Proceedings 29th international technical conference on coal utilization and fuel systems, p. 153–64. Clearwater, FL; 2004, April 18–22.

[56] Semadeni A. Potential forest resource contribution to energy supplies. Energy aspects of the forest industries. Proceedings of a seminar organized by the Timber Committee of the United Nations Economic Commission for Europe. Udine, Italy; 1978, November 13–17.

[57] Tillman D, Duong D. Biomass cofiring—basic issues and concerns. Proceedings 33rd international technical conference on clean coal and fuel systems. Clearwater, FL; 2008, June 1–5.

[58] Zulfiqar M, Moghtaderi B, Wall TF. Flow properties of biomass and coal blends. Fuel Processing Technology 2006;87(4):281–8.

[59] Berkowitz N. An introduction to coal technology. San Francisco: Academic Press; 1994.

[60] Demirbas A. Potential applications of renewable energy sources, biomass combustion problems in boiler power systems and combustion related environmental issues. Progress in Energy and Combustion Science 2005;31(2):171–92.

[61] Apaydin-Varol E, Putun E, Putun AE. Pyrolysis of different biomass samples for char production. Proceedings 33rd international technical conference on clean coal and fuel systems. Clearwater, FL; 2008, June 1–5.

[62] Biagini E, Tognotti L. Comparison of devolatilization/char oxidation and direct oxidation of solid fuels at low heating rate. Energy & Fuels 2006;3(1):986–92.

[63] Giuntoli J, Arvelakis S, Spliethoff H, de Jong W, Verkooijen AHM. Quantitative and kinetic thermogravimetric Fourier Transform Infrared (TG-FTIR) study of pyrolysis of agricultural residues: influence of different pretreatments. Energy & Fuels 2009;23(11): 5695–706.

[64] Tillman DA, Duong DNB, Miller BG, Bradley LC. Combustion effects of biomass cofiring in coal-fired boiler. Proceedings Power-Gen International. Las Vegas; 2009, December 8–10.

[65] Gaydon AG, Wolfhard HG. Flames: their structure, radiation, and temperature. London: Chapman and Hall; 1953.

[66] Tillman DA. Trace metals in combustion systems. San Diego: Academic Press; 1994.

[67] Bridgwater T. Biomass pyrolysis. Biomass and Bioenergy 2007;31(4), Update 26.

[68] Takahashi Y, Baba A, Yuasa H, Nakamura T, Miura R, Hirata M. R&D on coal and woody biomass co-firing technology for electric power utilities in Japan. Proceedings 29th international technical conference on coal utilization and fuel systems, p. 191–202. Clearwater, FL; 2004, April 18–22.

[69] Yilgin M, Pehlivan D. Volatiles and char combustion rates of demineralised lignite and wood blends. Applied Energy 2008;86(7–8):1179–86.

[70] Simone M, Biagini E, Galletti C, Tognotti L. Evaluation of global biomass devolatilization kinetics in a drop tube reactor with CFD aided experiments. Fuel 2009;88(10):1818–27.

[71] Bradley L, Miller SF. The effect of fuel composition on pyrolysis kinetics. Proceedings 34th international technical conference on clean coal and fuel systems. Clearwater, FL; 2009, May 31–June 4.

[72] Baxter LL. Ash deposit formation and deposit properties: a comprehensive summary of research conducted at Sandia's combustion research facility. SAND2000–8253. Livermore, CA: Sandia National Laboratories; 2000.

[73] Tillman DA. Biomass cofiring: the challenges before us. Presentation to EPRI workshop on biomass. Atlanta; 2009, Oct 2.

[74] Cronauer DC, Riley JT, Vorres KS, Crelling JC, Pisupati SV. editors. An examination of burning profiles as a tool to predict the combustion behavior of coals. Proceedings 211th ACS national meeting. New Orleans; 1996, March 24–28.

[75] Wagoner CL, Duzy AF. Burning profiles of solid fuels. Paper presented at the A.S.M.E. winter annual meeting and energy systems exposition. Pittsburgh; 1967, November 12–17.

[76] Duong D. Biomass cofiring and its effect on the combustion process. MS paper. Lehigh University Energy Systems Engineering Program (work sponsored by Foster Wheeler North America Corp.); 2010.

[77] Boylan D, Wilson S, Zemo B, Bush V. Switchgrass co-firing with coal for power generation. Proceedings 18th international Pittsburgh coal conference. Newcastle, NSW; 2001, December 3–7.

[78] Tillman D, Hughes E. Biomass cofiring guidelines. Palo Alto, CA: Electric Power Research Institute; 1997. Report TR-108952.

[79] Singer JG. Combustion: fossil power systems. Windsor, CT: Combustion Engineering; 1981. p. 13-2, 7-9.

[80] Battista Jr JJ, Hughes EE, Tillman DA. Biomass cofiring at Seward Station. Biomass and Bioenergy 2000;19(6):419–27.

[81] Tillman DA, Newell C, Hus P, Hughes E, Therkelsen K. Cofiring biomass in cyclone boilers using Powder River Basin coal: results of testing at the Michigan City Generating Station. Proceedings 8th biennial conference (Bioenergy '98), p. 285–93. Madison, WI; 1998, October 4–8.

[82] Prinzing DE, Hunt EF. Impacts of wood cofiring on coal pulverization at the Shawville Generating Station. Biomass usage for utility and industrial power: an Engineering Foundation conference. Snowbird, UT; 1996, April 28–May 3.

[83] Tillman D. Final Report: EPRI-USDOE cooperative agreement: cofiring biomass with coal. US Department of Energy, National Energy Technology Laboratory; 2001. USDOE Contract No. DE-FC22–96PC96252.

[84] Tillman DA, Payette K, Opalko A, O'Conner D. Opportunity fuel cofiring at Allegheny Energy. Palo Alto, CA: Electric Power Research Institute; 2004. Report TR-1004811.

[85] Tillman D, Hughes E. Wood cofiring in a cyclone boiler at TVA's Allen Fossil Plant. Palo Alto, CA: Electric Power Research Institute; 1997. Report TR-109378.

[86] Tillman, D, Stahl R, Bradshaw D, Chance R, Jett JR, Reardon L, et al. Cofiring wood waste and coal in cyclone boilers: test results and prospectus. Proceedings 2nd biomass conference of the Americas: Energy, environment, agriculture, and industry, p. 382–9. Portland, OR; 1995, August 21–24.

[87] Gurlyurtlu I, Crujeira AT, Abelha P, Cabrita I. Measurements of dioxin emissions during co-firing in a fluidised bed. Fuel 2007;86(14):2090–100.

[88] Tillman DA, Payette K, Banfield T, Plasynski S. Cofiring demonstration at Allegheny Energy Supply Company, LLC. Proceedings 19th international Pittsburgh Coal conference. Pittsburgh; 2002, September 23–27.

[89] Virginia City Hybrid Energy Center, *www.dom.com/about/stations/fossil/virginia-city-hybrid-energy-center.jsp*; 2010, accessed 11–10–11.

[90] Khan de Jong AAW, Jansens PJ, Spliethoff H. Biomass combustion in fluidized bed boilers: potential problems and remedies. Fuel Processing Technology 2008;90(1):21–50.

[91] Leckner B, Hansson KM, Tullin C, Borodulya AV, Dikalenko VI, Palchonok GI. Kinetics of fluidized bed combustion of wood pellets. Proceedings 15th international conference on fluidized bed combustion. Savannah; 1999.

[92] Zabetta CE, Barisic V, Moulton B. Foster Wheeler references and tools for biomass- and waste-fired CFBs. Proceedings 34th international technical conference on clean coal and fuel systems. Clearwater, FL; 2009, May 31–June 4.

[93] Boylan DM. Southern company tests of wood/coal cofiring in pulverized coal units. Proceedings EPRI conference on strategic benefits of biomass and waste fuels. p. 4-33–4-45. Washington, DC; 1993, March 30–April 1.

Chapter 5

Waste Fuel–Coal Blending

5.1. INTRODUCTION

Waste fuels can be defined as those combustible resources that are outside of the mainstream of fuels being utilized in the generation of electricity or the raising of process and space heat in industrial and commercial applications. These include residues or low-value products from other processes. Hazardous materials that can effectively be utilized with typical fossil fuels to reduce their impact on the environment are included as well.

For discussion purposes in this chapter, waste fuels include postconsumer materials such as tire-derived fuel, petroleum coke, and plastics and paper. This list is not meant to be exhaustive, but the issues discussed with these waste fuels will be applicable to all other types of solid and some liquid wastes.

5.2. TIRE-DERIVED FUEL

Used tires from automobiles, trucks, tractors, and other mobile equipment provide an energy resource of significant interest to many utilities. Tires—and tire-derived fuel (TDF)—have a high calorific value along with other favorable fuel characteristics. At the same time they present material preparation and handling issues for fuel users [1, 2]. For environmental reasons, they are increasingly difficult and costly to dispose of in landfills.

5.2.1. Overview

In 2007 there were approximately 2.5 million tons of scrap tires used as fuel, or about 54% of the total scrap tire market. In addition, there are an estimated 128 million scrap tires remaining in stockpiles throughout the United States, a reduction of more than 87% since 1990 [3]. Tires are not typically collected with other municipal solid waste and require special handling practices. As a result, many states have banned tires from disposal in landfills.

Also, tires are not easily compacted, do not decompose readily, consume space, and, due to their hollow shape, trap air or other gases. Stockpiled tires can catch fire, are difficult to extinguish, and produce air pollution and soil and groundwater pollution from residuals of the fire. Finally, tires can provide breeding grounds for mosquitoes and vermin when stored. A more complete description of TDF production and history is found in Tillman and Harding [4].

5.2.2. Typical Composition

Table 5.1 lists typical proximate, ultimate, and calorific value analyses of TDF samples [5]. As expected, those samples with steel have a lower heating value and higher ash content than those without steel. However, the calorific value, even for those samples with steel, is higher, in most cases, than that for coal. This is one of the attractive features of TDF: Because of its higher calorific content, it can replace coal with actually less mass being fed into the boiler. Even more pronounced is the low ash content of the samples without steel. Sulfur contents are typical of Eastern bituminous coals, but on a kg/GJ basis, the sulfur is lower than in most coals, potentially resulting in lower emissions.

Table 5.2 lists some of the major ash constituents in various TDF samples [6]. Of particular note is the very high zinc content of the ash resulting from the vulcanization processes. The emissions of TDF compared

TABLE 5.1 Typical Analyses of TDF

Parameter	TDF with Steel	TDF without Steel
Moisture	0.75	1.15
Volatile Matter	54.23	63.50
Fixed Carbon	21.83	29.28
Ash	23.19	6.22
Carbon	67.0	81.70
Hydrogen	5.81	7.18
Nitrogen	0.25	0.56
Sulfur	1.33	1.62
Oxygen	1.64	2.65
Ash	23.19	6.29
kJ/kg	31,058	36,681
Btu/lb	13,362	15,781

Source: Data from [5].

TABLE 5.2 Ash Constituents for TDF Samples

Parameter (wt% of ash)	TDF with Steel	TDF without Steel
Al_2O_3	1.12	7.85
SiO_2	4.13	20.95
TiO_2	0.26	7.40
Na_2O	0.16	0.81
K_2O	0.20	0.72
Fe_2O_3	80.10	23.05
MgO	0.38	0.99
CaO	1.80	3.61
ZnO	8.96	25.70
Other	0.21	1.62

Source: Data from [6].

to coal can be either the same or lower, with the exception of zinc emissions. It is also important to note that the iron content in the ash is high, even for wireless tires. This iron has the potential to flux slag formation in cyclone boilers, particularly if the iron is finely divided (e.g., not in bead wire).

The EPA Office of Air Quality Planning and Standards completed a study on trace element emissions from TDF burning facilities [7, 8]. The data were from 22 industrial facilities: 3 kilns (2 cement and 1 lime) and 19 boilers (utility, pulp and paper, and general industrial applications). In general, the results indicate that properly designed existing solid fuel combustors can supplement their normal fuels (coal, wood, and combinations of coal, wood, oil, petroleum coke, and sludge) with 10% to 20% TDF and still satisfy environmental compliance emission limits. In fact, dedicated tires-to-energy facilities indicate that it is possible to have emissions much lower than are produced by existing solid-fuel-fired boilers (on a heat input basis).

In a laboratory test program on controlled burning of TDF, with the exception of zinc emissions, potential emissions from TDF are not expected to be very much different from those from other conventional fossil fuels as long as combustion occurs in a well-designed, well-operated, and well-maintained combustion device [7]. Selected results from this study are shown in Table 5.3.

TABLE 5.3 Trace Element Emissions
from TDF Facility

	Emissions	
Element	g/MJ	lb/MMBtu
Lead	5.5E-7	1.3E-6
Cadmium	1.6E-6	3.7E-6
Chromium	2.0E-6	4.7E-6
Mercury	2.9E-7	6.7E-7
Copper	3.2E-6	7.5E-6
Manganese	6.9E-7	1.6E-6
Nickel	2.7E-6	6.3E-6
Tin	1.8E-6	4.2E-6
Beryllium	3.1E-5	7.3E-5
Zinc	6.0E-4	1.4E-3

Source: Data from [7].

5.2.3. Physical Characteristics

While the current options for waste tires are expanding, the most common uses are tire-derived fuel, civil engineering applications, manufactured products, and pyrolysis [9]. As a supplementary or blending fuel, TDF is used in cement kilns, lime kilns, paper mills, utility boilers, industrial boilers, iron foundries, copper smelters, and tire-derived fuel oil to supplement coal. Tire-derived fuel used as a blending fuel comes in three main forms. In some cases, whole tires have been injected into a boiler and burned. Others require that tire crumb be produced without any steel with particles less than about 6 mm size. The most common size for utilizing TDF is 25- to 50-mm squares with the steel removed. The economics of chipping to finer sizes and removing the steel are very important in determining the long-term potential for TDF usage.

Tire-derived fuel is broadly classified as a fuel feedstock derived from automobile, truck, off-road, and specialty tires. These tires are either used whole or chipped into pieces for blending with other, more conventional fuels such as coal. These tires may contain a significant amount of steel used in reinforcement of the sidewall and rim. Depending on production requirements, this steel may need to be removed from the tire.

The great advantage of TDF as a blending fuel is its high calorific value, low ash (if the steel is removed), and low to moderate sulfur content. This makes its usage in boilers attractive if it can be prepared and fed into the boiler in an economic manner.

5.2.4. Types of Tire-Derived Fuel

The three basic types of tire-derived fuels in use today are TDF with steel, TDF without steel, and crumb rubber. Tires come predominantly from passenger automobiles and trucks, but on a per-mass basis, tractor, sport utility vehicle, and commercial truck tires provide a significant amount of TDF [9].

Tire-Derived Fuel with Steel

TDF with steel implies utilizing the entire tire as is or shredded into sized pieces. There is essentially no processing done on these tires except for steel rim removal. This fuel has the highest ash content and the lowest calorific value of the TDF products, but it is the most economical from a processing perspective. However, the remaining steel can result in utilization problems such as increased slagging due to the additional iron present in the ash. Other utilization problems occur with materials handling. Steel wire, randomly liberated during the shredding process or in materials handling, can cause damage to coal handling belts, primary and secondary crushers, and other elements of the feed system. Further, steel "whiskers" on the tire chips can cause significant bridging in material handling systems—bridging that can be difficult to break up due to the interweaving of fuel particles.

Tire-Derived Fuel without Steel

In order to make the TDF fuel more compatible with energy production facilities, it is desirable to remove as much steel as possible. This is usually accomplished by removing the rim and steel bead from the tire sidewall and magnetically removing the steel. While most of the steel is removed, some still remains within the tire chip and may present handling issues. Manufacturing wireless chips requires additional machinery and additional cost. However, such chips are periodically produced at particle sizes of <15 to 25 mm (0.6–1 in.).

Crumb Rubber

Crumb rubber is TDF essentially without any steel. This is usually material chipped/ground to less than 6 mm (<$\frac{1}{4}$ in.) and passed through magnetic separators to remove any remaining steel. This material has the highest calorific value and lowest ash content, but it is the most expensive to process. It is mainly used in specialty applications but not for energy production.

5.2.5. Preparation and Handling Issues

In preparing tires for markets, scrap tires may be chopped, shredded, or ground. Shredded tires are typically cheaper to transport than whole tires. Costs to get tires to a market include the cost of transportation, labor, and shredding, which is approximately 50 cents to 90 cents per tire.

The two basic types of tire shredding equipment in commercial use are the solid cutter shredder and the replaceable cutter shredder [10, 11]. Both of these machines shear a whole tire to produce a chip with a fixed size in one dimension, but to control the other chip dimension, additional passes through the chipper may be required.

Tire-Derived Fuel Preparation

The solid cutter shredder is a low-speed shear shredder that lends itself to many shredding applications, including tires, municipal solid waste, oversized bulky waste, demolition and construction debris, and so on. Figure 5.1 shows the principle behind the solid cutter shredder. Two counterrotating shafts with intermeshing cutters are used to grab the whole tire and force it through the pinch points between opposing cutters.

The particle size of the tire chips is controlled by basically two factors: cutter width and number of hooks on each cutter. A cutter with a 25-mm cutter disc will produce a 25-mm wide chip, and the length dimension will be determined by the distance between successive hooks on the cutter perimeter. To get a controlled chip size, screening and multiple passes through the shredder may be required.

Advantages of the solid cutter shredder include

Flexibility: The solid cutter shredder is a multipurpose unit that can shred many types of materials.

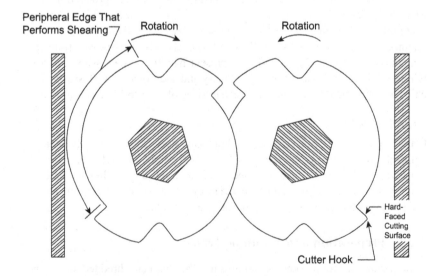

FIGURE 5.1   Solid cutter shredder shearing blades. *Source: [11].*

Simple design: The cutting box arrangement basically consists of two shafts and cutters that can be easily maintained in the field. Also, the cutters can be removed and resharpened to a distinct edge several times.

Variable chip size: Because the cutters are attached to each shaft, they can easily be changed to produce different chip sizes.

Bulk feeding capability: Many different feedstocks can be fed in bulk from grapple, front end loaders, conveyors, and so on, directly into the shredder hopper.

The two main disadvantages of the solid cutter shredder are that it was not specifically designed for tires and thus tends to tear the tire rather than shear or cut it; second, the sharpness of the cutter edge tends to deteriorate rapidly. This can be overcome with the use of harder steels, but they are more costly.

The second type of tire chipper is the replaceable cutter tire shredder. This is a low-speed shear shredder specifically designed for tires. The principle cutting feature is shown in Figure 5.2. Note the addition of star-shaped feed rolls that force the individual tires, one at a time, into the cutting zone. As the tires are fed into the cutting zone, the cutters engage them with a true shearing action.

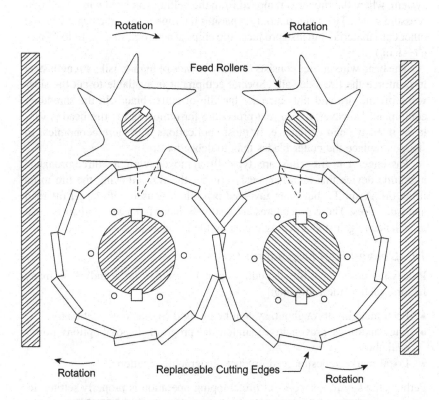

FIGURE 5.2 Replaceable cutter tire shredder system. *Source: [11].*

As with the solid cutter shredder, one dimension of the tire chip is controlled by the width of the cutter segments. However, because there are not any hooks, the other dimension may vary up to about 0.25 to 0.30 m in length. As a result, multiple passes are required to obtain a uniform chip size. The advantages of the replaceable cutter tire shredder are

Designed for tires: This type of shredder is specifically designed for tire shredding and thus produces a higher-quality cut and more uniform-sized chips.

Cutter blades: The removable cutter blades are made from a higher-quality steel alloy and can be replaced without having to replace an entire cutter assembly.

Some of the disadvantages of the replaceable cutter shredder are that it allows only one tire at a time into the feed rollers, and the chip size cannot be varied easily. Knife maintenance, however, is a critical issue for production of TDF in all systems.

Recently, alternative shredding/particle size–reduction equipment has been developed. Of particular interest is the development of the hydroblasting system, where the rubber is removed from the belting and bead wire with high-pressure water. The water not only separates the rubber from the bead wire and other cord material, but also produces tire chips at a particle size of 13 to 25 mm ($^1/_2$–1 in.).

The bead wire in tires consists of many strands of high-tensile-strength steel to reinforce the tire sidewalls. Special equipment is available to cut the side-walls off the tire and thus increase the life of cutter blades in the shredding equipment. However, the current procedure for removing the tire bead is very labor intensive and expensive. In most applications it is more economical to sharpen/replace the cutter blades than to debead a tire.

Passenger car tire rims are typically removed with a "three-pronged" hydraulic device that simultaneously crushes the rim, allowing the tire to be lifted off. A similar but more involved procedure is used with larger off-road and truck tires. These procedures are also very labor intensive, and costs must be balanced against sharpening/replacing the cutter blades.

Tire-Derived Fuel Preparation Costs

Processing costs for a tire shredding operation will vary depending on many factors, such as the following:

- Feed material, throughput, system being used (portable vs. stationary)
- Site conditions, preventative maintenance program, disposal options, power available
- Local markets, desired end product, geographic location

Perhaps the key to a successful tire chipping operation is properly setting the processing tipping fees, which must include tire processing, disposal, and profit. Fees must be evaluated on a "per-tire" versus "per-ton" basis; this will depend on the amount of processing to be done on the tires and the desired end product. In the mid-1990s, typical tipping fees were between $70 and $80 per ton, except in those areas where land disposal costs were extremely high [12]. Table 5.4 provides some estimates of the capital cost of different tire shredding equipment based on throughput and system configuration [13].

The use of TDF is, unfortunately, not without concerns. In most instances, the steel wires within the tire chip cause problems with large-scale fuel handling equipment. These wires intertwine and form large agglomerates that plug handling equipment and even tear conveyor belts.

The steel wire within a tire chip may also be a concern in the combustion of the TDF, as the additional iron may result in increases in corrosion rates and

TABLE 5.4 Capital Costs of Various Tire Shredding Systems

Shredder Type/ Manufacturer	Estimated Costs ($1000)	System Configuration	Estimated Throughput (kg/s)
Replaceable Cutters			
Columbus McKinnon	$500–$525 $435–$460	Portable Stationary	3.0–3.3: Rough Shred 2.0–2.5: 50 mm 1.0–1.3: 25 mm
Triple S/Dynamic	$475–$500 $400–$425	Portable Stationary	3.0–3.3: Rough Shred 2.0–2.5: 50 mm 1.0–1.3: 25 mm
Rotary Shear			
Eidal	$400–$425 $290–$315	Portable Stationary	2.5–3.0: Rough Shred 1.5–2.0: 50 mm 0.8–1.0: 25 mm
ERS	$500–$525 $425–$450	Portable Stationary	3.0–3.3: Rough Shred 2.0–2.5: 50 mm 1.0–1.3: 25 mm
Mac-Saturn	$400–$425 $340–$365	Portable Stationary	2.5–3.0: Rough Shred 1.5–2.0: 50 mm 0.8–1.0: 25 mm
Mitts & Merrill (Carthage)	$400–$425 $250–$275	Portable Stationary	2.0–2.5: Rough Shred 1.3–1.8: 50 mm 2.5–3: 25 mm
Shredding Systems	$450–$475 $375–$400	Portable Stationary	2.5–3.0: Rough Shred 1.5–2.0: 50 mm 0.8–1.0: 25 mm

Source: [13].

slagging on boiler waterwalls. Other constituents in the ash such as zinc or chromium may also be detrimental to overall boiler efficiency.

Finally, the overall cost of producing an acceptable product must be considered. While most tire chippers can produce a uniform-sized product in one dimension, the other dimension can vary significantly. Therefore, it becomes an economic decision as to whether multiple passes will be required through a chipper to produce a consistent, two-dimensional product. Screening, to verify size consistency, may have to be done at the end-user site to reduce problems associated with burning TDF.

5.2.6. Combustion Considerations

The use of tire-derived fuel as a blended feedstock in power generation has many advantages. First, because of the high calorific value of TDF (particularly the wire-free TDF), the cost of TDF in ¢/GJ (or $/MMBtu) is lower than that of any fossil fuel—with the possible exception of Powder River Basin (PRB) coal delivered locally or to a mine-mouth plant—and is competitive with even the lowest-cost biomass fuels. This, combined with the lower ash content of TDF (without steel), makes it a viable blending fuel. In addition, the sulfur content of TDF is less than most Eastern bituminous coals and comparable to medium-sulfur coals throughout the world.

By comparing the TDF fuel with nearly all other fossil fuels on a kg/GJ basis, the TDF is an attractive blending fuel for most boiler applications. The availability of a tire source to produce TDF within a local area and transportation costs to deliver the fuel to the boiler site need to be considered when specifying its use.

5.2.7. Case Studies

This section describes several test burns and applications of TDF in cyclone-fired boilers. Cyclone boilers are well suited for TDF applications because they easily handle chipped material and the ash is mainly taken out in fluid form. Thus, the additional steel is less of a problem in cyclones than in conventional pulverized coal boilers.

Case Study #1: Cofiring Tire-Derived Fuel at Allen Generating Station, TVA

Several TDF test burns have been completed at the TVA Allen Fossil Plant outside Memphis, Tennessee [14, 15]. The Allen Station consists of three cyclone boilers, each with the capacity to generate about 272 MW$_e$ of electricity. Early tests cofired up to 8% TDF with coal without significant problems. In addition, in the mid-1990s, a series of 11 tests were conducted to evaluate the impact of cofiring and trifiring coal, biomass, and TDF. Of these 11 tests, 3 were baseline (no biomass or TDF), 2 were 95% coal/5% TDF (mass basis),

TABLE 5.5 Summary of Allen TDF Cofiring Tests

Test No.	Coal (%)	Wood (%)	TDF (%)	Load (MW)	Efficiency (%)	NO_x (kg/J)	SO_2 (kg/J)	Opacity (%)
2	100	0	0	252	87.9	0.52	0.34	16.3
6	95	0	5	143	88.9	0.47	0.39	3.7
7	95	0	5	270	87.9	0.47	0.34	10.7
8	85	10	5	271	87.7	0.52	0.30	10.1
9	80	15	5	267	87.7	0.52	0.30	9.4
11	100	0	0	272	88.5	0.65	0.30	12.1

Source: Data from [14].

and 2 were trifired fuels (85%/10%/5% coal/biomass/TDF and 80%/15%/5% coal/biomass/TDF). These tests investigated the impact of cofiring and trifiring on the fuel yard and fuel handling system, the ability of the plant to achieve full capacity, the impact of alternate fuels on boiler efficiency and consequent net station heat rate, the impact on operating stability and operating temperatures, and the effect on emissions.

Table 5.5 provides a summary of the pertinent test results. As can be seen, the cofiring of TDF with coal at the 5% level reduced NO_x by about 8%, reduced opacity, and resulted in a slight increase in SO_2 emissions. Boiler efficiency was equal to or slightly greater during the cofiring tests than the coal-only tests. These short-term tests had the following results:

- Cofiring and trifiring did not result in any loss of boiler capacity or unit capacity.
- Cofiring and trifiring did not reduce boiler stability or operability and did not increase the excess air required for stable and effective operation.
- Cofiring and trifiring had impacts on boiler efficiency of less than 1% when the unit was firing at full load and not limited by opacity.
- Cofiring and trifiring had minimal impacts on estimated flame temperatures in the barrel.
- Cofiring and trifiring reduced NO_x emissions relative to all Western coal firing: 100% Western coal (Test 11) had a NO_x emission of 0.65 kg/J (1.5 lb/MMBtu) when firing at MCR and firing under conditions not restricted by opacity firing, while a blend of 85% coal/10% biomass/5% TDF resulted in a NO_x emission of 0.52 kg/J (1.2 lb/MMBtu).
- Cofiring TDF with Western coal did result in an increase in SO_2 emissions, but a blend of 85% coal/10% biomass/5% TDF achieved about the same level of SO_2 emissions as the coal alone.

- Opacity certainly did not increase and may have actually decreased during the cofiring and trifiring tests relative to baseline coal-only testing.
- Ash characteristics were essentially unchanged among the baseline, cofired, and trifired tests.

Case Study #2: Cofiring Tire-Derived Fuel at Baldwin Generating Station, Illinois Power

Illinois Power conducted an extensive program to test cofiring of TDF at their Baldwin Power Station, which has two cyclone-fired boilers [16–18]. More than 11,730 tonnes of TDF were burned during the course of the test program. This represented the ultimate disposal of over 1,293,000 tires. The objectives of these tests were to determine if TDF could be delivered to the boiler reliably, if the TDF could be burned reliably, if TDF posed hazards to the equipment, and if TDF cofiring affected the environmental operation of the plant. Cofiring levels were limited to less than 5% on an energy basis.

Because the cost to produce 25-mm tire chips is more expensive than 50-mm chips, the initial tests were to see if 50-mm chips could be successfully fed with the coal through the hammer mills. With only about 0.5 tonnes of chips fed, the screens on the hammer mill plugged. Later testing bypassed the hammer mill but still resulted in pluggage of the discharge chutes on the tripper. As a result, the majority of the testing was done with 25-mm chips.

Emission testing was done during the test program, but because of the small amount of cofiring, no appreciable differences were noted between the baseline and cofiring tests. Leaching tests were also completed on flyash and slag samples with no apparent differences between the baseline and cofiring tests. The only problem was the presence of unburned pieces of TDF and some wire in the slag. The recommended procedure was to purchase debeaded tire chips.

Carbon measurements made on the baseline ash and the ash from the cofiring tests were similar: 5.52% carbon-in-ash (baseline) and 5.50% carbon-in-ash (cofiring tests). The ESP efficiency varied from 95.6% to 96.2% during the baseline tests and 93.8% to 95.7% during the coal and TDF tests. Table 5.6 compares the mineral analyses of the flyash and slag during the baseline and cofiring tests.

The test program demonstrated that Baldwin Station could successfully cofire TDF as a supplementary fuel at a 2% level capacity. Budget estimates were also prepared for permanent cofiring. These showed an economic benefit of about $200,000/year to consumers.

Case Study #3: Cofiring Tire-Derived Fuel at Jennison Generating Station, NYSEG

New York State Electric and Gas (NYSEG) burned approximately 2,721 tonnes (300,000 tires) of TDF [19]. These tests were conducted at NYSEG's Jennison Station located in Bainbridge, New York. Jennison Station has the electrical

TABLE 5.6 Ash and Slag Analyses (wt%)

	Flyash		Slag	
Compound	Coal	Coal and TDF	Coal	Coal and TDF
SiO_2	50.86	50.15	52.51	53.11
Al_2O_3	20.00	18.71	19.66	19.47
TiO_2	1.31	0.84	0.78	0.70
Fe_2O_3	18.15	18.30	14.73	14.58
CaO	2.99	3.26	6.12	6.70
MgO	1.03	1.11	1.05	1.16
K_2O	2.76	2.62	1.80	1.80
Na_2O	1.28	1.27	0.68	0.72
P_2O_5	0.32	0.02	0.07	0.11
Undetermined	0.88	1.93	1.61	0.09
ZnO	—	0.60	0.10	1.14
Zn (ppm)	2000	483	—	724

Source: Data from [18].

generating capacity of 74 MW fired from four traveling chain grate stoker boilers that use approximately 590 tonnes of coal daily. The station has the capability of burning the equivalent of 4,500,000 scrap tires annually, using a blend of 25% tire chips with 75% coal. This equates to an energy savings of approximately 59,000 tonnes of coal.

The cofiring of the TDF required only minor changes in boiler operations. A noticeable amount of bead metal and some steel belt material melted into nuggets that stuck on or in between the stoker grate keys. Although the buildup of metal was visible, it did not interfere with the ability to efficiently fire the blend.

The ESP performance was not changed; there were no incidents of black smoke or increased emissions from the coal/TDF blends. There were no reports from neighbors of an odor from tires burning. Sulfur dioxide emissions decreased somewhat when burning the TDF/coal blends. Other measured constituents changed only slightly depending on the boiler load and TDF blends being used. In general, the emissions were about the same between the coal-only and the coal/TDF blends.

The only differences noted in the chemical analyses on the bottom ash and flyash samples were an increase in zinc in the flyash and an increase in metal from the tire chips in the bottom ash when firing the TDF blends compared to coal alone. Following these tests, magnetic separation equipment was installed at the ash pond to remove the ferrous metal from the bottom ash. This rendered the bottom ash acceptable for use by local municipalities as a traction agent in the winter. A scrap metal dealer recycled the metal.

The industrial hygiene data collected demonstrated that there were no detrimental as is human, health, or safety factors associated with the storage or handling of the tire chips at the plant. There were no noticeable odors from the tire chips, either inside or outside the plant, and the results of testing indicated there were no significant worker hazards associated with tire chip burning when compared to burning coal alone.

Case Study #4: Cofiring Tire-Derived Fuel at FirstEnergy Toronto Plant

In the early 1990s, Ohio Edison (now FirstEnergy) completed a four-day scrap tire test burn of cofiring whole scrap tires and pulverized coal [20, 21]. Testing was done at the Toronto Plant, located in Toronto, Ohio. This facility is a 162-MW_n plant with three generating units still in service. TDF testing was done in Unit 5, a 42-MW_n Babcock & Wilcox (B&W) pulverized coal-fired wet slag bottom boiler. The boiler was modified to accept whole scrap tires.

The environmental results from the test burn showed that no permit requirements were violated. Comparing the baseline to the 20% tire (energy basis) cofiring test showed lead emission rates were 5% lower and particulate emissions rates were 28% lower. Based on the coal bunkered the last two days of the test at the 20% cofiring rate, the sulfur dioxide emission rate was 13.7% lower than the calculated expected value. Ohio Edison obtained a modification of its operating permit to include up to 20% of total boiler heat input from tires. At this rate, over 3 million scrap tires could be converted to electricity each year at the Toronto facility.

Figure 5.3 shows the emissions of NO_x, SO_2, and particulates during these TDF tests. The reported values are the average of three different measurements. The figure clearly shows more than a 35% reduction in NO_x when 20% TDF (energy basis) is used compared to the baseline. In addition, there is relatively no change in the sulfur emissions or in the particulate emissions.

The mechanical results of the test burns showed that the equipment performed according to design. Tire feed rates varied from one tire every 34 seconds to one tire every 10 seconds during the test. To prove the mechanical aspects of the delivery system, a brief test was conducted with a feed rate of one tire every 7 seconds. However, with some minor mechanical changes and additional fine-tuning of the programmable logic controller for the delivery system, feed rates of one tire every 4 seconds are possible. A feed rate of one tire every 4 seconds corresponds to more than 40% energy input from tires. No

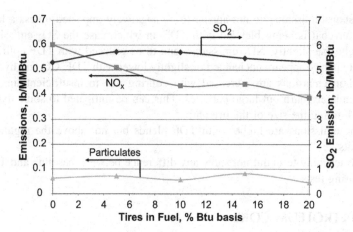

FIGURE 5.3 Emissions from FirstEnergy TDF program. *Source: Data from [20].*

boiler control problems were experienced, and boiler slag tapping was accomplished without incident.

5.2.8. Conclusions Regarding Tire-Derived Fuel as a Blend Fuel

Several conclusions can be drawn regarding the use of TDF cofired with coal in industrial and utility boilers:

- There is a sufficient supply of scrap tires in most areas of the United States to warrant the investigation of TDF cofiring in boilers.
- The boilers best suited for using TDF are cyclones, stokers, and fluidized bed boilers. Some unique situations exist with TDF in pulverized coal boilers. These boilers can use up to 20% to 25% TDF (energy basis).
- Tire chips greater than about 25 mm resulted in plugged hammer mills, coal transfer chutes, and so on, even in small cofiring amounts. Therefore, the TDF should be introduced after the hammer mills as the coal is proceeding directly to the bunkers.
- Steel in the tire chips (from the bead and steel belt) can cause problems in the handling of the TDF and in some instances in the burning of the fuel.
- For best results in obtaining a consistent TDF feed material, most utilities have gone to chipping the tires themselves. They can then control the size of the chip and the consistency of the product.
- TDF blends did not result in any major boiler operating problems. Full load was achievable at reasonable excess air levels. Unburned chips in the bottom slag have presented a problem if the chips are too large. Steel strings in the ash/slag have resulted in the need for a magnetic separator to remove them prior to selling the product.
- Boiler efficiencies were equal to or slightly greater with the TDF blends due to their higher calorific value and low moisture content.

- Gaseous emissions do not appear to be negatively impacted unless a low-sulfur coal is being blended with TDF, in which case the SO_2 emissions might rise slightly. NO_x emissions tend to be reduced with TDF cofiring, and opacity appears unchanged or slightly lower with TDF. CO and hydrocarbon emissions are increased with cofiring due to insufficient time to completely burn out large particles. This can be mitigated to some extent by limiting the size of the tire chip.
- Zinc emissions are higher with TDF blends but not above the regulated limits.
- Ash leaching tests did not show any difference between baseline and TDF cofiring tests.

5.3. PETROLEUM COKE

Petroleum coke is a solid by-product of petroleum refining and is used in the production of carbon electrodes for the aluminum industry, graphite electrodes for steel making, as fuel in power generation, and as fuel for cement kilns. In the United States, more than 1.5 million tonnes (1.68×10^6 tons) of petroleum coke are used by major utilities, as shown for 2000 in Table 5.7. The primary reason for this is the low cost of petroleum coke compared to coal.

Petroleum coke production processes are reviewed by Bryers [22, 23]; approaches include delayed coking and fluid coking. Delayed coking is a batch process where residual components of crude oil are heated to about 475°C to 520°C (890–970°F) in a furnace and then confined in a coke drum for thermal cracking reactions. The products of the coking process are gas, gasoline, gas oil, and coke.

In a typical delayed coking arrangement, two coke drums are used; one is being filled and reactions are proceeding while the produced coke is being removed by high-pressure water. The delayed coking process produces various types of petroleum coke, including needle coke, granular coke, sponge coke, and shot coke. Sponge and shot cokes are commonly found in the fuel market. Sponge coke is highly porous and anisotropic in nature; shot coke, which appears like an assembly of small balls, also exhibits anisotropic characteristics, but is less porous and much harder to crush. Shot coke often is the result of upset conditions in the delayed coker.

Fluid coke is produced in a fluidized bed reactor, where the heavy oil feedstock is sprayed onto a bed of fluidized coke. The oil feedstock is cracked by steam introduced into the bottom of the fluidized bed reactor. Vapor product is drawn off the top of the reactor, while the coke descends to the bottom of the reactor and is transported to a burner, where a portion of the coke is burned to operate the process. Fluid coke reactors are operated at about 510°C to 540°C (950–1000°F). Flexicoke is a variant of fluid coke, where a gasifier is added to the process to increase coke yields. Fluid coker installations tend to have yields that are lower than delayed coker installations, while flexicoker installations

TABLE 5.7 Petroleum Coke Consumption by U.S. Electric Utilities

Utility	Use (tonnes)	Use (tons)	Delivered Cost ($/GJ)	Delivered Cost ($/10⁶ Btu)
Central Illinois Public Service	23,400	26,000	0.86	0.91
Jacksonville Electric Authority	400,000	444,000	0.58	0.61
Lakeland Department of Water and Electric	1800	2000	0.41	0.43
Manitowoc Public Utilities	32,400	36,000	0.45	0.47
Michigan South Central Power	1800	2000	1.02	1.07
NIPSCO (NiSources)	156,800	174,000	0.62	0.65
Northern States Power	198,200	220,000	0.31	0.33
Ohio Edison Company	7200	8000	0.70	0.74
Owensboro, City of	8100	9000	0.51	0.54
Pennsylvania Power	183,000	203,000	0.70	0.74
San Antonio, City of	8100	9000	0.40	0.42
Tampa Electric (TECO)	190,000	211,000	0.48	0.51
Union Electric Company (Ameren)	111,700	124,000	0.58	0.61
Wisconsin Electric Power Company	132,400	147,000	0.66	0.70
Wisconsin Power & Light (Alliant Energy)	62,200	69,000	0.45	0.47
Total	**1,516,200**	**1,683,000**	**0.55**	**0.58**

Note: Totals may not add due to rounding.
Source: [3].

have yields that can be significantly greater than delayed coker installations. Fluid cokers produce layered and nonlayered cokes. Both delayed and fluid coke installations produce amorphous, incipient, and mesophase cokes, with the amorphous cokes having higher volatility and the mesophase cokes having the lowest volatility.

5.3.1. Fuel Characteristics of Petroleum Coke

Traditional analyses of fuels include proximate, ultimate, and ash elemental analysis, along with calorific value and, increasingly, trace metal concentrations. These values are presented in the following sections.

Proximate and Ultimate Analysis of Petroleum Coke

A significant body of literature exists concerning the traditional characteristics of petroleum coke [e.g., 22–26], and the influence of coking processes and conditions on such properties [e.g., 22, 23, 27]. Table 5.8 presents the typical characteristics of various petroleum cokes as a function of the coking method.

TABLE 5.8 Typical Fuel Characteristics of Petroleum Coke

Analysis	Delayed Coke(Sponge)	Shot Coke	Fluid Coke	Flexicoke
Proximate Analysis (wt%)				
Fixed Carbon	80.2	89.59	91.50	94.9
Volatiles	4.48	3.07	4.94	1.25
Ash	0.72	1.06	1.32	0.99
Moisture	7.60	6.29	2.24	2.86
Ultimate Analysis (wt%)				
Carbon	81.12	81.29	84.41	92.00
Hydrogen	3.60	3.17	2.12	0.30
Oxygen	0.04	0.93	0.82	0.00
Nitrogen	2.55	1.60	2.35	1.11
Sulfur	4.37	5.96	6.74	2.74
Ash	0.72	0.76	1.32	0.99
Moisture	7.60	5.69	2.24	2.86
Higher Heating Value				
MJ/kg	33.18	33.34	32.53	32.43
Btu/lb	14,298	14,364	14,017	13,972
HGI	54	39	35	55

Note: Some values in table depend on crude oil properties.
Source: [22, 23].

Note that flexicoke has the lowest volatile content and also the lowest calorific value. However, all of the petroleum cokes are high in calorific value. On occasion, off-specification petroleum coke is supplied to generating stations. A survey of tests conducted by others shows that petroleum cokes with 8% to 10% volatile matter are common; higher volatile matter (e.g., 10% to 14% volatiles) can also be encountered.

Ash Characteristics of Petroleum Coke

Petroleum coke is, inherently, a low-ash fuel, as shown in Table 5.9. However, the ash chemistry of petroleum coke remains significant due to issues of slagging and fouling and trace metal emissions. Bryers [22, 23] has reported general ash characteristics for petroleum cokes as a function of the coking method. These data are shown in Table 5.9. Note the high concentrations of vanadium and nickel in all but the fluid coke.

As a practical matter, the concentration of vanadium, nickel, and other inorganic matter in petroleum coke is largely a function of the source of the crude oil, as shown in Table 5.10. "Sweet" (low-sulfur) crude oils not only contain less ash, but can contain lower concentrations of vanadium, nickel, and other deleterious metals (see data from [22]). The vanadium content in the Venezuelan crude oil shown in Table 5.10 is particularly high, while the vanadium content in the Canadian crude oil is quite low. Mercury concentrations in petroleum coke are consistently and significantly lower than those associated with crude oils. This results from the thermal processes used in oil refining and the consequent volatilization of the mercury in the crude oil. By the time petroleum coke is formed, the mercury has been removed by the refining processes.

5.3.2. Petroleum Coke Issues

The use of petroleum coke, even with its low cost, is not without possible issues in the power plant. As a solid material similar to coal, it is handled in much the same manner as the predominant solid fuel during utilization. Two issues require special mention: pulverization and ash characteristics.

Handling and Pulverization

Being a solid fuel similar to coal, petroleum coke is usually blended with coal prior to pulverization. Depending on the origin and type of petroleum coke being blended and the coal it is being blended with, the pulverized petroleum coke may be found in the smaller sizes compared to the pulverized coal [22, 33, 34]. This might be advantageous, since the smaller particles require less reaction time, and with the low volatility of the petroleum coke, a smaller-size particle will reduce carbon-in-ash levels. Unfortunately this is not always the case; studies by Yu [35] indicated that the petroleum coke concentrates in the larger-size fractions during the comminution process. This can lead to an increase in carbon-in-ash levels relative to coal alone.

TABLE 5.9 Typical Ash Characteristics of Petroleum Coke

Analysis	Delayed Coke	Shot Coke	Fluid Coke	Flexicoke
Elemental Composition (wt%)				
SiO_2	10.1	13.8	23.6	1.6
Al_2O_3	6.9	5.9	9.4	0.5
TiO_2	0.2	0.3	0.4	0.1
Fe_2O_3	5.3	4.5	31.6	2.4
CaO	2.2	3.6	8.9	2.4
MgO	0.3	0.6	0.4	0.2
Na_2O	1.8	0.4	0.1	0.3
K_2O	0.3	0.3	1.2	0.3
SO_3	0.8	1.6	2.0	3.0
NiO	12.0	10.2	2.9	11.4
V_2O_5	58.2	57.0	19.7	74.5
Ash Fusion Temperatures (°C)				
Reducing				
Initial Deformation	1599	1436	1378	1538
Spherical	1599	1599	1386	1538
Hemispherical	1599	1599	1439	1538
Fluid	1599	1599	1474	1538
Oxidizing				
Initial Deformation	1374	1259	1095	749
Spherical	1425	1429	1155	790
Hemispherical	1431	1471	1183	853
Fluid	1433	1471	1224	1311

Source: [22].

Ash Species Mobility

With the stringent environmental regulations regarding solid fuel combustion flyash, the blending of petroleum coke with coal might have a negative impact on its use. Studies by Izquierdo [36] evaluated the leaching and mobility of the nine minor elements in slag samples and typical combustion flyash (silicon,

TABLE 5.10 Trace Metal Concentrations in Crude Oil from Various Locations (mg/kg Oil)

Metal	Typical	Libya	Venezuela	Alberta
As	0.0006–1.1	0.077	0.284	0.0024
Co	<1.25	0.032	0.178	0.0027
Cr	0.002–0.02	0.0023	0.430	—
Cu	N/A	0.19	0.21	—
Hg	0.09	—	0.027	0.084
Mn	0.4	0.79	0.21	0.048
Ni	14–68	49.1	117	—
Sb	0.002–0.8	0.055	0.303	—
Se	0.03–1	1.10	0.369	0.0094
V	15–590	8.2	1100	0.682
Zn	N/A	62.9	0.692	0.670

Source: [28–32].

aluminum, calcium, potassium, magnesium, sodium, iron, titanium, and sulfur), as well as 16 trace elements (As, B, Ba, Cd, Cl, Cr, Cu, F, Hg, Mo, Ni, Pb, Sb, Se, V, and Zn). These slag and ash samples came from a power plant in Spain where two levels of petroleum coke blending were completed: a low-percentage blend (4% m/m) and a high-percentage blend (24% m/m).

The results for the slag samples, even at the 24% petroleum coke addition level, suggested that the slag could be managed as an inert material because of the low mobility of all elements. This was not the case, however, for the flyash samples. While the mobility of most elements remained essentially unchanged, some did show an increase in leachability relative to the 100% coal test or between the 4% and 24% test. In particular, arsenic (As) and molybdenum (Mo) in the high petroleum coke blend resulted in leaching rates that exceeded the European standards for nonhazardous materials. Vanadium also showed a significant increase in mobility, but there were no regulations on vanadium. Based on these results, the amount of petroleum coke that can be blended may be a function of the ash properties and not boiler performance or emissions.

5.3.3. Petroleum Coke Utilization in Boilers

As shown earlier in Table 5.7, petroleum coke is used in several electric generating boilers as well as some large industrial plants. Because it has higher calorific value than most fossil fuels and a lower cost, petroleum coke is an excellent opportunity fuel to be blended with other fossil fuels. There are basically two general types of power-generating boilers: those that use crushed coal (typically <6 mm) and those that use pulverized coal with a particle size such that 70% passes a 200-mesh screen (74 μm). For illustrative purposes, this section summarizes two full-scale petroleum coke blend tests, one in a cyclone boiler and one in a pulverized coal boiler.

Petroleum Coke Use in a Cyclone Boiler

Cyclone boilers, comprising about 9% of the U.S. coal-fired capacity, were installed by electric utilities until about 1975 in order to use slagging coals typically found in the midwestern United States. Coal is crushed and introduced into a barrel, where combustion occurs in a highly intense environment (heat release rates are nominally 19 to 32 GJ/m³-hr [500,000–850,000 Btu/ft³-hr]). During combustion, a slag layer forms in the cyclone barrel and flows from the barrel to the furnace and then to a slag tank below the furnace. About 70% of the inorganic matter in the coal is removed as slag tapped from the barrels and the furnace, while the remaining 30% of the inorganic matter is removed as flyash.

Cyclone boilers are highly favorable for cofiring petroleum coke. Because cyclone boilers were developed for a crushed fuel rather than a pulverized fuel, and because they were designed to remove inorganic matter as slag, they have become known for fuel flexibility. At the same time, the requirement to have sufficient ash in the fuel to form a slag layer—typically >5% ash in the coal— limits the use of opportunity fuels in these combustion systems to cofiring applications.

Further, cyclones require significant volatile matter concentrations to promote ignition and combustion in the cyclone barrel. While petroleum coke contains significant concentrations of sulfur, the coal originally used in cyclone boilers typically was (and frequently remains) a high-sulfur coal. Consequently many of these boilers are equipped with scrubbers. Further, most cyclone boilers are concentrated in the midwestern United States and have ready access to petroleum coke from inland refineries and refineries with access to the Mississippi River transportation system.

While petroleum coke has been cofired in several cyclone units [37, 38], we use as an example the Bailly Generating Station (BGS), which is owned by Northern Indiana Public Service Co. (NIPSCO), a NiSources Company.

Petroleum Coke Cofiring at Bailly Generating Station

Petroleum coke was cofired in extensive tests at boiler #7 of the Bailly Generating Station as part of the biomass cofiring program developed through the Cooperative Agreement between USDOE and EPRI. These tests, conducted at the NIPSCO location, were designed to demonstrate the benefits and issues associated with firing petroleum coke with coal, urban wood waste with coal, and the triburn effect of firing petroleum coke and urban wood waste with coal. A four-month program was instituted during 1999, and 56 individual tests were performed.

During these tests, a baseline was developed, and then extensive cofiring and triburn testing was performed with petroleum coke percentages ranging from 10% to 25% on a mass basis, or 12% to 29% on a heat input basis. Urban wood waste percentages were 5%, 7.5%, and 10% on a mass basis or about 2.5%, 3.5%, and 5% on a calorific value basis [39–43]. This concept of burning petroleum coke with another opportunity fuel—a supplier of volatile matter as well as heat input—is not unique to the Bailly Generating Station. Owensboro Municipal Utilities also fired petroleum coke with TDF at its Elmer Smith Station Unit #1 [43].

Bailly Generating Station (BGS) Boiler #7 is a 160-MW$_e$ (net) cyclone fired unit. At MCR it generates some 151.3 kg/sec (1.2×10^6 lb/hr) of 165.5 bar/538°C/538°C (2400 psig/1000°F/1000°F) steam. At the time of testing, the plant fired a blend of 70% high-sulfur Illinois Basin coal and 30% low-sulfur Western bituminous Shoshone coal. For emissions control the BGS boilers are equipped with both electrostatic precipitators and a Pure Air wet scrubber from Air Products and Chemicals for SO_2 control. Consequently, the plant is well equipped to fire high-sulfur fossil fuels. During the testing, petroleum coke was fired directly with the blended Illinois Basin and Shoshone coals. During many tests, the petroleum coke was blended with urban wood waste for firing with the Illinois Basin and Western bituminous coals. Table 5.11 presents the salient fuel characteristics of the four fuels burned at the Bailly Generating Station.

Because 56 individual tests were conducted over a three-month period, a substantial body of data was obtained concerning the impact of petroleum coke cofiring on cyclone operations. The tests demonstrated that petroleum coke exerted a distinct benefit to boiler operations, improving efficiency and reducing the volumetric flow of fuel to the boiler. The technical and economic consequences of the improved boiler efficiency—up to 1.25% when cofiring 25% petroleum coke on a mass basis or 29% petroleum coke on a heat input basis—included reduced fuel flow to the bunkers and boiler. This, in turn, reduced the load on conveyors, fans, and other related mechanical systems. The improved efficiency resulted in a slightly reduced house load as a function of firing petroleum coke with coal.

These efficiency gains associated with cofiring petroleum coke occurred despite increases in unburned carbon (UBC) in the flyash. When cofiring petroleum coke with coal, the unburned carbon in the flyash increased from an average of 9.1% (baseline: 0% petroleum coke) to an average of 13.5% UBC with 10% petroleum coke cofiring (mass basis) and an average of 22.6% UBC when cofiring 20% petroleum coke (mass basis) [41]. However, these high UBC levels are somewhat misleading. With petroleum coke having <2% ash, and with 70% of the inorganic material leaving the boiler

TABLE 5.11 Characteristics of Fuels Burned at Bailly Generating Station

Parameter	Petroleum Coke	Illinois #6 Coal	Shoshone Coal	Urban Wood
Proximate Analysis (wt%)				
Fixed Carbon	78.27	41.95	42.15	12.50
Volatile Matter	13.90	34.43	37.56	52.56
Ash	1.34	9.66	5.63	4.08
Moisture	6.48	13.97	14.66	30.78
Higher Heating Value				
MJ/kg	33.31	25.87	25.37	13.48
Btu/lb	14,308	11,113	10,900	5788
Sulfur and Nitrogen				
Sulfur (%)	5.11	3.45	0.74	0.07
Sulfur (kg/GJ)	1.53	1.34	0.29	0.05
Sulfur (lb/10^6 Btu)	3.55	3.10	0.68	0.12
Fuel Nitrogen (%)	1.23	1.22	1.44	1.00
Fuel Nitrogen (kg/GJ)	0.37	0.47	0.57	0.75
Fuel Nitrogen (lb/10^6 Btu)	0.86	1.10	1.32	1.73

Source: [39].

as slag, there is less dilution of UBC with inorganic material in the flyash, and there is less flyash. The overall impact of flyash UBC on efficiency was negligible.

Because BGS employs a wet scrubber, SO_2 emissions were not evaluated. Focus was given to NO_x emissions, CO and hydrocarbon (THC) emissions, SO_3 emissions, and trace metal emissions. The NO_x emissions actually decreased with the addition of petroleum coke in the blend. The CO, THC, and SO_3 emissions showed only minor influences as a consequence of cofiring petroleum coke, as shown in Figure 5.4. Note that petroleum coke slightly increased CO emissions and slightly decreased THC emissions. Only the combination of petroleum coke and wood waste increased SO_3 emissions above noise levels, and the total SO_3 emissions under these conditions were <14 ppm_{vd}.

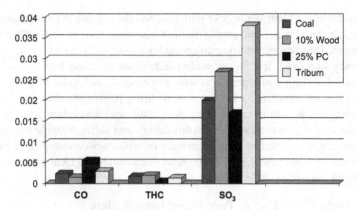

FIGURE 5.4 CO, hydrocarbon (THC), and SO$_3$ emission results from the Bailly Generating Station cofiring tests (values in ppm$_{vd}$ at 3% O$_2$). *Source: [33].*

Petroleum coke also reduced the trace metals found in the feed to BGS #7. Trace metal reductions occurred for arsenic, cadmium, chromium, lead, and mercury. Vanadium and nickel were the only metals increasing in concentration as a consequence of cofiring petroleum coke, as expected. Table 5.12 documents the concentrations of trace metals in the petroleum coke fired at BGS, relative to the coal and urban wood waste fired at that location. The concentrations of trace metals shown in the table indicate that, on a heat input basis, petroleum coke further reduces trace metal concentrations in the fuel fed.

TABLE 5.12 Concentrations of Trace Metals in Fuels Burned at Bailly Generating Station Tests

	Fuel (in mg/kg)			
Metal	Petroleum Coke	Low-Sulfur Coal	High-Sulfur Coal	Urban Wood Waste
Arsenic	0.337	1.402	2.194	2.145
Chromium	4.676	8.250	20.361	6.570
Lead	2.182	4.267	3.154	2.922
Mercury	0.016	0.020	0.033	0.013
Nickel	134.04	7.396	12.33	2.645
Vanadium	326.37	11.95	17.21	3.06

Source: [40].

The cofiring of petroleum coke with coal and the trifiring of petroleum coke with urban wood waste and coal at Bailly Generating Station boiler #7 clearly documented the benefits of using this opportunity fuel in cyclone boilers. Boiler efficiency improved, leading to modest reductions in house loads associated with fuel handling equipment and with forced draft and induced draft fans. NO_x emissions decreased as a function of firing petroleum coke and—most interestingly—as a function of firing petroleum coke and urban wood waste with the coal. Carbon monoxide, hydrocarbon, and sulfur trioxide emissions were largely unaffected by petroleum coke or urban wood waste. Trace metal emissions were decreased. The Bailly tests documented the clear advantage of petroleum coke in cyclone boilers.

Petroleum Coke Use in Pulverized Coal Boilers

Pulverized coal (PC) boilers are the most common large combustion systems for the generation of electricity in the United States and the industrialized economies of the world. PC boilers include both wall-fired and tangentially fired (T-fired) boilers; wall-fired boilers include both front-wall and opposed-wall configurations. Like cyclone boilers, they have the potential to use petroleum coke as a fuel provided that a sulfur dioxide scrubber has been installed on the unit. The vast majority of the petroleum coke fired in the United States is burned in PC boilers due to their dominance of the industry. Again, the low volatility in the petroleum coke limits its use in PC boilers; the typical cofiring percentage is on the order of 20% to 30% (calorific value basis).

Cofiring petroleum coke with coal in PC boilers has the economic potential associated with reducing fuel costs identified previously. However, there are certain limitations with PC boilers. Many of these units sell flyash as pozzolanic material under the American Society for Testing and Materials (ASTM) Specification C-618. ASTM C-618 limits the unburned carbon content in the flyash [43]. About 25% of the flyash generated in the United States is sold under ASTM Specification C-618. The ability to sell flyash—or the inability to sell it—can impact plant economics by $1 million to $3 million annually. Few cyclone boiler operators sell flyash.

Low-NO_x firing has been significantly advanced for PC boilers capitalizing on high volatile release at the base of the flame—and capitalizing on staged air–staged fuel combustion. While these concepts have been developed for cyclone firing, they have been sufficiently advanced for PC firing that several tangentially fired units using PRB coals have achieved NO_x emissions <64.5 g/GJ (0.15 lb/10^6 Btu) [44].

The constraints on PC cofiring of petroleum coke are more significant than in the case of cyclone firing. For all pulverized coal units, the Hardgrove grindability index (HGI) becomes critical because pulverizing petroleum coke to an appropriate size (e.g., 80% $<$ 200 mesh [74 μm]) may be uneconomical for certain petroleum cokes with low HGI values. Flame stability for wall-fired PC boilers becomes a critical issue and can limit the percentage of petroleum coke in the total fuel mix. Consequently the results of firing petroleum coke in PC boilers are mixed. An example of petroleum coke/coal blending is the Widows Creek Fossil Plant (WCF) of TVA; this example will highlight some of the potentials and limitations of petroleum coke utilization in large PC utility boilers.

Cofiring Petroleum Coke at Widows Creek Fossil Plant

Widows Creek Fossil Plant consists of two plants: A plant, with six smaller wall-fired boilers, and B plant, with two large T-fired boilers. Boiler #7 has a net generating capacity of 518 MW_e and boiler #8 has a net generating capacity of 531 MW_e. Typically they are operated at about 400 MW_e (net). These units generate 154.4 bar/538°C/538°C (2400 psig/1000°F/1000°F) steam. Both units are equipped with an SO_2 scrubber sufficient to handle the increased emissions resulting from petroleum coke cofiring [45]. Testing of cofiring petroleum coke was conducted on both boilers at 20% and 25% (mass basis, or 24% and 29%, calorific value basis), respectively, with a brief test at 35% in Unit #8. Characteristics of the petroleum coke and coal fired at the WCF cofiring tests are shown in Table 5.13.

The impact of the petroleum coke on efficiency was immediate and significant. Equations (5.1) and (5.2) document the impact on boiler efficiency for Units #7 and #8, respectively.

$$\eta_{\#7} = 99.5 + 0.029(\%\text{PetC}) - 0.453(\text{EO}_2\%) - 0.060(\text{T}_{ah}) \qquad (5.1)$$

and

$$\eta_{\#8} = 98.9 + 0.029(\%\text{PetC}) - 0.377(\text{EO}_2\%) - 0.060(\text{T}_{ah}) \qquad (5.2)$$

where %PetC is percent petroleum coke on a mass basis, $EO_2\%$ is percent excess O_2 at the air heater exit on a dry basis, and T_{ah} is the temperature of the air exiting the air heater (°C). The r^2 for these equations exceeds 0.99 [45]. The use of petroleum coke improved boiler efficiency; at 20% petroleum coke, boiler efficiency improved by over 0.5%, and net station heat rate decreased by >70kJ/kWh (65 Btu/kWh). This efficiency improvement could be attributed to reduced losses associated with hydrogen and moisture in the fuel. The efficiency improvements came despite increases in unburned carbon in the flyash. Table 5.14 summarizes the impacts of petroleum coke cofiring on unburned carbon in the flyash for both units. Given that flyash is 80% of the solid products of combustion from a PC boiler, these increases were significant. For plants selling flyash under ASTM C-618, these would be unacceptable results. For plants without flyash sales, these results can be accepted given the efficiency improvements.

Testing showed no appreciable impact on flame temperature, upper furnace temperature, or flame intensity when firing petroleum coke with high-sulfur coal. Flame intensity decreased when firing petroleum coke briefly with low-sulfur coal. Similarly, there were no operability issues; main steam and

TABLE 5.13 Fuel Characteristics for the Widows Creek Petroleum Coke Tests

Parameter	Petroleum Coke	Coal
Proximate Analysis (wt%)		
Volatile Matter	9.25	23.31
Fixed Carbon	84.66	56.81
Moisture	5.56	9.51
Ash	0.54	10.67
Ultimate Analysis (wt%)		
Carbon	84.4	66.4
Hydrogen	3.6	4.5
Oxygen	0.14	4.6
Nitrogen	0.97	1.25
Sulfur	4.8	3.07
Moisture	5.56	9.51
Ash	0.54	10.67
Higher Heating Value		
MJ/kg	32.84	26.94
Btu/lb	14,150	11,607

Source: [45].

TABLE 5.14 Unburned Carbon in Widows Creek Flyash

Boiler	Percent Petroleum Coke (Mass Basis)	Average %
7	0	NA
7	20	6.33
7	25	9.29
8	0	1.39
8	20	9.17

Source: [45].

TABLE 5.15 Average NO_x Emissions during Full Load Petroleum Coke Cofiring Tests at Widows Creek Fossil Plant

Boiler	Petroleum Coke (%)	NO_x Emissions (kg/GJ)	NO_x Emissions (lb/10^6 Btu)
7	0	0.14	0.32
7	20	0.17	0.39
7	25	0.17	0.40
8	0	0.11	0.25
8	20	0.12	0.28
8	35	0.13	0.31

Source: [45].

reheat steam temperatures did not decrease, and maintaining load was not an issue [45].

Environmental consequences of cofiring petroleum coke were also measured. SO_2 emissions were a function of scrubber performance, not fuel blend. SO_3 emissions increased modestly as a function of cofiring petroleum coke, although statistical analysis of the impact of petroleum coke on SO_3 emissions is inconclusive. Since SO_3 affected opacity, a portion of the WCF tests were performed with injection of hydrated lime as a means for SO_3 control. Hydrated lime injection proved successful [45].

NO_x emissions increased modestly during the cofiring of petroleum coke at WCF, as shown in Table 5.15. Given the high concentrations of char-bound nitrogen in petroleum, this result is not unexpected. The NO_x emissions associated with PC firing are fundamentally different from those associated with cyclone firing. The baseline NO_x emission level is much lower, and the impact of staging depends more heavily on fuel volatility due to the presence of the flame within the primary furnace.

Conclusions Regarding Cofiring Petroleum Coke in Pulverized Coal Boilers

The example just described demonstrates that petroleum coke can have certain economic advantages in PC boilers, particularly in terms of improving boiler efficiency and reducing fuel costs. Further, in many boilers, there will be no impact on main steam or reheat steam temperature and consequently no impact on turbine efficiency. However, the impact of petroleum coke on PC boilers is quite site-specific. Care must be taken to evaluate these impacts on any given location. The impact of petroleum coke cofiring on flyash may or may not be significant, depending on whether or not the flyash is sold as pozzolanic material.

The impact of petroleum coke cofiring on airborne emissions from PC boilers is not as favorable as for cyclone boilers. While there are numerous reports of units where NO_x emissions did not increase when petroleum coke was introduced with coal, there is potential for this outcome. That potential is consistent with the fuel characteristics of petroleum coke. Such potential may or may not be significant depending on whether a boiler has an associated selective catalytic reduction (SCR) system for NO_x control. Sulfur emissions also can increase, although this can be managed by scrubber technology.

5.3.4. Petroleum Coke Utilization in Other Systems

Petroleum coke can be fired in combination with coals or heavy oils in circulating fluidized bed boilers; alternatively, it can be fired as the sole fuel. Some examples of petroleum coke firing in fluidized bed boilers are shown in Table 5.16.

TABLE 5.16 Representative Circulating Fluidized Bed Boilers Firing Petroleum Coke or Petroleum Coke/Coal Blends

Owner	Steam Capacity (kg/sec)	Steam Capacity (10^3 lb/hr)	Main Steam Conditions (bar/°C)	Main Steam Conditions (psig/°F)
Purdue University	25.2	200	45/440	650/825
City of Manitowoc (100% pet coke)	25.2	200	67/485	975/905
Fort Howard Paper #3 (85% pet coke)	40.4	320	103/510	1500/950
Fort Howard Paper #5 (100% pet coke)	40.4	320	103/510	1500/950
Hyundai Oil, Korea (100% pet coke)	30.3	240	108/520	1565/970
University of N. Iowa (50% to 70% pet coke)	13.2	104.5	52/332	750/630
NISCO (2 reeheat boilers, 100% pet coke)	104.0	825	112/540/540	1625/1005/ 1005
JEA, Jacksonville, FL (2 × 300 MW$_e$ boilers firing coal and/or pet coke)	251.4	1993.6	172/538/538	2500/1000/ 1000

Source: [46–48].

Other examples of CFB boilers firing petroleum coke alone or in combination with other fuels include installations at the Nova Scotia Power Pt. Aconi Generating Station; Gulf Oil (now Chevron) in California; Oriental Chemical Industries in Korea; General Motors in Michigan; the Petrox refinery in Chile; a paper mill in Kattua, Finland; and numerous other sites. Most of these units fire a blend of petroleum coke and coal, with petroleum coke being the dominant fossil fuel.

The principle benefits associated with combusting petroleum coke in fluidized bed boilers include high boiler efficiencies and availabilities and control of airborne emissions. The principle issues associated with fluidized bed combustion of petroleum coke involve ash chemistry—management of limestone addition for optimized sulfur capture with minimum limestone cost—and management of vanadium–limestone interactions that cause agglomeration and fouling of heat transfer surfaces [49, 50].

Petroleum coke has also been fired in the Polk County, Florida, Clean Coal Technology Demonstration—in a 250-MW$_e$ (net) integrated gasification-combined cycle (IGCC). Typical blends are about 55% petroleum coke/45% coal. This installation employs a Texaco oxygen-blown gasifier to generate a synthesis gas that is then burned in a combustion turbine. The project has been quite successful in demonstrating the flexibility of petroleum coke in power generation. However, the Polk County project only highlights other applications for petroleum coke in the energy arena. The Polk County project is one of several gasification projects worldwide that employ petroleum coke as a part of the total feedstock. Similarly, petroleum coke has been fired successfully at the Wabash River Clean Coal IGCC project.

Petroleum coke has been studied extensively, and periodically employed, as an additive for the manufacture of metallurgical coke for the steel industry [51, 52]. This practice, ongoing for several decades, typically mixes 5% to 40% (weight basis) petroleum coke with coal in the coking ovens. The addition of petroleum coke reduces the reactivity of the metallurgical coke to CO_2 and improves its mechanical strength—making these cokes superior in performance to cokes made with coal alone [51]. Modest amounts of petroleum coke (e.g., ~3%) are sufficient to produce the desired improvements [51].

Petroleum coke has also been proposed in petroleum coke-water slurries [34, 53, 54] and petroleum coke-oil slurries [55]. Both of these slurries are considered to be replacements for heavy oil fired in existing boilers and furnaces. The coke-water slurries are more appropriate for cofiring than for single-fuel firing due to the lack of volatile matter in the fuel. The coke-oil slurries are proposed as a means for extending oil supplies and firing a high calorific value in boilers. The low ash content of the petroleum coke makes this option attractive, and the oil in the slurry provides sufficient volatile matter to support ignition and combustion.

Petroleum coke, then, is a highly flexible and useful blending fuel that can be used not only in conventional cyclone, PC, and fluidized bed boilers, but also in gasification systems, as a feedstock for metallurgical coke for the steel industry, and as a base fuel in coke-water slurries or coke-oil slurries. With increased production of this refinery by-product resulting from increased demand for gasoline, increased availability and use will follow.

5.4. WASTE PLASTICS AND PAPER

The disposal of municipal wastes is a world problem due to their partial biodegradability. Historically, these wastes have been landfilled like TDFs. However, with the increase in municipal waste production combined with the decrease in available landfill sites, economic alternatives are being sought. One of these methods is combustion of the municipal waste, especially waste plastics. Waste plastics are one of the most promising resources for fuel use as a blending fuel because of their high heat of combustion and increasing availability in local communities. Unlike paper and wood, plastics do not absorb much moisture, and the water content of plastics is lower than the water content of biomasses such as crops or kitchen wastes.

The methods for conversion of waste plastics into fuel depend on the types of plastics to be targeted and the properties of other wastes that might be used in the process. In general, the conversion of waste plastic into fuel requires feedstocks that are nonhazardous and combustible. In particular, each type of waste plastic conversion method has its own suitable feedstock. The composition of the plastics used as feedstock may be very different, and some plastic articles might contain undesirable substances (e.g., additives such as flame-retardants containing bromine and antimony compounds, or plastics containing nitrogen, halogens, sulfur, or any other hazardous substances) that pose potential risks to humans and to the environment.

The types of plastics and their compositions will condition the conversion process and will determine the pretreatment requirements and the combustion temperature for the conversion, as well as the energy consumption required, the fuel quality output, the flue gas composition (e.g., formation of hazardous flue gases such as NO_x and HCl), the flyash and bottom ash composition, and the potential of chemical corrosion of the equipment.

The major quality concerns when converting waste plastics into fuel resources are as follows [56]:

Smooth feeding to conversion equipment: Prior to their conversion into fuel resources, waste plastics are subject to various methods of pretreatment to facilitate the smooth and efficient treatment during the subsequent conversion process. Depending on their structures (e.g., rigid, films, sheets, or expanded (foamed) material), the pretreatment equipment used for each type of plastic (crushing or shredding) is often different.

Effective conversion into fuel products: In solid fuel production, thermoplastics act as binders that form pellets or briquettes by melting and
adhering to other nonmelting substances such as paper, wood, and thermosetting plastics. Although wood materials are formed into pellets using
a pelletizer, mixing plastics with wood or paper complicates the pellet
preparation process. Suitable heating is required to produce pellets from
thermoplastics and other combustible waste.

Well-controlled combustion and clean flue gas in fuel user facilities:
It is important to match the fuel type and its quality to the burner in
order to improve heat recovery efficiency. Contamination by nitrogen,
chlorine, and inorganic species, for instance, can affect the flue gas
composition and the amount of ash produced. Ash quality must also be in
compliance with local regulations when disposed of at the landfill.
Therefore, the fuel quality must be controlled in order to minimize its
environmental impact.

5.4.1. Waste Plastic Composition

Waste plastics are composed primarily of low-density polyethylene and high-
density polyethylene products. These two types make up nearly 60% of all
plastic production. Figure 5.5 denotes the distribution of waste plastics by type
[57]. Table 5.17 classifies various plastics according to the types of fuel they
can produce. As noted, thermoplastics consisting of carbon and hydrogen are
the most important feedstock for solid fuel production.

Because plastics are made from hydrocarbons, they tend to have very high
energy values, as noted in Table 5.18. These values compare well with fuel oil
and are considerably higher than most coals and TDF [57, 58].

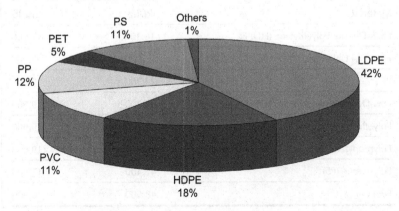

FIGURE 5.5 Plastic classification. *Source: [57].*

TABLE 5.17 Polymer as Feedstock for Fuel Production

Types of Polymer	Descriptions	Examples
Polymers consisting of carbon and hydrogen	Typical feedstock for fuel production due to high heat value and clean emissions	Polyethylene, polypropylene, polystyrene
Polymers containing oxygen	Lower heating value than above plastics	PET, phenolic resin, polyvinyl alcohol, polyoxymethylene
Polymers containing nitrogen or sulfur	These types produce NO_x and SO_x; flue gas cleaning required	*Nitrogen:* polyamide, polyurethane *Sulfur:* polyphenylene sulfide
Polymers containing halogens of chlorine, bromine, and fluorine	Source of hazardous and corrosive flue gas on thermal treatment and combustion	Polyvinyl chloride, polyvinylidene chloride, bromine-containing flame retardants, and fluorocarbon polymers

Source: [58].

TABLE 5.18 Heating Value of Some Common Plastics

	Heating Value	
Material	kJ/kg	Btu/lb
High-Density Polyethylene (HDPE)	46,300	19,900
Polyvinyl Chloride (PVC)	17,500	7,500
PET	22,700	9,800
Low-Density Polyethylene (LDPE)	46,300	19,900
Polyethylene (PE)	46,290	19,900
Polypropylene (PP)	46,170	19,850
Polystyrene (PS)	41,400	17,800
Fuel Oil	48,600	20,900

5.4.2. Waste Plastic and Paper Preparation

Refuse-derived paper and plastics densified fuel (RPF) is prepared from used paper, waste plastics, and other dry feedstocks [59]. Within the plastics, the thermoplastics play a key role as a binder in the briquetting or pelletizing process. Other components such as thermosetting plastics and other combustible wastes cannot form pellets or briquettes without a binding component. Approximately 15 wt% thermoplastics is the minimum required to be used as a binder to solidify the other components; however, more than 50 wt% could cause a failure in the pellet preparation. The components of RPFs are mainly sorted from industrial wastes and are sometimes also obtained from well-separated municipal waste. This type of solid fuel was set to be standardized in the Japanese Industrial Standards (JIS) in April 2002.

The plastic contents can be varied (within a range) to meet the needs of fuel users. The shape of the fuel will vary according to the production equipment (e.g., a screw extruder is often used to create cylindrical-shaped fuel with a variable diameter and length). In the production of solid fuel, the contamination of the targeted plastics with other plastics containing nitrogen, halogens (Cl, Br, F), sulfur, and other hazardous substances may cause air and soil pollution through flue gas emissions and ash disposal (e.g., inorganic components such as aluminum in the multilayer film of food packages produces flyash and bottom ash). Other contaminants such as hydrogen chloride might cause serious corrosion damage to the boiler.

Production Method

The solid fuel production process usually involves two steps: pretreatment and pellet production:

- Pretreatment includes coarse shredding and removal of noncombustible materials.
- Pellet production is comprised of secondary shredding and pelletization ($<200°C$).

Examples of these two types of production systems are presented. The first is a large-scale system with pretreatment for the separation of undesirable contamination such as metals and plastics containing chlorine; the other is a small-scale model without pretreatment equipment.

Large-Scale Model (3 ton/hour)

Industrial waste plastics, which have been separated and collected in factories, are ideal for use in solid fuel production. The fuel production facility consists of a waste unloading area, stockyard, pretreatment equipment, pelletizing equipment, and solid fuel storage. The pretreatment process includes crushing

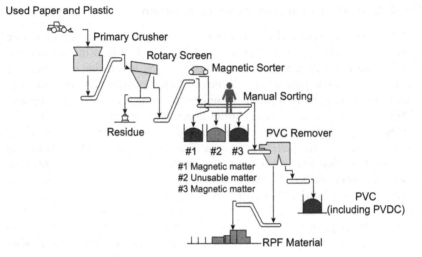

FIGURE 5.6 The pretreatment process. *Source: [56].*

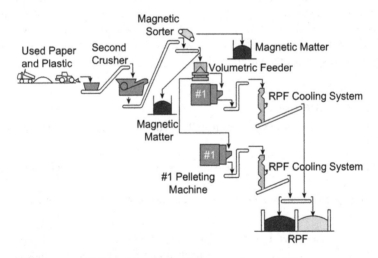

FIGURE 5.7 A pelletizing process *Source: [56].*

and sorting to remove unsuitable materials from the incoming wastes. A schematic diagram of the pretreatment process is shown in Figure 5.6.

After pretreatment, the mixture of paper and plastics is further processed in a secondary crusher and sorting process (conveyor and magnetic separator), and the resulting mixture is pelletized to produce solid fuel, as shown in Figure 5.7. The solid fuel is cooled to prevent natural ignition during storage, and it is further stored for shipping. The output of the process is usually solid fuel pellets

FIGURE 5.8   Heavy-duty machine to feed wastes. *Source: [56].*

of dimensions between 6 and 60 mm in diameter and 10 and 100 mm in length. The heating value of the pellets will change depending on the content of the plastics. A mixture of paper and plastics of a 1:1 weight ratio gives a heating value of approximately 7000 kcal/kg or higher.

Small-Scale Model (150 kg/hour)

The second system has a solid fuel production of 150-kg/h capacity. This facility does not have a pretreatment process, so the combustible wood, paper, and plastic wastes are directly fed into the crusher of the facility. This is carried out using a handling machine, as shown in Figure 5.8, where an operator must control the feed into the crusher to maintain a suitable ratio of each type of waste in order to produce the required fuel qualities, such as heating value. After crushing, the materials are transported through a pipe conveyor and are introduced into a twin-screw pelletizer. Figure 5.9 shows the entire process (the crusher, the pipe conveyor, and the pelletizer).

5.4.3. Waste Plastic Utilization

Experts agree that properly equipped, operated, and maintained incinerators or combustion facilities can meet the latest U.S. emissions standards while cofiring waste plastics [60]. In fact, plastics can be successfully burned in dedicated energy recovery facilities that achieve high combustion temperatures to eliminate dioxin and furan production.

One concern in utilizing waste plastics is in preventing melting of the material rather than combusting the material. If a melted plastic reaches a tube surface, it

FIGURE 5.9   Small-scale RPF production facility (150 kg/h capacity). *Source: [56].*

can stick and reduce heat transfer to the steam. In addition, the sticky surface permits solid ash particles to agglomerate and increase deposition [61].

5.5. HAZARDOUS WASTES

Hazardous wastes have long been burned both as a means of disposal and as a source of energy. Utilities such as Duke Power and Northern Indiana Public Service Co. have burned tar-laden dirt from manufactured gas plants in both pulverized coal and cyclone boilers. Many utilities are permitted to burn their own waste oils. Manufacturing industries have burned their own wastes as well. Over the past 30 years, hazardous wastes have entered the blending fuels arena, being supplied to cement kilns and industrial boilers. This practice has led to the promulgation of the Boiler and Industrial Furnace (BIF) regulations by the USEPA. Most significantly, this practice has been used by cement kilns and other industries as a means of reducing fuel costs as a component of product manufacturing costs.

Two types of hazardous wastes typically have been widely used in cofiring applications: spent solvents and other relatively light organic fluids, and waste oils from automotive and manufacturing sources. Both are widely used in boilers and kilns—particularly cement kilns—throughout the United States and the world.

5.5.1. Fuel Characteristics of Hazardous Wastes

The combustion of hazardous wastes with coal is governed by the BIF rules and by the Resource Conservation and Recovery Act (RCRA). Dellinger and colleagues [62] characterized representative hazardous wastes with respect to calorific value, as shown in Table 5.19.

TABLE 5.19 Calorific Values for Selected Combustible Hazardous Wastes

Compound	Formula	kJ/kg	Btu/lb
Acetonitrile	C_2H_3N	30,850	13,260
Tetrachloroethylene	C_2Cl_4	4981	2140
Acrylonitrile	C_3H_3N	33,195	14,270
Methane	CH_4	55,581	23,880
Pyridine	C_5H_5N	32,776	14,090
Dichloromethane	CH_2Cl_2	7116	3060
Carbon Tetrachloride	CCl_4	1005	432
Hexachlorobutadiene	C_4Cl_6	8874	3820
Benzene	C_6H_6	41,986	18,050
Monochlorobenzene	C_6H_5Cl	27,628	11,880
1,2-Dichlorobenzene	$C_6H_4Cl_2$	19,130	8230
1,2,4-Trichlorobenzene	$C_6H_3Cl_3$	14,232	6120
1,2,3,4-Tetrachlorobenzene	$C_6H_2Cl_4$	10,925	4700
Hexachlorobenzene	C_6Cl_6	7493	3220
Nitrobenzene	$C_6H_5NO_2$	25,158	10,820
Aniline	C_6H_7N	36,544	15,710
Hexachloroethane	C_2Cl_6	1926	828
Chloroform	$CHCl_3$	3140	1350
1,1,1-Trichloroethane	$C_2H_3Cl_3$	8330	3580

Source: [61].

5.5.2. Combustion of Hazardous Wastes in Rotary Kilns

Cement kilns are the most common users of hazardous wastes as blending fuels. Cement kiln energy recovery is an ideal process for managing certain organic hazardous wastes. The burning of wastes or hazardous wastes as supplemental fuel in the cement and other industries is not new. In the United States alone, nearly 30 cement kilns use hazardous waste as a supplemental fuel, thus saving the equivalent of 636×10^6 liters (168×10^6 gallons) of oil or 900×10^3 tonnes (1×10^6 tons) of coal [63].

Waste fuels are used in the cement industry worldwide, and emissions have been thoroughly investigated in light of government and environmental regulations. Because of the potential toxic effects on human life from some of the pollutants from cement kilns burning fossil fuels combined with hazardous waste, very stringent regulations have been enacted. For example, Table 5.20 lists the maximum achievable control technology (MACT) limits for cement kilns in the United States [64].

Test results have shown that virtually all of the organics, originating from fossil fuel, oil, or waste fuels are destroyed in the process because the material reaches temperatures exceeding 1340°C (2450°F) and nearly 1930°C (3500°F) in the gas phase in the cement kiln [64–73]. This high temperature must be maintained for several minutes in order to form the clinker minerals that give the cement its properties.

Although cement kilns are used to destroy wastes by combustion, they are not hazardous waste incinerators. They do have some common characteristics, but they also have distinct differences. From a combustion viewpoint, at least three key differences exist [74]:

1. When wastes are fed at the clinker discharge end of the kiln, the volatilized material experiences a time-temperature history much more severe than in most incinerators (2200–3000°F for 4–12 seconds compared to 1800–2400°F for 2–6 seconds, respectively).

TABLE 5.20 MACT Standards for New and Existing Cement Kilns

Pollutant	Existing Kilns	New Kilns
Dioxins/Furans	0.2 ng/dscm TEQ*	0.2 ng/dscm TEQ*
Particulate Matter	0.030 gr/dscf	0.030 gr/dscf
Mercury	72 μg/dscm	72 μg/dscm
Semivolatile Metals (Cd, Pb)	670 μg/dscm	670 μg/dscm
Low-Volatile Metals (As, Be, Cr, Sb)	63 μg/dscm	63 μg/dscm
HCl + Cl$_2$	120 ppm$_v$	120 ppm$_v$
CO	100 ppm$_v$	100 ppm$_v$
Total Hydrocarbons	Main: 20 ppm$_v$ Bypass: 10 ppm$_v$	Main: 20 ppm$_v$ Bypass: 10 ppm$_v$

*TEQ: Toxicity equivalents
Source: [63].

2. The waste-to-oxygen ratios in the exit gases are generally lower in cement kilns than in incinerators (% O_2 is between 2% and 6% in kilns compared to 4% and 12% in incinerators).

3. The raw meal preheat zones of the process serve as a "low-temperature afterburner" with a high surface-to-volume ratio.

It is the effects of these design and operating parameters that minimize the emissions of toxic combustion by-products.

5.5.3. Waste Oil Utilization

Waste oil is a unique hazardous waste with a long history of utilization. Typical sources of waste oil include automotive oils, machinery cutting and cooling oils, and other sources of lubricants. The opportunities to use this material as an opportunity fuel are worldwide. U.S. waste oil production and consumption exceeds 4.2×10^9 l/yr (1.1×10^9 gal/yr), of which 67% is burned as fuel and another 4% is rerefined [74]. A significant quantity is generated in Canada annually as well. Blundell [74] reports that $200 \times 10^6 - 250 \times 10^6$ l/yr ($53 - 66 \times 10^6$ gal/yr) of waste oil is generated in the Ontario province alone; of this 15% is burned in cement kilns, 7% is burned in small furnaces, and 27% is refined again.

In the United Kingdom, 447,000 tonnes of waste oil are generated annually, of which 380,000 tonnes are used—largely as fuel [75]. Significant attention has been given to this waste disposal problem/energy resource opportunity in such other locations as Bulgaria [76], New Zealand [77], Spain [78], and throughout the European Union. States from California [79] to Vermont [80] are paying particular attention to waste oil, its use as a blending fuel, and its proper disposal.

The general fuel characteristics of waste oils are shown in Table 5.21. Note the differences between mineral oil and synthetic automotive oil. Note also the broad range in properties, particularly as associated with mineral oil.

Typical trace metal concentrations have also been measured in waste oils, as shown in Table 5.22. Note that there are significant differences between typical concentrations in the United States and in New Zealand.

There are three basic uses of waste oil as a blending fuel: in small space heaters and boilers, in larger boilers, and in cement kilns. Of these, cement kilns are the most prominent due to their continued search for low-cost alternatives to coal, oil, and traditional energy sources. In New Zealand, for example, two cement kilns dominate the use of all waste oil in that country. Typical emissions from the combustion of waste oil in various applications are shown in Table 5.23. Note that SO_2 is not shown in this table because of its dependency on the sulfur content of the incoming fuel.

Given the typical emissions associated with firing waste oils, it is useful to consider case studies of firing waste oils in cement kilns [82–84]. Therefore, an example of cofiring waste oils in cement kilns in Germany is presented.

TABLE 5.21 Representative Properties of Waste Oils

Property	Mineral	Synthetic Automobile
Water Content (wt%)	9.65–64.1	1.0–15.0
Flash Point (°C)	160–180	140–180
Density (g/cm^3)	0.89–0.95	0.90–0.92
Sediments (wt %)	1.38–10.5	0.8–23.25
Higher Heating Value (MJ/kg)	18.90–39.2	36.0–43.8
Higher Heating Value (Btu/lb)	8,130–16,860	15,480–18,830
Cl (wt%)	0.07–0.25	0.05–0.13
S (wt%)	0.69–1.10	0.42–1.32
N (wt%)	1.60–1.95	0.70–2.35

Source: [77].

TABLE 5.22 Typical Trace Metal Concentrations in Waste Oils from the United States and New Zealand (in mg/kg)

Metal	United States	New Zealand
Lead	1100	82
Arsenic	12	8
Cadmium	1	0.8
Chromium	6	2.6
Zinc	800	249

Source: [76].

Waste Oil in Cement Kiln in Germany

A study utilizing data from cement plants in Germany was done to estimate the emissions of various metals as a function of waste material [83]. There are 76 cement kilns in operation, of which 40 are permitted to use alternate fuels such as tires, waste oil, waste wood, and so forth. A "typical" cement kiln consisting of a raw mill section, a preheater-rotary kiln section, and a cement mill section was used to describe cement production in Germany. Using partitioning factors based on information from operating kilns, a mass balance model was developed for this "typical" kiln. Using this information, elemental distributions were calculated for cadmium, lead, and zinc when using waste oils at the

TABLE 5.23 Selected Emission Factors for the Combustion of Waste Oil

Emission	Small Boilers	Space Heaters	Atomizing Burners
Particulates	7.68 (64)	0.34 (2.88)	7.92 (66)
PM-10	6.12 (51)	ND	6.84 (57)
NO_x	2.28 (19)	1.32 (11)	1.92 (16)
CO	0.6 (5)	0.20 (1.7)	0.25 (2.1)
TOC	0.12 (1)	0.12 (1)	0.12 (1)
HCl	7.92 (66)	ND	ND
Lead	6.6 (55)	0.05 (.41)	6.0 (50)
Arsenic	.013 (0.11)	3×10^{-4} (2.5×10^{-3})	7.2×10^{-3} (6×10^{-2})
Beryllium	ND	ND	2.2×10^{-4} (1.8×10^{-3})
Cadmium	1.1×10^{-3} (9.3×10^{-3})	1.8×10^{-5} (1.5×10^{-4})	1.4×10^{-3} (1.2×10^{-2})
Chromium	2.4×10^{-3} (2×10^{-2})	2.3×10^{-2} (1.9×10^{-1})	2.2×10^{-2} (1.8×10^{-1})
Nickel	1.3×10^{-3} (1.1×10^{-2})	6×10^{-3} (5×10^{-2})	1.9×10^{-2} (1.6×10^{-1})

Note: In kg/m^3 and $lb/10^3$ gal of waste oil burned.
Source: [81].

maximum allowable rate of 30%. The results are shown in Table 5.24. The results show that nearly all the trace metals exit with the clinker, and destruction and removal efficiency (DRE) numbers for the three metals are estimated to be 99.96% for lead, 99.95% for zinc, and 99.94% for cadmium.

5.6. CONCLUSIONS

A wide variety of waste fuels are currently being utilized as blending fuels with conventional fossil fuels in energy production. The use of these waste fuels provides significant opportunity to reduce industry annual operating costs due to lower fuel prices. Care must be taken when proposing to cofire waste fuels, since the actual cost of a waste fuel must reflect any additional preparation, comminution, emissions, and efficiency costs related to the waste fuel cofiring. Nonetheless, many solid waste fuels are successfully, economically, and environmentally being used in the industrial and power generation markets.

TABLE 5.24 Metal Concentrations for Typical Cement Kilns in Germany

	Metal (in g/t of clinker)		
	Cadmium	Lead	Zinc
Inlet Streams			
Waste Oil	0.0635	4.77	31.77
Coal	0.0261	10.43	7.39
Raw Meal	0.31	23.25	72.85
Outlet Streams			
Clean Gas	0.000256	0.0152	0.058
Clinker	0.399	38.43	111.94
Destruction and Removal Efficiency (%)	99.94	99.96	99.95

Source: [74].

REFERENCES

[1] McGowin CR. Alternate fuel cofiring with coal in utility boilers. Proceedings EPRI conference on waste tires as a utility fuel; 1991. EPRI GS-7538.

[2] Winslow J, Ekmann J, Smouse S, Ramezan M, Harding NS. Cofiring of coal and waste. International Energy Association report; 1996. IEACR/90.

[3] Harding NS. Cofiring tire-derived fuel with coal, Proceedings 27th international technical conference on coal utilization and fuel systems. Clearwater, FL; 2002.

[4] Tillman DA, Harding NS. Fuels of opportunity: characteristics and uses in combustion systems. London: Elsevier; 2004.

[5] Granger JE, Clark GA. Fuel characterization of coal/shredded tire blends. Proceedings EPRI conference on waste tires as a utility fuel; 1991. EPRI GS-7538.

[6] Skolnik L. Tire cords. In: Kirk and Othmer, editors. Concise encyclopedia of chemical technology, vol. 24. New York: John Wiley & Sons; 1985.

[7] Reisman JI, Lemieux PM. Air emissions from scrap tire combustion. EPA report; 1997. No. EPA-600/R-97-115.

[8] Tillman DA. Trace metals in combustion systems. San Diego: Academic Press; 1994. p. 140–41.

[9] Goodyear Tire and Rubber Company, Scrap tire recovery; 2002.

[10] American Recycler. Equipment spotlight—tire shredders; 2002.

[11] Bakkom TK, Felker MR. Tire shredding equipment. Proceedings EPRI conference on waste tires as a utility fuel; 1991. EPRI GS-7538.

[12] U.S. Scrap tire market. Scrap tire news reports. Washington, DC: Scrap Tire Management Council; 2002.

[13] Bakkom TK, Felker MR. Tire shredding equipment. Proceedings EPRI conference of waste tires as a utility fuel; 1994. EPRI GS-7538.

[14] Tillman DA. Report of the cofiring combustion testing at the Allen Fossil plant using Utah bituminous coal as the base fuel. Final report to Tennessee Valley Authority and Electric Power Research Institute, prepared by Foster Wheeler Environmental Corporation; 1996.

[15] Weinhold JF. TDF cofiring tests in a cyclone boiler, strategic benefits of biomass and waste fuels. Washington, DC: Electric Power Research Institute; 1993.

[16] Costello PA, Waldron RG, Witts WH. Tire-derived fuel and thermal treatment waste incineration—commercial operation in coal-fired cyclone units. ASME FACT Division, joint PowerGen conference; 1996.

[17] Stopek DJ, et al. Testing of tire-derived fuel in a cyclone-fired utility boiler, Proceedings EPRI conference on waste tires as a utility fuel; 1991. EPRI GS-7538.

[18] Stopek DJ, Licklider PL, Millis AK, Diewald DJ. Tire derived fuel (TDF) cofiring in a cyclone boiler—at Baldwin Station, strategic benefits of biomass and waste fuels. Washington, DC: Electric Power Research Institute; 1993.

[19] Murphy PM, Tesla MR. Co-firing tire derived fuel in a stoker fired boiler: strategic benefits of biomass and waste fuels. Washington, DC: Electric Power Research Institute; 1993.

[20] Gillen JE, Szempruch AJ. Ohio Edison tires-to-energy project, strategic benefits of biomass and waste fuels. Washington, DC: Electric Power Research Institute; 1993.

[21] Horvath M. Results of the Ohio Edison whole-tire burn test, Proceedings EPRI conference on waste tires as a utility fuel; 1991. EPRI GS-7538.

[22] Bryers R. Utilization of petroleum coke and petroleum coke/coal blends as a means of steam raising. In: Coal—blending and switching of low-sulfur Western coals. New York: ASME; 1994. p. 185–206.

[23] Bryers RW. Utilization of petroleum coke and petroleum coke/coal blends as a means of steam raising. Fuel Processing Technology 1995;44:121–41.

[24] Heintz EA. Review: the characterization of petroleum coke. Carbon 1996;34(6):699–709.

[25] Heintz EA. Effect of calcination rate on petroleum coke properties. Carbon 1995;33 (6):817–20.

[26] Lapades DN, editor. Encyclopedia of energy. New York: McGraw-Hill; 1976.

[27] Silva AC, McGreavy C, Sugaya MF. Coke bed structure in a delayed coker. Carbon 2000;38:2061–8.

[28] Edwards LO. Trace metals and stationary conventional combustion sources, vol. 1. Austin, TX: Radian Corporation; 1980. Technical report prepared for USEPA, Contract No. 68-02-2608.

[29] Tillman DA. Trace metals in combustion systems. San Diego: Academic Press; 1994.

[30] Thompson WE, Harrision JW. Survey of projects concerning conventional combustion environmental assessments. Research Triangle Park, NC: Research Triangle Institute; 1978. USEPA Report 600/7-78-139.

[31] Golightly DW. Atomic Emission Spectroscopy. In: Geifer B, Taylor J, editors. Survey of various approaches to the chemical analysis of environmentally important materials. Washington, DC: National Bureau of Standards; 1973.

[32] Ondov JM, Zoller WH, Olmez I. Elemental concentrations in the National Bureau of Standards' environmental coal and fly ash standard reference materials. Analytical Chemistry 1975;47(7):1102–9.

[33] Milenkova KS, Borrego AG, Alvarez D, Menendez R, Petersen HI, Rosenberg P, et al. Coal blending with petroleum coke in a pulverized-fuel power plant. Energy & Fuels 2005;19:453–8.

[34] Xu R, Hea Q, Caia J, Pan Y, Shen J, Hu B. Effects of chemicals and blending petroleum coke on the properties of low-rank Indonesian coal water mixtures. Fuel Processing Technology 2008;89:249–53.

[35] Yu J, Kulaots I, Sabanegh N, Gao Y, Hurt RH, Suuberg ES, et al. Utilization of fly ash coming from a CFBC boiler co-firing coal and petroleum coke in Portland cement. Energy & Fuels 2000;14(3):591–6.

[36] Izquierdo M, Font O, Moreno N, Querol X, Huggins FE, Alvarez E, et al. Influence of a modification of the petcoke/coal ratio on the leachability of fly ash and slag produced from a large PCC power plant. Environmental Science & Technology 2007;41:5330–53.

[37] Tillman DA. Petroleum coke as a supplementary fuel for cyclone boilers: characteristics and test results. Proceedings international joint Power-Gen conference. Phoenix; 2002, June 24–28. Paper 2002–26157.

[38] Letheby K. Utility perspectives on opportunity fuels. Proceedings 27th international technical conference on coal utilization and fuel systems. Clearwater, FL; 2002, March 4–7.

[39] Tillman, DA. Final Report: EPRI-USDOE cooperative agreement: cofiring biomass with coal. Palo Alto, CA: Electric Power Research Institute; 2001. Contract No. DE-FC22-96PC96252.

[40] Hus PJ, Tillman DA. Cofiring multiple opportunity fuels with coal at Bailly Generating Station. Biomass & Bioenergy 2000;19(6):385–94.

[41] Tillman DA. Biomass cofiring: field test results. Palo Alto, CA: Electric Power Research Institute; 1999. Report TR-113903.

[42] Tillman DA. Petroleum coke as a supplementary fuel for cyclone boilers. Proceedings 27th international technical conference on coal utilization and fuel systems. Clearwater, FL; 2002, March 4–7.

[43] Hower JC, Robertson JD, Roberts JM. Petrology and minor element chemistry of combustion by-products from the co-combustion of coal, tire-derived fuel, and petroleum coke at a Western Kentucky cyclone-fired unit. Fuel Processing Technology 2001;74:125–42.

[44] Pearce R, Grusha J. Tangential low NO$_x$ system at Reliant Energy's limestone unit #2 cuts lignite, PRB, and pet coke NO$_x$. Proceedings EPRI-DOE-EPA combined power plant/air pollution control symposium. Clearwater, FL; 2001, August 20–24.

[45] Tillman DA. Results of testing Widows Creek fossil plant boilers 7 and 8 cofiring petroleum coke with coal. Prepared for the Tennessee Valley Authority, Chattanooga. Sacramento: Foster Wheeler Environmental Corporation; 1995.

[46] Abdulally IF, Reed KA. Experience update of firing waste fuels in Foster Wheeler's circulating fluidized bed boilers. Proceedings of Power-Gen Asia. Hong Kong; 1994, August 23–25.

[47] Castro AL, Chelian PK. Application of Foster Wheeler CFB boilers to burn vacuum residuals for low-cost power with low emissions. Proceedings Mexico power. Monterrey; 1996, October 8–10.

[48] Dyr RA, Compaan AL, Hebb JL, Darling SL. The JEA Northside repowering project: low-cost power and low emissions with CFB repowering. Proceedings PowerGen conference. Orlando; 2000, November 14–16.

[49] Anthony EJ, Iribarne AP, Iribarne JV, Talbot R, Jia L, Granatstein DL. Fouling in a 160 MW$_e$ FBC boiler firing coal and petroleum coke. Fuel 2001;80:1009–14.

[50] Conn RE. Laboratory techniques for evaluating ash agglomeration potential in petroleum coke fired circulating fluidized bed combustors. Fuel Processing Technology 1995; 44:95–103.

[51] Alverez R, Pis JJ, Diez MA, Barriocanal C, Canga CS, Menendez JA. A semi-industrial scale study of petroleum coke as an additive in cokemaking. Fuel Processing Technology 1998; 55:129–41.

[52] Zubkova VV. The effect of coal charge density, heating velocity and petroleum coke on the structure of cokes heated to 1800°C. Fuel 1999;78:1327–32.

[53] Prasad M, Mall BK, Mukherjee A, Basu SK, Verma SK, Narasimhan KS. Rheology of petroleum coke-water slurry. Proceedings 23rd international technical conference on coal utilization and fuel systems. Clearwater, FL; 1998, March 9-13. p. 1109–16.

[54] Vitolo S, Belli R, Mazzanti M, Quattroni G. Rheology of coal-water mixtures containing petroleum coke. Fuel 1996;75(3):259–61.

[55] Xu HZX, Weiyi S, Xinyu C, Qiang Y, Zhenyu H, Junhu Z, et al. Study on flow resistance and heat transfer for petroleum coke-residual oil slurry in a pipe. Proceedings 23rd international technical conference on coal utilization and fuel systems. Clearwater, FL; 1998, March 9–13. p. 1119–28.

[56] United Nations Environmental Programme. Converting waste plastics into a resource compendium of technologies. Division of Technology, Industry and Economics. Osaka/ Shiga, Japan: International Environmental Technology Centre; 2009.

[57] Eulalio AC, Capiati NJ, Barbosa SE. Municipal plastic waste: alternatives for recycling with profit. Society of Plastics Engineers technical conference, Orlando; 2000, May 7–11.

[58] Renewables and alternate fuels, Table 4. Washington, DC: U.S. Energy Information Adminsitration, 2007.

[59] Boavida D, Abelha P, Gulyurtlu I, Cabrita I. Co-combustion of coal and non-recyclable paper and plastic waste in a fluidised bed reactor. Fuel 2003;82:1931–8.

[60] Piasecki B, Rainey D, Fletcher K. Is combustion of plastics desirable? American Scientist 1998;86(4):364.

[61] Vamvuka D, Salpigidou N, Kastanaki E, Sfakiotakis S. Possibility of using paper sludge in co-firing applications. Fuel 2009;88:637–43.

[62] Dellinger B, Torres JL, Rubey WA, Hall DL, Graham JL. Determination of the thermal decomposition properties of 20 selected hazardous organic compounds. Cincinnati: Industrial Environmental Research Laboratory, Office of Research and Development, U.S. Environmental Protection Agency; 1984. EPA-600/2-84-138.

[63] Questions and answers about cement kilns, their operation and their role in processing organic wastes. Houston: Southdown, Inc; 1992. Brochure.

[64] Lighty JS, Veranth JM. The role of research in practical incineration systems—a look at the past and the future. Proceedings 27th symposium international on combustion. Pittsburgh: The Combustion Institute; 1998.

[65] Sun B, Sarofim AF, Eddings EG, Paustenbach DJ. Reducing PCDD/PCDF formation and emissions from a hazardous waste combustion facility—technological identification, implementation, and achievement. 21st international symposium on halogenated environmental organic pollutants and POPS, Kyongju, Korea; 2001.

[66] Tillman DA, Seeker WR, Pershing DW, DiAntonio D. Developing incineration process designs and remediation projects from treatability studies. Remediation Journal 1991;1(3): 251–73.

[67] Eddings EE, Lighty JS, Kozinski JA. Determination of metal behavior during the incineration of a contaminated montmorillonite clay. Environmental Science & Technology 1994;28: 1791–800.

[68] McClennen WH, Lighty JS, Summit GD, Gallagher B, Hillary JM. Investigation of incineration characteristics of waste water treatment plant sludge. Combustion Science & Technology 1994;101:483–503.

[69] Holbert CH, Lighty JS. Trace metals behavior during the thermal treatment of paper-mill sludge. Waste Management Journal 1999;18:423–31.

[70] Pershing DW, Lighty JS, Silcox GD, Heap MP, Owens WD. Solid waste incineration in rotary kilns. Combustion Science & Technology 1993;93:245–76.

[71] Senior C, Sarofim AF, Eddings EE. Behavior and measurement of mercury in cement kilns. IEEE-IAS/PCA 45th cement industry technical conference. Dallas; 2003.

[72] Eddings EE, Pershing DW. Hydrocarbon emissions due to raw materials in the manufacture of Portland cement. Proceedings 89th annual meeting of the Air & Waste Management Association. Nashville; 1996, June 23–28.

[73] Hansen E, Pershing DW, Sarofim AF, Heap MP, Owens WD. An evaluation of dioxin and furan emissions from a cement kiln cofiring waste. Waste combustion in boilers and industrial furnaces. Air & Waste Management Association; 1995.

[74] Blundell G. Used motor oil forum background paper: provincial and state policies on used motor oil management. Proceedings policy forum: used motor oil. Recycling Council of Ontario; 1998, May 26.

[75] Environmental Resources Management. Waste oil recycling. London: Department of Trade and Industry; 2002.

[76] Metodiev M. Utilization of waste oil-derivatives. Annual of the University of Mining and Geology, "St. Ivan Rilski." Mining and Mineral Processing 2002;II(44–45):147–9. Sofia, Bulgaria.

[77] Woodward-Clyde. Final report: assessment of the effects of combustion of waste oil, and health effects associated with the use of waste oil as a dust suppressant. Prepared for the Ministry for the Environment, Wellington, NZ; 2000.

[78] Regueira LN, Anon JR, Proupin J, Labarta C. Recovering energy from used synthetic automobile oils through cogeneration. Energy & Fuels 2001;15:691–5.

[79] Myers P. Hazardous waste generation trends in California: 1997–2001. Sacramento: State of California, Department of Toxic Substances Control, Permitting Division; 2003.

[80] Air Pollution Control Division and Hazardous Materials Management Division. Vermont used oil analysis and waste oil furnace emissions study. Waterbury: Vermont Agency of Natural Resources; 1996.

[81] USEPA. Waste oil combustion. 1996; AP-42.

[82] Lamb CW. Detailed determination of organic emissions from a preheater cement kiln co-fired with liquid hazardous wastes. Proceedings 3rd international congress on toxic combustion by-products. Cambridge, MA; 1993, June.

[83] Achternbosch M, Brautigam KR. Co-incineration of wastes in cement kilns—mass balances of selected heavy metals. IT3 conference. Philadelphia; 2001, May.

[84] Benestad C. Incineration of hazardous waste in cement kilns. Waste Management & Research 1989;7(4):351–61.

Chapter 6

Environmental Aspects of Fuel Blending

6.1. INTRODUCTION

Since the 1960s, people have come to understand that industrial growth and energy production were the cause of potentially harmful pollutants being released into the environment. Studies to understand emissions, sources, and the effects on human health and the environment led to stringent legislation. Controls were for air emissions, waterway discharges, and solids disposal. Environmental control is mainly driven by government legislation and regulations at the local, national, and international levels. For fired systems, the constituents that have the most emphasis are particulates, NO_x, SO_2, and mercury [1–3]. Consequently, the regulatory climate, at the local, national, and international levels, has influenced the blending of different types of coals such as Powder River Basin (PRB) subbituminous coal with Eastern bituminous coal and the cofiring of different fuels such as biomass with coals.

6.2. REGULATORY CLIMATE AS IT INFLUENCES BLENDING AND COFIRING

The Federal Clean Air Act (CAA) is the key driver for all air pollution control legislation in the United States. The original act was enacted in 1963 and has experienced five significant amendment cycles in 1965, 1967, 1970, 1977, and 1990. Prior to the 1990 amendment, regulatory elements included the National Ambient Air Quality Standards (NAAQS), the New Source Performance Standards (NSPS), and the New Source Review (NSR). The 1990 amendment additionally addressed SO_2 and NO_x and control of air toxics. In an effort to reduce SO_2 emissions, the blending of high-sulfur bituminous coals with low-sulfur subbituminous—particularly PRB—coals became common practice.

In addition, with the careful selection of certain subbituminous coals, NO_x reductions were observed. This was caused by two factors: decreased nitrogen concentration and increased reactivity of the lower-rank coal. When cofiring biomass with coal, inherently low-sulfur concentration in most biomass fuels decreases the SO_2 generated. In addition, NO_x formation has been observed to also decrease in many cofiring demonstrations. Certain biomass fuels are inherently lower in nitrogen concentration, and most biomass fuels are very reactive [4]. If the system is designed properly and the biomass is combusted in the proper locations, NO_x emissions can be reduced.

6.3. BLENDING FOR ENVIRONMENTAL AND ECONOMIC REASONS

The practice of blending different fuels is commonly employed throughout the industry. Blending, if done correctly, can be beneficial both environmentally and economically. The initial driving force for using PRB subbituminous coals was largely in blends with bituminous coals to achieve SO_2 regulations compliance. PRB coal is also significantly cheaper than bituminous coals. Blending of the fuels also resulted in NO_x emission reductions. Biomass cofiring with coal began in both Europe and the United States as a means of reducing emissions. For Europe, the practice of cofiring became common because of the economic subsidies designed to reduce CO_2 emissions. However, not all environmental consequences of fuel blending are favorable. For instance, the cofiring of biomass with coal can affect the salability of the flyash, resulting in increased landfills.

6.4. AREAS OF CONCERN

Typically, several emission constituents of concern are addressed through legislation. These include particulates, sulfur dioxide (SO_2), nitrogen oxides (NO_x), mercury, and fossil carbon dioxide (CO_2).

6.4.1. Particulates

Solid and liquid matter of organic or inorganic composition that is suspended in the flue gas or atmosphere is known as particulate. Particulates come from the noncombustible, ash-forming mineral matter in the fuel that is released from the combustion process and is carried by the flue gas. Some examples of particulates formed during a reaction include solid ash particles, unburned carbon, and condensable volatile organic compounds (VOC). Most particle sizes range from 1 to 100 μm.

The two subsets that are typically referred to most often include PM_{10} and $PM_{2.5}$. PM_{10} is particulate matter that is 10 μm and finer, whereas $PM_{2.5}$ is particles that are 2.5 μm and finer. Gases such as SO_2, NO_x, and VOCs can transform in the atmosphere to fine particulates by chemical reactions—for example, sulfuric acid, nitric acid, and photochemical smog. Of the two subsets, $PM_{2.5}$ is considered to have more deleterious health effects. The effects of particulate emissions include impaired visibility, soiling of the environment, and respiratory problems [2, 5].

6.4.2. Sulfur Dioxide

Sulfur dioxide (SO_2) in fossil fuels will be present as either organic or inorganic compounds. Inorganic sulfur will mainly be present as pyrite for coals, whereas organic sulfur exists either as aromatic rings or in aliphatic functional groups [6]. SO_2 is formed as a combustion product as the sulfur oxidizes. Fuel sulfur will mostly convert to SO_2, with small quantities converting to SO_3. Sulfur oxides are typically related to irritation of the human respiratory system, reduced visibility, materials corrosion, and effects on vegetation [2]. Once in the atmosphere, SO_3 will typically convert to H_2SO_4, which has been identified as a health hazard and contributes to acid rain formation. The concerns over sulfur emission were a key driver for the increased use of PRB coal in the United States [5, 7, 8].

6.4.3. Nitrogen Oxides

Nitrogen oxides (NO_x) are primary pollutants emitted in combustion processes. They contribute to acid rain, ozone formation, visibility degradation, and human health concerns. NO_x emissions are typically formed from two sources: nitrogen compounds in the fuel and molecular nitrogen in the air, which is necessary for the combustion process [2]. The three mechanisms by which NO_x is produced are thermal, prompt, and fuel.

Thermal NO_x is formed by high-temperature reactions of nitrogen with oxygen, which is known as the Zeldovich mechanism. NO_x formation increases exponentially with temperature. Prompt NO_x is formed by several relatively fast reactions between nitrogen, oxygen, and hydrocarbon radicals. This is sometimes referred to as Fenimore NO_x. Prompt NO_x becomes important at lower temperatures but is much less important at higher temperatures, which tend to be dominated by thermal NO_x. Fuel NO_x is formed by the oxidation of organonitrogen compounds found in the fuel [9]. The blending of highly reactive fuels such as PRB coal and certain biomass has shown, both at lab and full scale, the benefits of NO_x reduction. Highly reactive fuels promote early release of nitrogen in the fuel mass.

6.4.4. Mercury

Natural processes such as volcanic activity and weathering of rocks release mercury into the environment. Human activities such as mining, combustion of solid fuels, production of products such as chlor-alkali, and other activities also add mercury into the atmosphere. When released to the atmosphere, mercury circulates between air, water, and soil until it comes to rest in sediments or landfills. Mercury pollution has caused severe health effects in several chemical incidents; the most notable form is methylmercury [10].

In some countries, coal combustion can be a significant source of mercury in the atmosphere. However, the contribution from coal combustion in most developed nations is relatively minor. In order to address the concerns of mercury emission, the United States promulgated several regulations, including the Clean Air Mercury Rule (CAMR), the Energy Policy Act, Global Climate Change, and the Clear Skies Initiatives (CSI). CAMR was overturned in early 2008 [10].

6.4.5. Fossil CO_2

With concerns over global warming, the reduction of CO_2 has been of increasing importance. According to the Environmental Integrity Project, approximately 40% of all carbon dioxide emission comes from power plants. CO_2 emissions are a function of the carbon content in the fuel. Other forms of carbon include unburned carbon in the flyash, carbon monoxide (CO), and hydrocarbon emissions from the combustion process [5]. The combustion or cofiring of biomass with coal is a means by which to reduce fossil CO_2, since the combustion of biomass is considered CO_2 neutral.

6.5. ASH MANAGEMENT FOR POWER PLANTS

The management of ash for power plants is important. Proper ash management directly impacts daily operation and revenue for a plant. The following subsections describe the two common types: bottom ash and flyash.

6.5.1. Bottom Ash

There are several uses and applications for bottom ash. These include filler material for structural applications, aggregate in road bases and subbases, pavement, feedstock in the production of cement, aggregate in lightweight concrete products, and snow and ice traction control material. Even though the use of bottom ash is not as significant as the use for flyash, management of bottom ash is still important for power plants [11].

6.5.2. Flyash

Flyash is a pozzolanic material that in the presence of water and calcium hydroxide produces cementitious compounds. Flyash is a filler in hot mix asphalt applications and improves the fluidity of flowable fill and grout because of its spherical shape and particle size distribution. In general, flyash has several application; some of them include raw material in concrete products and grout, feedstock in the production of cement, fill material for structural applications, and soil modification or stabilization. Other applications include ingredients of flowable fill, components in pavement, and mineral filler in asphalt [11]. Due to its many uses, flyash is a salable by-product. It contributes to a plant's revenue and therefore must be managed appropriately.

6.6. BLENDING FOR EMISSION BENEFITS

The blending and cofiring of different fuels have played key roles in reducing the emission of various constituents for power plants. The blending of PRB coal with bituminous coal has been a means for reducing sulfur oxides and, in many cases, nitrogen oxide emission, which is contributed by the reactivity of the PRB coal. The cofiring of biomass fuels with other solid fuels reduces sulfur emission, nitrogen oxide emission (depending on the design and particular biomass fuel), and fossil carbon dioxide emission. Blending has proven to be an effective method of reducing emission constituents.

6.6.1. Blending PRB Coal with Other Solid Fuels

The National Environmental Policy Act (NEPA) of 1970 called for a significant reduction in sulfur dioxide emissions. In response to the act, existing power plants were adapted and new power plants were built to burn low-sulfur subbitiminous coal, particularly from the Powder River Basin [5]. On a tonnage basis, the use of PRB coal is almost as much as that of bituminous coal (Figure 6.1).

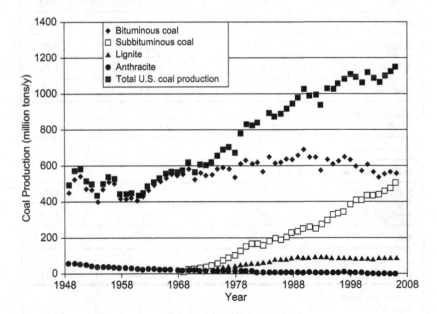

FIGURE 6.1　U.S. coal production by type. *Source: [5].*

6.6.2. Emission Aspects

The blending of Powder River Basin coals with other solid fuels has been proven to reduce certain emission constituents. Table 6.1 show the typical fuel analyses for Black Thunder (PRB) subbituminous coal, Long Fork Central

TABLE 6.1 Representative Fuel Analyses for Selected Solid Fuels

	Black Thunder Subbituminous Coal	Central Appalachian Bituminous Coal (Long Fork)	Petroleum Coke (Shot Coke)
Proximate Analysis (wt% as received)			
Fixed Carbon	34.94	50.09	89.59
Volatile Matter	30.72	31.23	3.07
Ash	5.19	11.52	1.06
Moisture	29.15	7.16	6.29
Ultimate Analysis (wt% as received)			
Carbon	51.30	66.93	81.02
Hydrogen	2.87	4.43	3.16
Oxygen	10.46	7.55	0.94
Nitrogen	0.68	1.34	1.59
Sulfur	0.35	1.07	5.95
Ash	5.19	11.52	1.06
Moisture	29.15	7.16	6.29
Higher Heating Value (Btu/lb)	8,888	12,114	14,364
Mercury (ppm$_w$)	0.12	0.08	0.02–0.09*
Pollutant Measures (lb/10^6 Btu)			
Sulfur (as SO_2)	0.79	1.77	8.28
Nitrogen	0.77	1.11	1.11
Ash	5.84	9.51	0.74
Carbon (as CO_2) [lb/10^6 Btu]	115.44	110.50	112.81

Dependent on the crude oil.
Source: [12].

Appalachian bituminous coal, and shot petroleum coke. It also compares some pollutant measures, such as sulfur, nitrogen, and ash loading; in addition, it calculates carbon dioxide loading. Table 6.2 details the same parameters for selected blends of PRB coal with Central Appalachian coal and with petroleum

TABLE 6.2 Selected Blending Scenarios for Black Thunder Subbituminous Coal with Central Appalachian Bituminous Coal, and Black Thunder Subbituminous Coal with Petroleum Coke

	20% PRB–80% Central Appalachian	40% PRB–60% Central Appalachian	20% PRB–80% Petroleum Coke	40% PRB–60% Petroleum Coke
Proximate Analysis (wt% as received)				
Fixed Carbon	47.06	44.03	78.66	67.73
Volatile Matter	31.13	31.03	8.60	14.13
Ash	10.25	8.99	1.89	2.71
Moisture	11.56	15.96	10.86	15.43
Ultimate Analysis (wt% as received)				
Carbon	63.80	60.68	75.08	69.13
Hydrogen	4.12	3.81	3.10	3.04
Oxygen	8.13	8.71	2.84	4.75
Nitrogen	1.21	1.08	1.41	1.23
Sulfur	0.93	0.78	4.83	3.71
Ash	10.25	8.99	1.89	2.71
Moisture	11.56	15.96	10.86	15.43
Higher Heating Value (Btu/lb)	11,469	10,824	13,269	12,174
Pollutant Measures (lb/10^6 Btu)				
Sulfur (as SO_2)	1.61	1.44	7.28	6.10
Nitrogen	1.05	0.99	1.06	1.01
Ash	8.94	8.30	1.42	2.23
Carbon (as CO_2 lb/10^6 Btu)	111.27	112.12	113.16	113.58

coke. Environmental benefits of blending PRB coal with other solid fuels can be observed.

SO_2

The use of PRB coal, when blending with higher-sulfur solid fuels, can significantly reduce sulfur concentration of the fuel mass and consequently SO_2 emissions. The reduction of SO_2 emissions has been a key driver for the increased use of PRB coal over the years. Table 6.1 shows typical fuel analyses for several solid fuels. Note the sulfur content and SO_2 loading for the PRB coal compared to those for the other fuels.

NO_x

The inherent fuel nitrogen concentration of PRB coal is typically lower compared to bituminous coals and petroleum cokes. Therefore, the formation and emission of nitrogen oxides can be reduced when blending or combusting PRB coals. In addition, evolution patterns and reactivity of PRB coals have proven to have positive effects on NO_x formation. Figures 6.2 and 6.3 depict the volatile evolution patterns for Black Thunder subbituminous coal and Pittsburgh Seam bituminous coal, respectively. For the Pittsburgh Seam coal, volatile nitrogen evolution lags behind volatile carbon evolution. These results are supported by the fact that bituminous coals, generally, generate more NO_x compared to PRB coals in utility and industrial boilers.

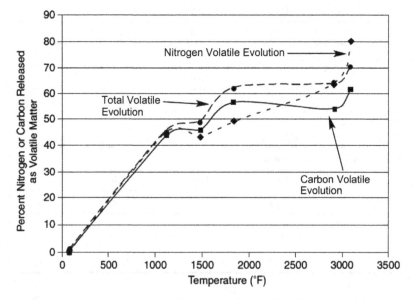

FIGURE 6.2 Volatile matter evolution patterns for Black Thunder subbituminous coal.

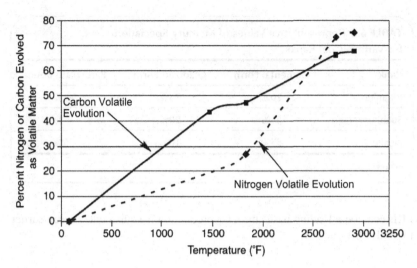

FIGURE 6.3 Volatile matter evolution patterns for Pittsburgh Seam bituminous coal.

Particulates and Ash Management

As shown in Table 6.1, ash loading can be lower or higher when compared to other solid fuels. PRB coals generally are much lower in ash concentration than bituminous coals, but the increased mass flow of fuel can cause increased ash generation. Petroleum cokes have even lower ash concentrations and generally higher carbon concentrations compared to those of the bituminous and PRB coals. Consequently, if PRB coal and petroleum coke are blended, ash management may become an issue. This is due to the fact that both fuels are inherently low in ash, so dilution effects are not present; unburned carbon in flyash may increase past acceptable levels, which will affect the salability of the flyash.

Mercury

Mercury concentrations in PRB coals typically are higher than those in the bituminous coals and petroleum cokes. In addition, the speciation of mercury for the different fuels has an impact on the amount of mercury that can be captured in postcombustion equipment. Table 6.3 lists representative speciation values of mercury for lignite coal, subbituminous coal, and bituminous coal. Particulate bound and oxidized forms of mercury can be more easily captured in postcombustion equipment. If PRB coal is blended with a fuel that has higher concentrations of chlorine or other halogens, oxidization of elemental mercury will occur.

Fossil CO_2

The blending of PRB coal with other solid fuels does not address the concerns over fossil CO_2 production. Even though the carbon concentration is lower for

TABLE 6.3 Representative Values of Mercury Speciation for Various Coal Ranks

Rank	Elemental Form	Oxidized Form	Particulate Bound
Bituminous	20%	35%	45%
Subbituminous	65%	20%	15%
Lignite	85%	10%	5%

Source: [13].

PRB coal, on a loading basis, the CO_2 production is equivalent, as seen earlier in Tables 6.1 and 6.2.

6.6.3. Selected Case Studies

This section discusses several blending case studies. The blending of PRB coal with other solid fuels has been and is still very successful within the utility sector.

DTE Energy, Monroe Power Plant

To maintain a competitive economic position and meet environmental constraints, Monroe Power Plant converted to blending PRB coal with Central Appalachian bituminous coal. Monroe Power Plant is a 3100-MW$_e$ (net) station with four boilers that nominally burn between 60% and 70% PRB coal with an Eastern bituminous coal. It typically consumes between 25,000 and 30,000 tons of coal on a daily basis, equating to approximately 8 to 10 million tons on an annual basis.

Testing showed that the units could consume southern PRB coals, but consuming northern PRB coals resulted in unacceptable problems. Severe slagging problems were experienced when iron concentrations in the Eastern bituminous coal became excessive. Blending is performed to achieve several effects, including reduced fuel cost, reduced SO_2 emissions, reduced NO_x emissions, and several other parameters [7, 8, 14].

In order to blend at high percentages of PRB coal without experiencing significant slagging and fouling problems, the plant invested in a digital fuel tracking system and an on-line coal analyzer. The tracking system enabled operators to make operational changes proactively to accommodate changes in fuel conditions. Digital fuel tracking combined with on-line coal analysis monitored fuel quality for the plant in the fuel yard and provided the operators information about the coal blend being burned, what they will see in one hour, two hours, and four hours, and what is on the main conveyor belt to the plant.

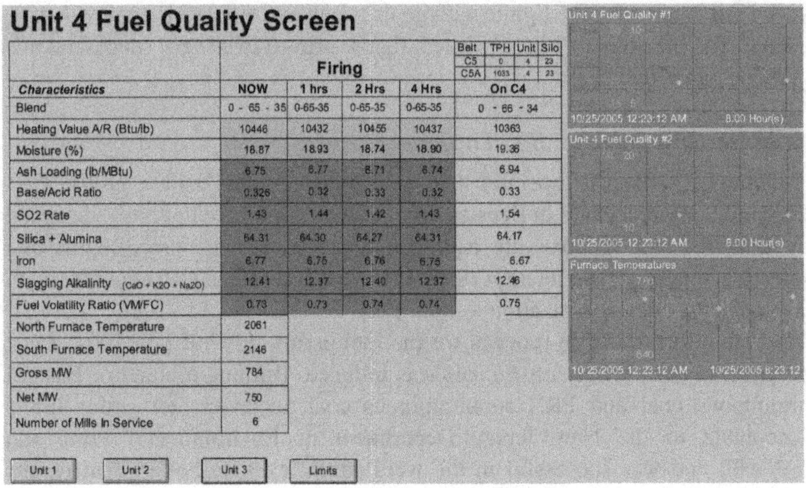

Unit 4 Fuel Quality Screen

						Belt	TPH	Unit	Silo
		Firing				C5	0	4	23
						C5A	1635	4	23
Characteristics	NOW	1 hrs	2 Hrs	4 Hrs	On C4				
Blend	0 - 65 - 35	0-65-35	0-65-35	0-65-35	0 - 66 - 34				
Heating Value A/R (Btu/lb)	10446	10432	10455	10437	10363				
Moisture (%)	16.87	18.93	18.74	18.90	19.36				
Ash Loading (lb/MBtu)	6.75	6.77	6.71	6.74	6.94				
Base/Acid Ratio	0.326	0.32	0.33	0.32	0.33				
SO2 Rate	1.43	1.44	1.42	1.43	1.54				
Silica + Alumina	64.31	64.30	64.27	64.31	64.17				
Iron	6.77	6.76	6.76	6.76	6.67				
Slagging Alkalinity (CaO + K2O + Na2O)	12.41	12.37	12.40	12.37	12.46				
Fuel Volatility Ratio (VM/FC)	0.73	0.73	0.74	0.74	0.75				
North Furnace Temperature	2061								
South Furnace Temperature	2146								
Gross MW	784								
Net MW	750								
Number of Mills In Service	6								

Unit 1   Unit 2   Unit 3   Limits

FIGURE 6.4   Monroe Power Plant Unit 4 fuel quality screen. *Source: [14].*

Figure 6.4 shows some of the parameters and analyses that an operator typically sees at the Monroe Power Plant; this particular screen is for Unit 4 [14]. In addition to the fuel quality, the screen provides information on SO_2 loading, which has consequences for sulfur emission and precipitator performance. Additionally, the measure of base/acid ratio (B/A) provides the operators information for not only slagging and fouling problems but also potential of opacity problems.

The blending of high percentages of PRB coal with Eastern bituminous coal has been and is still successful at the Monroe Power Plant, benefiting both environmentally and economically.

Luminant, Big Brown Plant

Luminant's Big Brown Plant was the first large-scaled, lignite-fueled power plant in the state of Texas. There are two supercritical boilers with a total capacity of 1190 MW_e (net). PRB coal was introduced to the plant to improve fuel quality and reduce emissions. Big Brown originally burned locally mined lignite for 28 years, but production was decreased quickly. Consequently, in order to prepare for the end of the mine's life, the blending of PRB coal with lignite was introduced in 2000.

The units at Big Brown, prior to blending with PRB coal, were capacity limited due to lower coal quality. Unlike most units that experience slagging and fouling problems when blending PRB coals, Big Brown's mix of PRB coal with lignite reduced fuel ash concentration. The blend also increased volatile matter content, lowered sulfur content, and increased the calorific value when compared to 100% lignite. Consequently, the plant typically burns a blend of

42% PRB coal with 58% lignite with goals to achieve 60% PRB coal with 40% lignite. The blending of PRB coal with lignite proved to be very successful and reduced emissions [15].

Atlantic City Electric, B.L. England Station

Atlantic City Electric operates two coal-fired cyclone boilers at the B.L. England Station, located in Beesley's Point, New Jersey. Both units—designed by Babcock and Wilcox—are equipped with electrostatic precipitators and selective noncatalytic reduction (SNCR) systems. Unit 2 is also equipped with a wet flue gas desulphurization system [16].

In an effort to obtain renewal for the fuel permit on Unit 1 in 2001, a test program to burn lower-sulfur coals was initiated. Burning a blend of Eastern bituminous coal and PRB subbituminous coal achieved fuel sulfur limits acceptable to the New Jersey Department of Environmental Protection (NJDEP) and was successful in the wet-bottom, cyclone boiler. Test results demonstrated that consistent fuel blend properties are required to minimize impacts on operations and emissions.

Higher moisture concentration, increased coal fines, and differences in ash characteristics had significant impact on boiler operation. Ash characteristics were particularly important under reducing conditions to ensure proper slag tapping while operating with the overfire air system for NO_x emissions control. A blend of 30% PRB coal provided the necessary regulatory improvement without adversely impacting operations. Higher percentages of PRB coal would require extensive modifications to both the boiler and electrostatic precipitators [16].

6.7. COFIRING BIOMASS WITH COAL

Cofiring biomass with coal is well understood, and the experience is vast. It has several environmental benefits, including SO_x, NO_x, and particulates emissions reduction. The following subsection briefly discusses these areas and also looks at several cofiring tests.

6.7.1. Emission Aspects

Table 6.4 shows selected pollutant measures when cofiring various blends of woody biomass with Central Appalachian bituminous coal. Sulfur and ash loading decreases as the fraction of woody material increases in the blend; NO_x emissions have been observed to decrease if the system is designed properly. The following subsections further discuss these aspects.

SO_2

Almost all of the sulfur in a fuel will convert to SO_2 given enough oxygen availability and residence time. The benefits of sulfur reductions are easily

TABLE 6.4 Impact of Cofiring Woody Material with Central
Appalachian Bituminous Coal on Emission Constituents

Parameter	0% Wood Cofiring	5% Wood Cofiring (Heat Input Basis)	10% Wood Cofiring (Heat Input Basis)	20% Wood Cofiring (Heat Input Basis)
Mass Percentage of Wood	0	11.25	21.12	37.59
Blend Heating Value (Btu/lb)	12,114	11,317	10,618	9,450
lb $SO_2/10^6$ Btu	0.88	0.85	0.81	0.73
lb Fuel/10^6 Btu	83	88	94	106
lb $H_2O/10^6$ Btu	5.91	9.79	13.67	21.43
lb Ash/10^6 Btu	9.51	9.26	9.02	8.53
lb N/10^6 Btu	1.11	1.11	1.11	1.12

Source: [17].

obtained through cofiring biomass with coal. Since most biomass fuels have inherently low sulfur concentrations, SO_x reductions have been observed by many cofiring applications.

NO_x

Volatile nitrogen can form reactive species such as amine species (e.g., NH_3, HCN, and tar nitrogen), which can react in the presence of oxygen, forming NO, some N_2O, and a small portion of N_2 [18]. The principle of NO_x reduction is to design a system and control the combustion process in order to essentially form more N_2 than other forms of nitrogen species. In a pulverized coal (PC) fired application, one of these techniques involves initiating the combustion process in a fuel-rich region.

NO$_x$ reductions have been observed in several demonstration tests. These tests demonstrated the benefits of separate injection of the biomass at the center of the fireball. The presence of the biomass promoted early ignition of the total fuel mass. The early ignition of the biomass at the center of the fireball promoted the combustion of the coal flame. More nitrogen could be released into a fuel-rich region, promoting the formation of N_2.

Particulates

When cofiring biomass with coal or other solid fuels that have higher ash concentrations, ash concentration will decrease slightly as the fraction of biomass in the fuel blend increases (refer to Table 6.4). Sometimes, fines are generated when cofiring biomass with coal. The generation of fine particles that do not experience complete combustion in the boiler has created concerns for certain postcombustion equipment. Fines can cause baghouse fires, so when cofiring biomass, fire suppression systems should be installed.

Mercury

Mercury concentrations are inherently low in biomass fuels. Therefore, the cofiring of biomass with other solid fuels will typically decrease mercury emission. This has been observed in both demonstration and experimental tests.

Ash Management

Ash management can become a problem when cofiring biomass with coal. Under ASTM C-618, flyash generated from the cofiring of biomass cannot be used as pozzolanic material. The unburned carbon concentration also has been shown to increase in some projects when significant percentages of biomass are combusted. Plants must consider these factors when evaluating the potential of biomass cofiring.

Fossil CO_2

Biomass is considered a CO_2 neutral fuel. During the life of the plant, photosynthetic processes occur, consuming CO_2 from the atmosphere. Once combusted, the same amount of CO_2 is considered to be released, so the net CO_2 production is essentially zero. Therefore, fossil CO_2 is decreased when biomass cofiring is employed. Wood waste substitution for coal reduces fossil CO_2 emissions by 1.05 ton fossil CO_2/ton wood burned [19–21].

6.7.2. Cofiring in Europe

The cofiring of biomass with coal is common practice in the European market [22]. Regulatory and economic drivers exist and promote the use of biomass in utility boilers. The combustion of biomass is commonly used to meet regulatory demands for increasing the use of renewable energy. It can be base loaded, and the capital cost is lower than that of other renewable energy sources such as wind or solar.

6.7.3. Selected Case Studies

The following subsections briefly cover selected biomass cofiring experiences.

Albright Generating Station

Albright Generating Station #3 is a 140-MW_e (gross) tangentially fired (T-fired) four-cornered furnace located in Albright, West Virginia. Sawdust was cofired

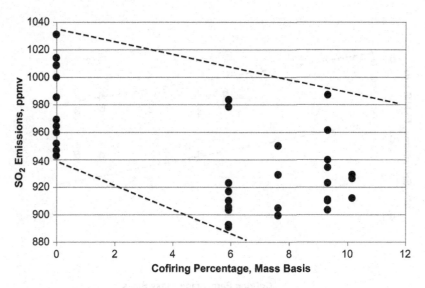

FIGURE 6.5 SO₂ emissions as a function of cofiring percentage. *Source: [23].*

with coal from May 30, 2001, to July 27, 2001. Albright used the separate injection cofiring technology. Sawdust was injected in opposing corners, right at the center of the fireball. The percentage of cofiring ranged from 0% to 10% on a mass basis [4, 23, 24].

Opacity and CO for the demonstration showed that up to 10% mass basis, minimal impacts were observed. SO_2 emission decreased the most, since the sawdust had essentially no sulfur (Figure 6.5). Significant NO_x reductions were observed during the demonstration test. Figure 6.6 shows the NO_x concentrations versus the cofiring percentage. For this particular test and fuel blend, NO_x reductions were ≥3% for every additional percent sawdust (heat input basis) added to the blend [23, 25].

Unburned carbon (UBC) concentrations were measured for both the bottom ash and the flyash. UBC concentrations did not vary as a function of cofiring percentage. Mercury emission was also reduced by the cofiring of sawdust with coal.

Seward Generating Station

Seward Generating Station is located in western Pennsylvania. The plant has three boilers (#12, #14, and #15). Boiler #12 is a Babcock & Wilcox (B&W) front wall-fired boiler with a design capacity of 136,080 Kg/hr (300,000 lb/hr), with main steam conditions of 4654 kPa (675 psig) and 446°C (835°F). It is equipped with two pressurized EL-56 ball and race pulverizers [26, 27]. Biomass cofiring tests were performed in December 1996 and July 1997. The program involved designing and installing a system for separate injection of the biomass [27, 28].

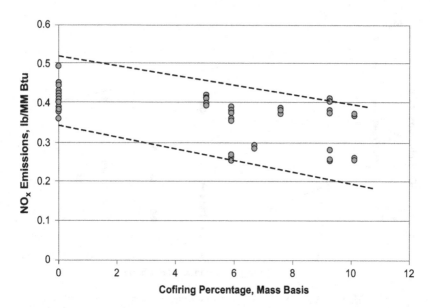

FIGURE 6.6   NOₓ emissions versus cofiring percentage for the Albright demonstration test. *Source: [2].*

Cofiring with biomass had an impact on boiler efficiency. At 10% to 12% biomass (mass basis), boiler efficiency loss was about 0.5%. The CO emission level stayed constant and did not exceed 20 ppm$_v$. Opacity concentrations were observed to slightly increase. Because of the inherently low concentration of sulfur in the sawdust, sulfur dioxide emissions were reduced. NOₓ emissions were also reduced by the cofiring technique (Figure 6.7). The impact on unburned carbon concentration was minimal [27].

6.7.4. Cofiring with Waste

Waste materials discussed in this section include opportunity fuels such as reclaimed and reprocessed wastes from energy and other industries. These fuels include anthracite culm, gob, slack, and other coal mining wastes. Included in this category are fines from coal processing plants such as fines previously impounded in ponds. Coal fines and impoundments are found throughout the coal-using industry. Substantial quantities of coal products have promoted the development of several opportunity fuels, such as coal-water slurries (CWS) and direct combustion of waste coal [12, 29, 31].

6.7.5. Emission Aspects

Coal-water slurries were developed to address concerns for environmental and safety issues in coal mining and coal beneficiation, cleanup of waste coal ponds

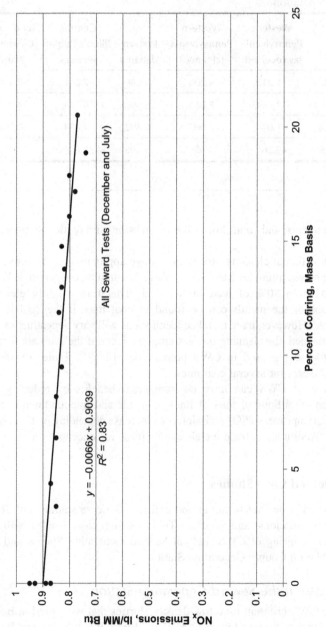

FIGURE 6.7 NO$_x$ reduction versus mass percentage of biomass in the fuel. *Source: [27].*

Within the figure:

All Seward Tests (December and July)

Percent Cofiring, Mass Basis

NO$_x$ Emissions, lb/MM Btu

$y = -0.0066x + 0.9039$
$R^2 = 0.83$

TABLE 6.5 Pollutant Measures for Several Sources of Coal Fines in CWS Preparation

Parameters	Western Pennsylvania (as received)	Western Pennsylvania (cleaned)	Eastern Alabama	Central Illinois Public Services	Rend Lake Cleaning Plant
lb S/10^6 Btu	1.312	4.006	3.280	7.882	7.485
lb SO$_2$/10^6 Btu	2.624	8.013	6.560	15.764	14.971
lb N/10^6 Btu	1.073	0.907	0.492	2.074	0.741
lb Ash/10^6 Btu	49.26	6.652	15.498	31.632	4.447

Source: [12].

and impoundments, and control of airborne emission with particular emphasis on NO$_x$.

Conventional coal cleaning processes remove approximately 11% to 25% of mercury and selenium; 26% to 50% of arsenic, cadmium, copper, nickel, and zinc; and 51% to 70% of lead in the coal. Therefore, slightly elevated concentrations of the metals can be found in coal fines if the coal is not rebeneficiated. However, trace metal concentrations will vary depending on the coal being cleaned, the cleaning process employed, and if the fines are further cleaned before being used in CWS preparation [12, 30]. Table 6.5 shows pollutant measures for several coal fines.

Production of CWS can have environmental benefits by reducing the accumulation of billions of tons of fines in ponds and impoundments. NO$_x$ emissions can improve, so SO$_2$ emissions can increase depending on the source of the fines. Additionally, trace metals can increase in concentration.

6.7.6. Selected Case Studies

Since the focus is on fuel blending and cofiring, the case studies will detail cofiring demonstrations and studies. Two cofiring case studies will be discussed: the cofiring of CWS fuels at Seward Generating Station and the cofiring at Marion County Generating Station.

Cofiring CWS at the Seward Generating Station

The Seward CWS program used low-density slurries that were fired in Boiler #14 and Boiler #15. Boiler #14 is a 32-MW$_e$ wall-fired B&W boiler, and Boiler #15 is a 147-MW$_e$ tangentially fired boiler. Tests were conducted during the summer of 1997, with the plant operating at 130 to 136 MW$_e$. Cofiring

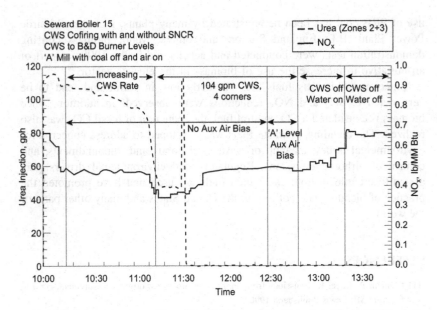

FIGURE 6.8 NO_x emissions while cofiring 393 l/m coal-water slurries with the SNCR system shut off. *Source: [12].*

percentages achieved were as high as 35% on a heat input basis. Significant improvements were seen in NO_x formation, which was observed in both reduced emission and reduced urea consumption. Figure 6.8 shows that by firing 104 gpm of CWS, the SNCR system could be turned off, while NO_x emissions could be maintained at 0.5 lb/10^6 Btu. SO_2 emissions could be managed during the tests [12].

Cofiring CWS at the Marion County Generating Station

A CWS was cofired in Boiler #3—cyclone boiler—of the Marion County Generating Station. Boiler #3 is a 33-MW_e unit that is equipped with 2×7-ft -diameter cyclone barrels. Demonstration tests showed that operationally a cofiring limit of 20%, on a heat input basis could be achieved. Above 20%, opacity and carbon conversion efficiency degraded. Flyash carryover also increased. NO_x reductions were only modestly observed and were maximized at approximately 34%, on a heat input basis. The lack of O_2 used in firing the system attributed to the low NO_x reductions [12].

6.8. CONCLUSIONS

Environmental and regulatory drivers have been key factors in promoting many fuel blending and cofiring practices. The increased use of PRB coal over the years was a means to reduce sulfur dioxide emissions. Successful use of PRB coal has been demonstrated by many plants, including Monroe Power Plant, B.L. England Station, and others. Major biomass cofiring demonstration tests were conducted and accelerated in the 1990s as part of an initiative to increase the use of biomass in the electric sector. Since most biomass fuels are inherently low in sulfur and tend to be very reactive, SO_2 and NO_x reductions were observed. In addition, since biomass is considered a CO_2 neutral fuel, the reduction of fossil CO_2 was also realized. The blending of waste fuels was developed to address concerns for environmental issues, cleanup of waste coal ponds and impoundments, and control of airborne emissions. Regulatory and environmental drivers play a significant role in how and which fuels are used and have promoted the practice of blending and cofiring in the United States and many other parts of the world.

REFERENCES

[1] Clement R, Kagel R. Emissions from combustion processes: origin, measurement, control. Chelsea, MI: Lewis Publishers; 1990.

[2] Kitto JB, Stultz SC, editors. Steam: its generation and use, 42nd ed. Barberton, OH: Babcock & Wilcox; 2005.

[3] Miller BG. Coal energy systems. Boston: Academic Press; 2005.

[4] Tillman DA, Duong DNB, Miller BG, Bradley LC. Combustion effects of biomass cofiring in coal-fired boiler. Proceedings Power-Gen International. Las Vegas; 2009, December 8–10.

[5] Miller BG, Tillman DA, editors. Combustion engineering issues for solid fuel systems. Boston: Academic Press; 2008.

[6] Johnsson JE, Glarborg P. Sulfur chemistry in combustion I: sulfur in fuels and combustion chemistry. Pollutants from combustion: formation and impact on atmospheric chemistry. Amsterdam: Kluwer Academic; 2000. p. 263–82.

[7] Duong D, Miller B, Tillman D. Characterizing blends of PRB and Central Appalachian coals for fuel optimization purposes. Proceedings 31st international technical conference on coal utilization and fuel systems. Clearwater, FL; 2006, May 21–25.

[8] McLenon, KR. Fuel quality: a shift supervisor perspective. Proceedings electric power conference. Chicago; 2007, May 1–3.

[9] Baukal Jr CE. Industrial combustion pollution and control. New York: Marcel Dekker; 2004. p. 247–52, 327–35, 379–82.

[10] Sloss LL. Economics of mercury control. IEA Clean Coal Centre; 2008. Report CCC/134.

[11] U.S. Environmental Protection Agency. Wastes - Resource conservation - Reduce, reuse, recycle - Industrial materials recycling, *www.epa.gov/epawaste/conserve/rrr/imr/ccps/flyash. htm*; 2011, accessed 6-21-11.

[12] Tillman DA, Harding NS. Fuels of opportunity: characteristics and uses in combustion systems. Amsterdam: Elsevier; 2004.

[13] Yang H, Xu Z, Fan M, Bland AE, Judkins RR. Adsorbents for capturing mercury in coal-fired boiler flue gas. PubMed 2007;146(1–2):1–11.

[14] Tillman DA, Dobrzanski A, Duong D, Dosch J, Taylor K, Kinnick R, et al. Optimizing blends of Powder River Basin subbituminous coal and bituminous coal. Proceedings PRB coal user's group. Atlanta; 2006, May 2–4.

[15] Peltier R. Luminant's Big Brown Plant wins for continuous improvement and safety programs. Power Magazine, *www.powermag.com/coal/Luminants-Big-Brown-Plant-wins-for-continuous-improvement-and-safety-programs_117_p3.html*; 2008, accessed 01-06-11.

[16] Russell JM, Gillespie MB, Gibson WC, Bhamidipati VN, Mahr D. Considerations for low-sulfur coal blending at B. L. England Station. Proceedings Power-Gen International. Las Vegas; 2003, December 9–11.

[17] Tillman DA, Duong DNB. Fuel selection for cofiring biomass in pulverized coal and cyclone fired boilers. Proceedings 34th international technical conference on clean coal and fuel systems. Clearwater, FL; 2009, May 31–June 4.

[18] Konttinen J, Hupa M, Kallio S, Winter F, Samuelsson J. NO formation tendency characterization for biomass fuels. Proceedings 18th international conference on fluidized bed combustion. Toronto; 2005, May 22–25.

[19] Hus PJ, Tillman DA. Cofiring multiple opportunity fuels with coal at Bailly Generating Station. Biomass & Bioenergy 2000;19(6):385–94.

[20] Tillman D, Hughes E. Opportunity fuel cofiring at Allegheny Energy. Palo Alto, CA.: Electric Power Research Institute; 2004. Report TR-1004811.

[21] Tillman DA. Biomass cofiring: the technology, the experience, the combustion consequences. Biomass & Bioenergy 2000;19(6):365–84.

[22] Zabetta EC, Barisic V, Peltola K, Hotta A. Foster Wheeler experience with biomass and waste in CFBs. Proceedings 33rd international technical conference on clean coal and fuel systems. Clearwater, FL; 2008, June 1–5.

[23] Tillman D, Hughes E. Annual report on biomass cofiring program. Palo Alto, CA: Electric Power Research Institute; 2001. Report TR-1004601.

[24] Tillman DA, Payette K, Banfield T, Plasynski S. Cofiring demonstration at Allegheny Energy Supply Company, LLC. Proceedings 19th international Pittsburgh coal conference. Pittsburgh; 2002, September 23–27.

[25] Tillman DA, Miller BG, Johnson DK, Clifford DJ. Structure, reactivity, and nitrogen evolution characteristics of a suite of solid fuels. Proceedings 29th international technical conference on coal utilization and fuel systems. Clearwater, FL; 2004, April 18–22.

[26] Tillman D, Hughes E. Biomass cofiring guidelines. Palo Alto, CA: Electric Power Research Institute; 1997. Report TR-108952.

[27] Tillman D, Battista J, Hughes E. Cofiring wood waste with coal at the Seward Generating Station. Proceedings 23rd international technical conference on coal utilization and fuel systems. Clearwater, FL; 1998, March 9–13.

[28] Tillman D. Biomass cofiring: field test results. Palo Alto, CA: Electric Power Research Institute; 1999. Report TR-113903.

[29] Wiltsee GA. Strategic analysis of biomass and waste fuels for electric power generation. Palo Alto, CA: Electric Power Research Institute; 1993. Report TR-102773.

[30] Tillman DA. Trace metals in combustion systems. San Diego: Academic Press; 1994.

[31] Tillman DA. The combustion of solid fuels and wastes. San Diego: Academic Press; 1991.

Chapter 7

Modeling and Fuel Blending

7.1. INTRODUCTION

Engineers have relied on theory and experiments for design and analysis. Through modeling, a few tools can aid in the design and analysis process. Models are used primarily to increase the understanding of a physical process. They can also aid in the redesign of a system. Modeling can be performed through either numerical analysis or physical modeling. Each technique has different insights with different benefits. If used properly, modeling can and has been very successful at providing design and operational ingenuity.

Physical or cold flow models enable visualization of a process or a system. They have traditionally been used in analyzing fuel bunkers or silos, coal or fuel pipes, burner design, boiler simulations, evaluations of gas flows through air pollution equipment (e.g., scrubbers, selective catalytic reduction systems), and other processes. Physical and CADD models have been and continue to be used extensively in the design process. Once the design of a system such as a steam boiler reaches a certain stage, the boiler is modeled in order to validate the design and layout. Physical models are still used in many applications like the ones just listed. The physical scaling down of processes and components has been very successful, but physical modeling is expensive and time consuming. These aspects have been major drivers for the development and advancement of computerized numerical modeling.

Numerical modeling, which includes computational fluid dynamic (CFD) modeling, began in the 1960s when computers with significant computational capability were first commercially available [1]. Over the years, the sophistication and resolution of these tools increased, and improvements continue to be made today. Numerical modeling has several benefits, including as a way to obtain necessary information quickly at a reduced cost and for design validation or examination of interactions between systems. Therefore, if a system does not function as anticipated, modeling can aid in the troubleshooting process—determining the problem and then helping to devise a solution.

Numerical modeling can be used to look at flue gas and steam-water flow to evaluate boiler behavior or evaluate design modifications. Applications can include predicting temperature distribution in a furnace, burner design, nitrogen oxide (NO_x) emissions, boiler furnace circulation systems, and many others. At the same time, the many approaches to numerical modeling have limitations. Typically these tools are used to calculate outcomes in steady-state operations. Further, many, such as Gibbs Free Energy-based thermodynamic models, assume all reactions go to equilibrium [2, 3].

This chapter provides an overview of the different approaches to modeling. It also briefly looks at the various modeling applications utilized in the electric utility industry and in basic process industries.

7.2. THE PURPOSES OF MODELING

Models—both computer-based and physical—increase the understanding of physical and chemical processes ranging from fuel handling to combustion and emissions management. Consequently, modeling aids in the design and redesign of a system or component. Models can also enable visualization of a process, thus enhancing understanding, particularly during design and fabrication phases. Engineers have used various modeling techniques for decades in order to validate the design and layout of different processes. Models are used to examine the interaction between systems and to troubleshoot processes.

The advantages of modeling are many and can mitigate mistakes while a project is in the design stage. Modeling can be accomplished through either numerical or physical techniques. Depending on the process or equipment and the extent of detail necessary, numerical modeling may be sufficient. At times, however, physical models or both physical and numerical modeling may be critical to the success and understanding of a process.

7.3. SPECIFIC APPLICATIONS OF MODELING

Design modifications as a result of modeling are seen in burner applications, boiler design and layout, selective catalyic reduction (SCR) configurations, and many others. Foster Wheeler and other boiler manufacturers have used computer modeling to further the understanding of biomass cofiring and its impact on burner performance and the heat transfer profile in the boiler. Prior to the implementation and construction of an SCR, it is common to use physical cold flow models to understand the mixing behavior of the flue gas and ammonia injection system upstream of the catalyst (Figure 7.1). This enables a designer to properly incorporate the necessary equipment to maximize mixing patterns and, subsequently, NO_x reduction. Another application looks at the flow pattern through pulse jet fabric filters and a spray dryer absorber (Figure 7.2). Both computer and physical modeling are used in many other applications to help understand the necessary process.

FIGURE 7.1 Physical model of an SCR. *(Photo by Air Flow Sciences. Used with permission.)*

7.3.1. Modeling to Reduce the Use of Physical Tests and Costs

When models simulate a particular process with adequate accuracy, testing at full scale can be minimized. Full-scale tests can be expensive and time consuming. Models—particularly computer models—help to enable upfront research to be conducted, thus minimizing the cost associated with full-scale tests. Some of the numerical models that are now considered computer models include the very popular computational fluid dynamics models, stochastic models, linear programming models, thermodynamic models, and chemical percolation models. All of these are designed to simulate physical flow activities, and many also address combustion, gasification, and other chemical reactions.

FIGURE 7.2   Physical model of a pulse-jet fabric filter and spray dryer absorber. *(Photo by Air Flow Sciences. Used with permission.)*

The increased power of computers over the years has enabled the capacity and ability of models to be more robust and dynamic. Computation time has decreased drastically. With the ability to perform three-dimensional modeling, the need for physical models has decreased. However, in certain applications, the need and importance of physical models still exist (e.g., the installation of SCR).

7.3.2. Methods of Modeling

Several modeling techniques are available and exist for different applications. These include CFD models, zonal models, thermodynamic models, discrete elemental models, and physical or cold flow models. The term *numerical model* defines a mathematical description of a physical process using numerical approaches [1]. Numerical models are used for several reasons. They can be used as an analytical approach to solve equations that describe a system that may not exist. They may also be used when many iterations of a calculation are necessary. The time spent on simulation and computation is drastically reduced. The level of detail of the model depends on the required accuracy and resources available to use the model.

The two basic approaches to mathematical modeling are modeling the behavior of a system and modeling the fundamental physics and chemistry of a system in order to determine its behavior. For modeling the behavior of a system, these models include network flow models and heat exchanger heat transfer correlations. For modeling the fundamental physics and chemistry of a system to determine behavior, these models include CFD and chemical reaction models [1].

Applying numerical modeling to engineering systems has some limitations. Numerical models can only be applied where there is sufficient understanding of the chemistry and physics [1]. Numerical modeling typically defines activities of a system in steady state, not in transient conditions. If the mathematical description is not available or is too complex, numerical modeling may not be appropriate for the analysis. Computer capability continues to limit the detail of a numerical system, so simplified descriptions are typically developed in order to make the analysis manageable. In addition, the level of precision and accuracy depends on the accuracy of the inlet condition and other boundary conditions.

Furthermore, the level of detail is influenced by the limitations of turbulent flow modeling. Currently, numerical approaches are not capable of modeling the full detail of temporal and spatial fluctuations associated with the turbulence in a system. Within these limitations, numerical models provide insights into the orders of magnitude of outcomes, the directions of change (if modeled), and the orders of magnitude of change. Following are summaries of various modeling methods. Subsequently, a few will be examined in more detail.

Physical (Cold Flow) Models

A physical model can be described as a small-scale replica of an apparatus [1]. Numerous such models have been built to evaluate bunkers and silos [4], for coal and fuel pipes for pneumatic transport [5], for duct work [6], and for pollution control systems [6]. Such models also have been constructed to evaluate burners and their installations [7]. The level of detail depends on the need and circumstances. When modeling silo or bunker discharge systems, considerable detail is required if the question is the flow of solids—particularly dissimilar solids (Figure 7.3). In applications such as the installation of SCR or the design of burners, physical models are highly desirable. They enable the observer to better understand the flow patterns and physical phenomena of the process. The use of numerical models—in certain applications—cannot replace physical models.

Numerical Zonal Models

Zonal models are typically one-dimensional models and provide a first approximation to system behavior. Over the years, the technique has been augmented by or even replaced with more sophisticated numerical simulations (e.g., CFD). Zonal models are fundamental tools available for the analysis of various systems. In these models or simulations, a few key assumptions must be made. The first assumption basically ignores any two-way coupling between thermophysical and hydrodynamic characteristics; uniform mixing is assumed within each of the defined zones [8]. Plug flow is assumed to occur between zones.

FIGURE 7.3 A cold flow model (1:1 scale) of the discharge of the bunkers at the Allen Fossil Plant, constructed to determine whether a blend of coal, wood chips, and tire chips would flow successfully to the stock feeders and cyclone burners. (The blend did flow properly.) This model was designed by N. S. Harding and built by Reaction Engineering International. *(Photo by D. Tillman.)*

Properties that are outputs from one zone are used in the following zone. Zonal models enable simplification of the system, at the same time providing useful insight into the problems or system of interest while minimizing the use or cost of complex numerical simulations. Chemical reactions can be accounted for by a network of zones or reactors. The furnace or process can be divided into a defined series of zone/reactors interconnected within one another. The results obtained from one zone are transferred to the subsequent zone. An advantage to zonal modeling is its ability to represent complex phenomena within reasonable computation time frames. However, two-way coupling of properties is not considered; uniform flow is assumed to exist.

The Monroe Power Plant provides an example of the use of zonal modeling. The plant has four supercritical boilers, each capable of producing approximately 775 to 795 MW_{net}. In an effort to quantify the effects of combustion on other equipment or processes, four zonal models were built to look at the problems associated with incomplete combustion, different blending scenarios, and many more. The models provided several important implications. Blended coals have more complete combustion than either of the parent coals.

This phenomenon is supported by analyzing the pyrolysis activation energies. In the rate-determining step of char oxidation, activation energies showed that the subbituminous PRB coal drives the blend char reactivity. This implied that blend char oxidation is essentially as reactive as char oxidation for PRB, thus providing more complete combustion. The models also provided other technical implications/conclusions. Besides having the ability to show where the fuel burned, the models represented how well the fuel burned, implications for electrostatic precipitator performance, boiler efficiency, economic savings potential, and the like [9].

At the Monroe Power Plant, zonal modeling was used to quantify where different combustion reactions occurred as a function of coal type in a blend and coal particle size [7]. The modeling provided a basis for evaluating the cost of varying mill performance [10]. Figures 7.4 and 7.5 illustrate the application of zonal modeling at the Monroe Power Plant [7, 9, 10]. Application of zonal modeling to the combustion processes has been conducted for the four boilers at the Monroe Power Plant. The assumption of first in, first out enables a simplified approach to studying combustion behavior when blending significant percentages of PRB coal with Central Appalachian coal [9].

Thermodynamic Models

Thermodynamic models can be used to predict various properties such as enthalpy or phase equilibrium. Model categories include equations of state, activity coefficient, empirical, or special system specific. Model selection can depend on parameters such as process species and compositions, pressure and temperature ranges, availability of data, and other aspects [2, 3, 11]. These models can help to understand system behavior. For power generation applications, thermodynamic models can help to predict and understand flame temperature, furnace temperature profiles, and other parameters. When modeling using thermodynamic equations, all reactions that are being considered must go to equilibrium.

Gibbs free energy is a thermodynamic potential that is similar to enthalpy and internal energy and that depends on initial and final states of the system. In the application of a combustion process where the reactants and products are in equilibrium with the surroundings, the equation depicts the maximum value of

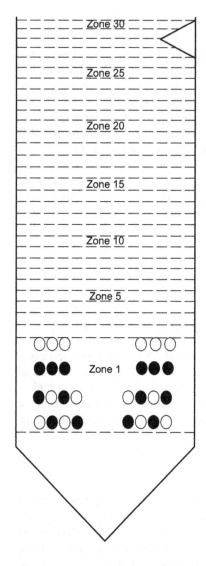

FIGURE 7.4 The 30 zones that are applied to waterwell sections of Monroe Power Plant boilers. *Source: [7].*

reversible work that can be obtained. Free energy is typically used to determine the temperature of burning a fuel while considering the effects of dissociation [1]. Gibbs calculations are at the heart of thermodynamic modeling.

Discrete Element Analysis Models

The discrete element method (DEM) is a family of numerical methods computing the motion of a large number of micrometer-scale-size or larger particles. The term *discrete element method* is most commonly associated with the definition where it allows for finite displacements and rotations of discrete bodies and

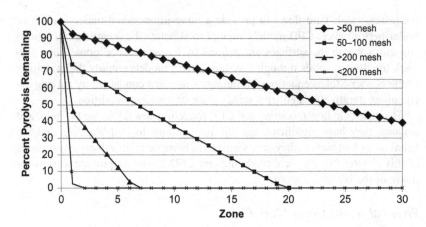

FIGURE 7.5 Completion of pyrolysis in the Monroe Power Plant boilers' waterwall sections as a function of coal particle size. *Source: [7].*

recognizes any new contacts automatically as calculations progress [12]. DEM is typically used in addressing engineering problems that involve granular and discontinuous materials. Applications include granular and particulate materials, silo flow, sediment transport, jointed rocks, and others. Discrete element analysis is quite similar to stochastic modeling as described by Scavuzzo [4].

Discrete element modeling generally consists of a group of spheres that have various forces such as friction, collision, cohesion, damping, gravity, and molecular acting on them. Well-defined input parameters can provide some of the most accurate results. The drawback with DEM models is that they are complex and processor intensive, so they can be impractical for real-time calculations. Applications to real-time fuel blending optimization or operator advisory screens become very difficult [13]. The stochastic models are event-driven, probability-based models. Engineering Consultants Group (ECG) uses AccuTrack—a stochastic model—to predict flow patterns in a bunker. The results closely match those of the DEM results, with a high degree of accuracy, but only require a fraction of the run time when using DEM [13].

Empirically derived models are based on measured data obtained through testing of full-scale boilers. Testing at full scale is becoming expensive, so the availability of such data is limited. However, if conducted properly, models that are derived from full-scale boilers are more accurate and are valuable in further understanding the products of combustion.

Computational Fluid Dynamics Models

Computational fluid dynamics modeling is based on the principles of fluid mechanics, utilizing numerical methods and algorithms to solve problems that involve fluid flows. Models can integrate chemical reactions—combustion processes—with fluid flows to provide a three-dimensional understanding of boiler performance. CFD models attempt to simulate the interaction of liquids and gases where the surfaces are defined by boundary conditions. They also track the flow of solids through a system. These models employ the principles of the Navier-Stokes equations. Simulations are then conducted by solving the equations iteratively as either a steady-state or transient condition.

Because cold flow modeling and CFD modeling now dominate the field, and because they have significant specific applications to fuel blending and the behavior of fuel blends, they are examined in more detail in subsequent sections of this chapter [6]. Physical modeling and CFD modeling provide a basis for many of the blend decisions.

Physical (Cold Flow) Modeling

Physical or cold flow modeling is perhaps the oldest form of modeling available, and, with the exception of thermal processes, it has certain unique features that are of considerable benefit to engineers and scientists. One example is physical modeling of thermal processes in pilot scale furnaces and related facilities, which is used at The Energy Institute of Pennsylvania State University; the test facilities at the University of Utah; EERC in North Dakota; Southern Research Co. in Birmingham, Alabama; the National Energy Technology Center in Pittsburgh, Pennsylvania; Sandia National Laboratories in Livermore, California; and numerous other locations. These test facilities have specific purposes and are outside the purview of this discussion.

Physical modeling is used for many applications, including the flow of fuels through bunkers and silos (refer, for example, to Figure 7.3), the flow of fuels through pneumatic pipes, the flow of fuels through burners, the flow of air through windboxes of boilers, and the flow of gases through ductwork and air pollution control systems. It is a remarkably flexible engineering tool, capable of evaluating both steady-state and non-steady-state conditions—including transient and upset conditions.

7.4. PRINCIPLES OF PHYSICAL MODELING

In cold flow modeling, a physical representation of the full-scale application is constructed. Scales can be very small; in the evaluation of the flow of blends of dissimilar solids through bunkers, ECG began by building a cold flow model of the bunker system at a scale of 1:50 [4] (Figure 7.6). The bunker discharge model for the Allen Fossil Plant of TVA, which was shown in Figure 7.3, was built at a scale of 1:1. The modified burners for the Monroe Power Plant were modeled at a 1:2 scale. Typically, however, models are built at scales of 1:8 to 1:16, with 1:12 being the most common.

In order to validate computer models of coal flow through bunkers, scale models of actual bunker designs have been developed by ECG. The model uses

FIGURE 7.6   The 1:50 model of bunker discharge built by ECG for use in modeling the bunkers at Conemaugh Generating Station. For the flow of dissimilar granular solid materials, ECG tested its model with kitty litter and Miracle Gro plant food. *Source: [4]. (Photo by ECG. Used with permission.)*

several materials that closely match coal characteristics. To further validate both physical and computer models of coal flow through bunkers, ECG employs radio frequency identification devices (RFIDs) in full-scale bunkers. Field tests have been conducted at the Conemaugh Station to monitor flow patterns and flow rates [13].

In constructing the physical model, the level of detail is a function of the purpose of the model. In all cases, the flow pattern and Reynolds number regime—laminar or turbulent—are matched to the full-scale application. Matching the Reynolds number regime is sufficient if the boundary layer is not significant and if general conditions are adequate for modeling purposes. Reynolds number matching becomes more significant as lift or drag characteristics are important to the modeling or if the boundary layer is significant [6].

Gretta and Grieco [14] state that in evaluating air pollution control equipment, physical modeling depends on three similarities between the model and the prototype:

- Geometric similarity means that the model and the prototype must be identical in shape and that the dimensions of the two must be carefully related by the scale factor.
- Kinematic similarity requires that the flow streamlines have similar patterns between the model and the prototype.
- Dynamic similarity requires that the relative influences of system forces in the prototype and the model be very close.

FIGURE 7.7 Schematic representation of the air/coal flow loop constructed by Air Flow Sciences for EPRI to evaluate pneumatic transport issues, including roping. *(Photo by Air Flow Sciences. Used with permission.)*

If possible, frictional forces from the wall should be as similar as possible [14]. At the same time, walls are typically plexiglass in order to give the engineers the ability to see and measure events in the physical model.

In constructing physical models, flow patterns become essential to the modeling effort. When modeling the flow of fuel through pneumatic pipes, for example, the configuration of elbows is of great concern. Air Flow Sciences has built a detailed loop to evaluate the pneumatic flow of coal and fuels through pipes (Figures 7.7 through 7.9).

Physical models are particularly adept at evaluating particle layout in transport systems—coal pipes, postcombustion ductwork, and postcombustion systems [6]. However, the use of this modeling approach requires more schedule time and more cost than numerical modeling such as CFD [6].

7.4.1. Some Applications of Physical Modeling

Physical or cold flow models are typically used in understanding the flow patterns of flue gas when SCR installations are considered. This has been demonstrated at plants such as Xcel Energy King Station, Ameren Coffeen 2, First Energy Mansfield 1 & 2, and others [15]. Through the use of cold flow modeling, computer models can help to validate the design of the SCR. Where dissimilar fuels are blended together, producing different forms of flyash, this can be particularly useful.

Physical models were particularly useful in developing modified burners for the Monroe Power Plant from 2004 to 2006. The burner configuration at Monroe is nearly unique, with two coal pipes feeding into one and then into the

FIGURE 7.8   Overview of the air/coal flow loop constructed at the Air Flow Sciences facility in Livonia, Michigan. *(Photo by Air Flow Sciences. Used with permission.)*

large burners. The modifications converted the burners from impeller-based to swirl-stabilized. As such, rope breakers became significant. Physical modeling helped in the basic design of the rope breakers and also in their placement [7].

Numerous other examples can be cited. We have already seen bunker flow models and coal roping models, but windbox modeling is also a common application of physical modeling (Figure 7.10). Windbox modeling is useful in support of fuel blending and the management of air to individual burners. Physical modeling is returning to prominence as engineers seek to understand non-steady-state and transient conditions and the physical interactions of dissimilar materials blended together.

7.4.2. Computational Fluid Dynamics Modeling

The most common and most popular numerical modeling technique, and the one used here for comparison to physical modeling, is computational fluid dynamics, which became quite popular in the 1970s, 1980s, and 1990s with the increasing power of computers. CFD modeling is based on the principles of fluid mechanics, utilizing numerical methods and algorithms to solve problems that involve fluid flows. Models attempt to simulate the interaction of liquids and gases where the surfaces are defined by boundary conditions. CFD models are based on the

FIGURE 7.9 Detail of the Air Flow Sciences cold flow physical model built for EPRI to test air/ coal flow in pipes, including roping phenomena. *(Photo by Air Flow Sciences. Used with permission.)*

principles of the Navier–Stokes equations. Simulations are then conducted by solving the equations iteratively as steady-state or transient. Firms such as Fluent offer CFD software as a platform for modelers and analysts.

7.5. THE BASIC APPROACH OF COMPUTATIONAL FLUID DYNAMICS MODELING

In CFD modeling, a computational three-dimensional grid of many cells is constructed to model the reactor or transport system (e.g., the ductwork, windbox, or furnace and boiler). In many applications, more than a million cells are used. Further, the three-dimensional grid does not employ cells of a uniform size; in areas where reactions or flows are more critical, cells may be smaller [6]. Once the grid is constructed, all relevant equations are installed in each cell. Equations for gases are well defined and include conservation of mass, momentum, and energy. They are complex differential equations and require the power of computers. The equations are then solved for the given cell or location, and bordering cells receive the products of those equations to be solved subsequently [6]. CFD modeling is performed at steady-state conditions. Alternative scenarios can be evaluated, but transient conditions are difficult to

FIGURE 7.10   Windbox modeling at Air Flow Sciences. *(Photo by Air Flow Sciences. Used with permission.)*

manage. In CFD modeling, the 1:1 scale is almost always used. There is no need for scaling, and Reynolds numbers are matched identically.

When thermal processes are involved (e.g., combustion chemistry or gasification chemistry), the equations include the kinetics of the chemical reactions: drying, pyrolysis, gas-phase oxidation, char oxidation, and gasification reactions. From these, temperatures for heat transfer and products of combustion can be calculated. Figure 7.11 shows the ability of CFD modeling to consider blends of dissimilar materials; this figure is a trajectory for coal and wood in a cyclone boiler. Figure 7.11 illustrates that CFD modeling can be used for blends of dissimilar materials. It also illustrates the use of different-sized cells. Figure 7.12 illustrates the use of CFD modeling to calculate combustion temperatures in cyclone combustion. Both Figures 7.11 and 7.12 rely on CFD modeling by Reaction Engineering International.

7.5.1. Computational Fluid Dynamics Modeling of Combustion Processes

Combustion processes for various systems must be well understood because they have consequences for components such as burners, boiler tube metals and spacing, postcombustion systems, and others. Modeling of the combustion

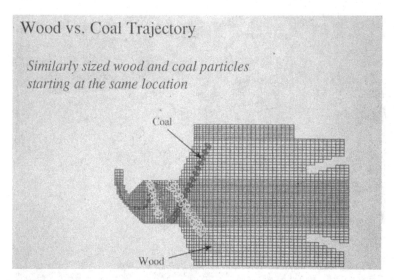

FIGURE 7.11 Calculating the trajectories of wood and coal particles in a cyclone using computational fluid dynamics modeling by Reaction Engineering International. *Source: [16].*

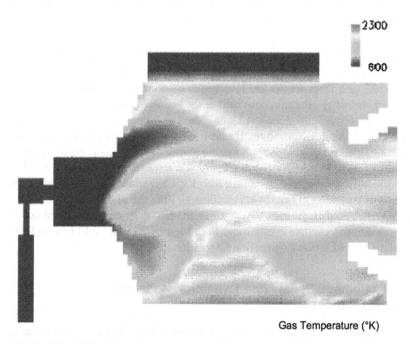

FIGURE 7.12 Cyclone barrel temperatures calculated for the Willow Island cyclones by Reaction Engineering International, using a blend of 90% coal–10% wood waste. *Source: [17].*

processes helps to provide the basis for understanding and system development. Modeling is a common practice when the behavior and characteristics of a combustion system must be predicted. The results are not absolute but help to provide insight. CFD modeling is commonly used to predict and/or verify the combustion behavior of firing systems. For example, models can help to depict conditions both before and after new burners are retrofitted to a boiler. Other selective combustion behaviors can also be modeled through CFD analysis. Creative Power Solutions (CPS) analyzed the aerodynamic and chemical processes of Unit 2 of the Four Corners Power Plant in Fruitland, New Mexico [18]. Plant data were collected and used to verify the computational fluid dynamics models.

Two studies of two different operating conditions were investigated. In both cases, the boiler was under steady-state, full-load conditions, where the total fuel mass flow rate is constant. The combustion process was modeled in two main processes: The Eulerian method modeled the continuous phase, while a Lagrangian method modeled the coal particulate and ash. An iterative solver coupled the two approaches. Both main heat release and oxidation of nitrogen were implemented.

The models considered work by DeSoete, Fenimore, and Smoot et al., for the consumption of HCN. They also considered the oxidation of N_2 through both the prompt (Fenimore) and/or thermal (Zeldovich) mechanisms. The relationship between CO and the amount of unburned carbon in the solid particles was depicted. High carbon contents were located in the corner between the back and side wall of the furnace. CO was also observed to have concentrations in the same locations for both the numerical and the experimental results. This information can provide several insights into the combustion conditions.

CFD modeling of the combustion process for bubbling fluidized bed boilers was developed by Åbo Akademi [12]. The Åbo Akademi Furnace Model attempts to model turbulent two-phase flow of a bubbling bed. It addresses the issues of fuel blending and, in particular, the cofiring of biomass. It predicts the solid fuel and gaseous combustion processes, focusing on devolatilization mechanisms, the chemical composition of the released volatiles, and the conversion of biomass particles. The model currently ignores complex axial and lateral mixing of fuel, bed material, and fluidizing air. When combustion of biomass mixtures is simulated, fuel-specific characteristics have to be considered.

The particles model of biomass conversion considers drying, devolatilization, char-carbon conversion, and ash particle formation. Once the carbon has been burned off, the ash particles remain, consisting of mainly inorganic material. Interaction with the surrounding gas is not currently considered in this model. When addressing biomass mixing, the chemistry and mathematical complexity increases. Mixing is also not currently considered in the model.

Advanced CFD models have provided insights into evaluating the performance of the boiler and the components, have aided in the development of conceptual designs and retrofits, and have helped with boiler operational improvements and changes such as oxycombustion [19]. This has been demonstrated through other software packages such as SmartBurn, which has helped to reduce NO_x emissions in tangentially fired, wall-fired, and cyclone-fired boilers [20].

CFD modeling to evaluate combustion processes is one of the more advanced techniques utilized. Some of the applications include temperature mapping, flame shaping, and evaluation of emissions formation and control. CFD has been used to study variables of combustion. Changes in boiler configuration can be studied in order to have a better understanding. Several universities and commercial companies are developing the use of three-dimensional models.

Sheng et al. [21] used computational fluid dynamics modeling to look at the synergistic effects of fuel blending in pulverized coal (PC) power plants. The study focused on combustion when blending Australian black coals in a pilot-scale furnace. Two different approaches were used to model the combustion flow: a two-mixture fraction model and a single-coal approach. In the two-mixture approach, the combustion of the two component coal blends was tracked by considering two mixture fractions. The model was able to—with a reasonable degree of accuracy—predict the ignition location and NO_x emission behavior for the blends. The synergistic effects of the chemical interactions between the parent fuels were considered. Conversely, the single-coal approach does not adequately represent the combustion behavior of fuel blends [21]. Arenillas et al. [22] have also considered the application of modeling the combustion processes of blending as a series of binary coal blends. The use of drop tube reactors was also employed.

7.5.2. Products of Combustion Modeling

Understanding products of combustion is becoming even more important as the regulatory climate pushes for more stringent emission reductions. By studying and analyzing how different products of combustion are formed and behave with different firing configurations and fuels, the reduction of emission constituents can be better achieved. The use of accurate models, along with detailed measurements, is an essential aspect.

The modeling of products of combustion can be performed through CFD modeling, zonal modeling, thermodynamically driven processes, and empirically derived techniques. CFD modeling is a common approach and is relatively successful. Empirically derived models provide one of the more accurate representations of the process, but they can become system/unit specific.

The use of CFD modeling to look at products of combustion is a common approach utilized throughout the industry. Modeling and analysis have been conducted by various companies and universities. Reaction Engineering

International (REI) performed modeling studies of the Southern Research Institute Combustion Research Facility (CFR) unit in order to look at the effects of oxy-fuel firing [23]. Three case studies were performed. The first case, which served as a baseline, was a traditional air-firing configuration. The next two cases looked at the effects of temperature and NO_x formation with oxy-fuel firing. Case #2 introduced flue gas recycle (FGR) into the center of the burner as the coal carrier. Case #3 removed most of the coal carrying FGR and introduced it around the outside of the burner quarl.

In the baseline case study, the coal ignites just as it leaves the quarl of the burners, which is typical of single-register, pulverized coal burners. For cases #1 and #2, the temperature profiles past the overfire oxygen region were very similar. However, for case #3, the temperature profile was higher due to the delayed combustion process. Combustion is stratified and continues into the overfire oxygen region. NO_x concentrations, when using FGR, were significantly higher when compared with the air-firing case. When FGR was introduced into the burner center, NO_x concentration was notably high near the burner due to the high velocities. This is also important to consider when blending fuels, since the combustion behavior will be different from burning a single fuel [23].

Another study conducted CFD modeling of the combustion process and emission from pilot-scale and full-scale PC boilers. A fairly good correlation exists between the modeling results and experimental results for predicting NO_x emissions. Experimental data were derived from a 50-kW test furnace that burned 500 to 600 kg of coal at various conditions. The model was then compared against data obtained for 550-MW opposed wall-fired and 575-MW tangentially fired boilers [24].

Some of the challenges when modeling products of combustion include the detection limits of the measurement instrument, the limitations of operational data that are available, and the complexity of the chemistry surrounding each compound. In order to model with increased accuracy and to represent the data accurately, laboratory-scale and full-scale testing is necessary. The challenges surrounding the modeling of products of combustion can be significant, depending on the degree of accuracy that is sought and the compound being modeled.

The detection limits of measurement instruments directly impact the degree of accuracy of the data and consequently the model results. Additionally, this places a limit on the constituents or compounds that can be measured. Measurement of species such as mercury is significantly affected by detection limits; commonly, the degree of accuracy in mercury measurements is questioned.

The use of operational data to validate a model is of critical importance. However, many plants will only measure selected products of combustion on a regular basis. Obtaining operational data over long periods of time becomes difficult. The limitations of operational data create challenges in modeling the products of combustion.

The formation of products of combustion is a set of complex reactions that are still not completely understood. The complexity of these reactions makes the modeling of products of combustion challenging. The degree of accuracy for each compound varies and depends on the depth of understanding and the research conducted. This complexity is further enhanced when blends of fuels are considered. The interaction between fuels exists and does not behave as the weighted average of the parent fuels. Duong et al. [9] studied these interactions with the modeling of the Monroe Power Plant. Models must consider these interactions when blends of fuels are considered.

7.5.3. Other Applications of Computational Fluid Dynamics Modeling

Other models include the mechanistic analysis of CFBs by Jian et al. These models were developed from one-dimensional analysis considering parameters such as the effect of fuel particles' size, combustion and heat transfer characteristics, and several other factors [25].

Multiphase reactive jet flows appear in engineering applications such as the oxygen-based steel-making process, water- and/or steam-diluted gas turbine combustors for NO_x control, fire-suppression systems, and others. Jet reactors' low behaviors are complex. They are operated at high temperatures—typically the 1500- to 3000 K range—and need to be contained within heavy metal enclosures [26]. CFD in combination with experimental measurements is used to further understand multiphase and multispecies reacting jet systems and have been ongoing for the last three decades [27].

CFD models are applied to the study of oxy-fuel combustion to look at combustion, heat transfer, and pollutant formation characteristics. Various coal combustion techniques were experimentally investigated using the pilot-scale CANMET Vertical Combustor Research Facility (VCRF). The pilot-scale flames were simulated using CFD. The model provided insight on the observed variation in NO_x production through several test runs: the increase in the O_2-enriched air, the drop at higher swirl settings, and the small reduction in recycled flue gas. The model results compared well with measured data in all test cases [27].

7.6. MODELING FOR BLENDING PURPOSES

When blending fuels of distinctly different characteristics, some aspects can be averaged, while others cannot be treated this way. The interaction or synergies between fuels is apparent and has been demonstrated. Therefore, it becomes important that there is an understanding of not only the parent fuels but also the characteristics associated with the fuel blends.

7.6.1. The Traditional Approach to Blending Analysis

Traditionally, when fuel blending is considered, most analysts will simply calculate the weighted average of the various fuel parameters. However, several studies have shown that the characteristics of blended fuel do not necessarily behave as the weighted average of the parent fuels. This has been demonstrated by Tillman, Miller, Duong, and several others [7, 10]. Consequently, models that attempt to understand and predict fuel behaviors and characteristics must consider the phenomena relating to fuel blending. Certain fuel parameters can utilize an averaging technique, while others cannot.

7.6.2. The Detailed Analytical Approach to Blending

To model with increased accuracy and detail, the effects of blending different fuels—particularly when blending fuels of distinctly different behaviors such as coals of various ranks or coals with biomass—must be studied and analyzed. Blending analysis can be conducted through techniques such as thermogravimetric analysis (TGA) for both the fuels and the char, drop tube reactor (DTR) analysis, and others. Reaction kinetics can be obtained through these techniques, which can be used to understand the blends relative to the parent fuels.

7.7. LIMITATIONS OF MODELING

Physical and computer models are critical to further the understanding of various aspects of blend combustion, but it must be recognized that there are limitations. The limitations of modeling depend on many factors, including the depth of research, the complexity of the chemistry for the particular compound, the limitations of operational data, and so on. Further study and research are necessary in order to improve the accuracy of models.

The literature on modeling is extensive, with just a few additional references representing much research [28–32]. The use of models is important to the design and understanding of various processes within electric utilities. They provide a basis by which a particular process can be studied and visualized. Numerical and physical models have been in use for many years. However, some of the limitations associated with modeling merit continued research, and this is particularly necessary when blending of fuels is considered.

7.8. CONCLUSIONS

For fuel blending, numerous approaches to modeling exist including physical modeling and various forms of computer modeling such as CFD. Each of the approaches has its uses and limitations; the engineer or scientist must choose the approach to be taken based on the needs of the project.

Physical modeling provides a basis for understanding the flow of materials and gases through a system, and the influence of blending on such flows. It provides a means for understanding both steady-state and transient conditions. It also provides tools for training. Physical modeling is more expensive and time consuming than computer/numerical modeling. Further, it cannot address chemical interactions directly.

Computer/numerical modeling, such as CFD modeling, can be achieved more rapidly, and less expensively, than physical (cold-flow) modeling. Computational modeling can address the chemistry of blending, including chemical reactions and temperature consequences. However, computer/numerical modeling is best used to define steady-state conditions, plus it does not address transient conditions well.

REFERENCES

[1] Kitto JB, Stultz SC. Steam: its generation and use. Barberton, OH: Babcock & Wilcox, A McDermott Company; 2005.

[2] Gordon S, et al. Computer program for calculation of complex chemical equilibrium compositions, rocket performance, incident and reflected shocks, and Chapman-Jouget detonations. Washington, DC: National Aeronautics and Space Administration; 1971.

[3] Gordon S, McBride BJ. Computer program for calculation of complex chemical equilibrium compositions, rocket performance, incident and reflected shocks, and Chapman-Jouget detonations. Interim Revision. Cleveland: NASA Lewis Research Center; 1976. NASA SP-273. (Updated with Interim Revisions in 1989 and 1993.)

[4] Scavuzzo J. Utilization of AccuTrack for monitoring fuel handling and switching. Proceedings electric power conference; 2009.

[5] Mudry R. Personal communication with Tillman DA. October 3, 2011.

[6] Linfield KW, Mudry R. Pros and cons of CFD and physical flow modeling: a white paper. Livonia, MI: Air Flow Sciences Corporation; 2008.

[7] Tillman DA, Dobrzanski A, Duong D, Dezsi P. Fuel blending with PRB coals for combustion optimization: a tutorial. Coal Technology Association conference. Clearwater, FL; 2006.

[8] Adewumi M. Module engineering applications (I)—compositional modeling of gas-condensate reservoirs: the zero-dimensional approach, *www.e-education.psu.edu/png520/m20_p4.html*; 2008, accessed 3-12-10.

[9] Duong D, Miller B, Tillman D. Characterizing blends of PRB and Central Appalachian coals for fuel optimization purposes. Proceedings 31st international technical conference on coal utilization and fuel systems. Clearwater, FL; 2006, May 21–25.

[10] Duong D, McLenon K. The application of advanced fuel characterization to power plant operations. Proceedings electric power conference. Baltimore; 2008, May 6–8.

[11] Edwards JE. Process modelling selection of thermodynamic models. Thornaby, UK: P&I Design Ltd White Paper; 2001.

[12] Mueller C, Brink A, Hupa M. Numerical simulation of the combustion behavior of different biomasses in a bubbling fluidized bed boiler. Proceedings 18th international conference on fluidized bed combustion. Toronto; 2005, May 22–25.

[13] Santucci M, Scavuzzo J, Hoffman J. Some computer applications for combustion engineering with solid fuels. In: Bruce GM, Tillman DA, editors. Combustion engineering issues for solid fuel systems. Boston: Academic Press; 2008. p. 393–421.

[14] Gretta WJ, Grieco GJ. Consideration of scale in physical flow modeling of air pollution control equipment. International joint Power-Gen conference; 1995.

[15] FERCo. SCR Cold Flow Modeling Plant List, *www.ferco.com/Files/SCR-CFM.pdf*; 2010; accessed 9-20-11.

[16] Tillman DA. Cofiring alternate fuels in coal-fired cyclone boilers. Proceedings American Flame Research Committee international symposium: combustion in industry—status and needs into the 21st century. Baltimore; 1996, September 30–October 2.

[17] Tillman D, Payette K, Banfield T. Cofiring biomass at Allegheny Energy: conclusions of an extended demonstration program. Proceedings 20th Pittsburgh coal conference. Pittsburgh; 2003, September 15–19.

[18] Piffaretti S, Abdon A, Engelbrecht EG, Orth M, Toqan M. Validation of a CFD based modeling approach to predict coal combustion using detailed measurements within a pulverized coal boiler. Proceedings 33rd international technical conference on coal utilization and fuel systems. Clearwater, FL; 2008, June 1–5.

[19] Griendl L, Hohenwarter U, Karl J. CFD aided design of an oxycoal test burner with high recirculation rates. Proceedings 33rd international technical conference on coal utilization and fuel systems. Clearwater, FL; 2008, June 1–5.

[20] Lu P, Ma Z, Ohl C. Leverage CFD modeling in SmartBurn coal combustion improvement. Proceedings 32nd international technical conference on coal utilization and fuel systems. Clearwater, FL; 2007, June 10–15.

[21] Sheng C, Moghtaderi B, Gupta R, Wall TF. A computational fluid dynamics based study of the combustion characteristics of coal blends in pulverized coal-fired furnace. Fuel 2004;83(11–12): 1543–52.

[22] Arenillas A, Backreedy RI, Jones JM, Pis JJ, Pourkashanian M, Rubiera F, et al. Modelling of NO formation in the combustion of coal blends. Fuel 2002;81(5):627–36.

[23] Fry A, Davis K, Wang D. CFD Modeling of single burner oxy-coal combustion retrofit concepts. Proceedings 33rd international technical conference on coal utilization and fuel systems. Clearwater, FL; 2008, June 1–5.

[24] Korytnyi E, Saveliev R, Perelman M, Chudnovsky B, Bar-Ziv E. Computational fluid dynamic simulations of coal-fired utility boilers: an engineering tool. Fuel 2008; 88(1):9–18.

[25] Jian Z, Suo Y, Cheng F, Zhang Y, Gu X. Modeling study of a new circulating fluidized bi-bed boiler combustion system. Journal of Thermal Science 1999;8(3):207–13.

[26] Dahikar SK, Joshi JB, Shah MS, Kalsi AS, RamaPrasad CS, Shukla DS. Experimental and computational fluid dynamic study of reacting gas jet in liquid: flow pattern and heat transfer. Chemical Engineering Science 2010;65(2):827–49.

[27] Chui EH, Douglas MA, Tan Y. Modeling of oxy-fuel combustion for a Western Canadian subbituminous coal. Fuel 2003;82(10):1201–10.

[28] Behjat Y, Shahhosseini S, Hashemabadi SH. CFD Modeling of hydrodynamic and heat transfer in fluidized bed reactors. International Communications in Heat and Mass Transfer 2008;35(3):357–68.

[29] Chilka A, Orsino S. Multi-zonal model for accurate and faster combustion simulations. Proceedings 32nd international technical conference on coal utilization and fuel systems. Clearwater, FL; 2007, June 10–15.

[30] Shadle LJ, Shamsi A, Zhang G-Q, Archer D. Solids mixing in a spouted, fluidized bed, cold flow model. Proceedings 15th international conference on fluidized bed combustion. Savannah; 1999, May 16–19.

[31] Kaneko Y, Shiojima T, Horio M. DEM Simulation of fluidized beds for gas phase olefin polymerization. Chemical Engineering Sciences 1999;54:5809–21.

[32] Bicanic N. Fragmentation and discrete element methods. In: Milne I, Ritchie RO, Karihaloo B, editors. Comprehensive structural integrity, 3rd ed.; 2003. p. 427–57.

Chapter 8

Institutional Issues Associated with Coal Blending

8.1. INTRODUCTION

Institutional issues are both economic and noneconomic in nature, and they significantly influence all blending projects. Blending is a technical, an environmental, and an economic practice. This chapter deals with institutional, economic, and permitting considerations. Institutional concerns include (but are not limited to) sources of supply for various fuels, such as Powder River Basin (PRB) coals, Central Appalachian coals, Illinois Basin coals, various petroleum cokes, biomass, and wastes; fuel prices for all individual fuels; equipment modifications and their costs; transportation systems, including availability and costs; operating requirements, including personnel actions and their costs; and externalities (e.g., taking care of wastes, supporting other local industries such as sawmills).

The blending and firing of solid fuels for both electric utility and industrial applications are occurring in a changing environment. Emissions have been reduced dramatically through both capital investments and operating practices. Among those practices is the blending of bituminous coals or petroleum cokes with low-emission solid fuels such as PRB coals and Adaro coal from Indonesia. Figures 8.1 and 8.2 highlight the decrease in SO_2 and NO_x emissions [1]. Particulate emissions—including PM_{10}—have decreased by more than 75% (mass basis) over the past 50 years. Today, less than 8% of the particulates emitted to the atmosphere result from electricity generation, and less than 6% result from industrial activity. These reductions have come not only from the use of natural gas but, more importantly, from changing practices in firing coal and other solid fuels. Coal and solid fuels still are used to generate about half of the electricity consumed in the United States. The emissions management shown in Figures 8.1 and 8.2 is paralleled by other economies (e.g., Europe).

The U.S. Energy Information Agency (EIA) estimates that over 0.4×10^{15} Btu (0.42 exajoules or EJ) of waste were consumed for energy purposes in the United States in 2006 [2]. This is not an insignificant amount of energy being utilized in various industries, including the power industry. The combination of

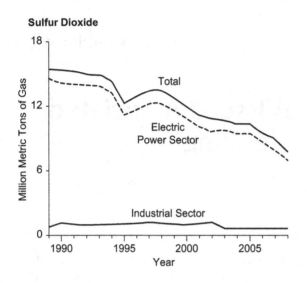

Sulfur Dioxide

FIGURE 8.1 Sulfur dioxide emissions in the United States, 1989–2008. *Source: [1].*

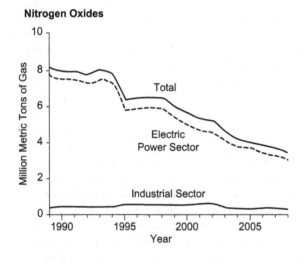

Nitrogen Oxides

FIGURE 8.2 Nitrogen oxide emissions in the United States, 1989–2008. *Source: [1].*

wood and biomass supplies 3.4×10^{15} Btu/yr (3.6 EJ/yr) to the U.S. economy and, with wastes, supplies 3.8×10^{15} Btu/yr (4.0 EJ/yr) to the U.S. economy, as shown in Figure 8.3 [1]. This makes up 50% of the renewable energy, including hydroelectric power [1]. The vast majority of the wood, biomass, and waste is used in industrial applications (e.g., pulp and paper manufacturing using spent pulping liquor as a fuel) rather than in electricity generation in utility boilers.

In addition, the EIA estimates the utilization of petcoke for energy applications at over 1 million tons in 2009, equal to about 30×10^{12} Btu/yr, with about 13% (~4×10^{12} Btu/yr) being used by independent power producers [4].

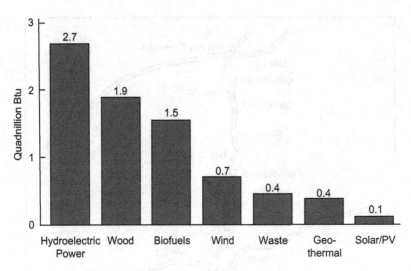

FIGURE 8.3 2009 use of renewable energy in the United States. *Source: [1].*

Petroleum coke is increasingly popular as utilities install scrubbers on power plants. It can be used as a supplemental fuel in pulverized coal (PC) or cyclone boilers, limited in application only by its low volatility; alternatively, it can be the primary fuel—in some cases the only fuel—in circulating fluidized bed boilers. Petroleum coke is being produced in increasing quantities as more heavy crude oils are being refined and as the product slate from oil refineries is changed [5].

Powder River Basin coals from both Wyoming and Montana were originally viewed as opportunity fuels. The 1970s saw the significant opening of these fields and the tremendous deposits that exist there (Figure 8.4). Original concerns were the amount of land to be disturbed by mining, the needs for water associated with mine land reclamation, and the slagging and fouling properties of these coals [6]. They have since become mainstream fuels—to the point that they support generation of approximately 20% of all electricity used in the United States. Most of this coal is produced at mines along a 117-mile stretch of railroad known as the Joint Line in Wyoming, which stretches from Donkey Creek in the north to Shawney Junction in the south [7]. This stretch of railroad is now largely triple tracked, yet it is operating at >95% capacity (Figure 8.5). PRB coals are, however, blend fuels for the most part. All coals are, in fact, blend fuels.

8.2. INSTITUTIONAL ISSUES ASSOCIATED WITH FUEL BLENDING

Institutional and economic issues associated with fuel blending have to be considered in the broadest of terms: fuel supply, transportation, utilization, and

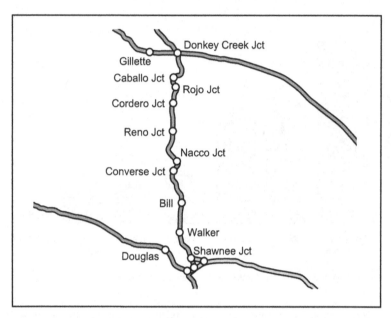

FIGURE 8.4   The Joint Line and the Southern Powder River Basin, which contains the largest mines in the basin. For the BNSF, the Joint Line is the Orin Subdivision. *(Map by Pentrex, Inc. Used with permission.)*

FIGURE 8.5   Coal hauling on the triple-tracked Joint Line. The line is always used at or very near full capacity. *Source: [7]. (Photograph by Pentrex, Inc. Used with permission.)*

waste management. They have to be considered in terms of externalities— impacts of the fuels and fuel blending on the community at large. Externalities can be beneficial (e.g., job creation, infrastructure improvement) or detrimental (e.g., waste disposal considerations). In all such considerations we have to make a distinction between the price of a fuel or blend and the cost of that fuel or blend.

Waste fuels, biomass, and petroleum coke are used when they are the low-cost option for industrial or power generation purposes. These fuels are used when they are lower in cost than the base or design fuels. Such alternate fuels are making their way into the industrial and utility markets as blending fuels, cofired with a manufacturing or utility plant's base fuel.

Dai [4] provides an overview of many examples of blending different fuels with coal in a variety of boilers. He notes that blends require careful planning regarding issues such as preparation, handling, storage, and so forth. Other work in Turkey [8] also demonstrates that blending of lignite coals with alternate fuels requires a good understanding of the physical as well as the chemical properties of both the primary coal and the secondary fuel to ensure a successful program.

Blending with PRB coal, the most common single type of solid fuel blending, is a complex decision with many components. A simple decision to enter into, or increase, the blending of PRB and bituminous coal includes a detailed evaluation concerning regulatory compliance; attention to coal acquisition and delivery; capital modifications to the coal yard, the mills, the burners, the boiler, and the postcombustion controls; and operating and maintenance changes ranging from housekeeping to boiler operations and firing practices.

This chapter focuses on the use of blending for economic and technical advantage and the economic and institutional concerns associated with the practice of fuel blending. Blending also carries with it nontechnical, noneconomic, institutional issues, including concerns such as regulatory and permitting processes, changes to air and water permits, and many more.

Johnson [9] cites numerous changes in the permitting environment, particularly for older coal-fired power plants. The compromise between the late senator Edwin Muskie of Maine and the late senator Jennings Randolph of West Virginia is no longer operative. This compromise grandfathered older coal-fired power plants, protecting them from the most stringent requirements of the Clean Air Act (CAA) of 1970 and its successors. Resulting from the CAA was the so-called "Wepco decision," where the regulatory environment could significantly impact fuel blending—coal-coal blending and selected coal–opportunity fuel blending. The recently promulgated Cross State Air Pollution Regulations (CSAPR) affect both new and older boilers.

Institutional issues also include dealing with the community and the work force on issues ranging from job security, plant performance, community relations, and governmental relations. In the process of blending, if it means changing fuels, institutional issues may include dealing with interest groups seeking either to close the plant down or to impose other permit conditions. This chapter deals with the myriad of institutional concerns surrounding fuel blending—and changing fuels to accomplish blending. Coal–coal blends do not always cause the same level of institutional impact that can be experienced by coal–petroleum coke blends, coal–biomass (cofiring) blends, or coal–waste blends. Yet many issues remain, even for the coal–coal blends. This chapter attempts to sort through the issues and provide examples of how they are dealt with.

8.3. ECONOMIC CONSIDERATIONS ASSOCIATED WITH BLENDING

Economics is always the underlying force behind any decision to blend coals or blend alternate fuels with coals. However, it is not merely the cost of the second or third fuel that must be addressed. With each new fuel, there are many hidden, but real, costs that must be evaluated before making a final decision to blend fuels. This section provides some guidelines as to what additional institutional economic issues must be reviewed and the associated costs included when the overall economic impact of fuel blending is considered. The economic issues, then, include the fuels themselves, production of the fuels, transportation of the fuels, and—finally—the use of these solid sources of energy.

8.3.1. Fuels Availability

Fuel markets fluctuate considerably. A fuel that has been determined to be a viable, economic blending fuel one year may not be as economic in future years as the demand for fuels increases and decreases (Figure 8.6).

Whether the blending fuel is another coal, biomass, petcoke, or waste, an economic evaluation must be completed to assess the viability of the blended fuel. As an example, an eastern U.S. utility was contemplating blending some biomass with its current coal to reduce its annual fuel costs. The first study completed was to determine the availability of biomass sources within a 50-mile (80-km) radius of the plant [10]. This distance is generally accepted as the farthest distance a biomass fuel can economically be transported (depending on initial cost) to a plant for blending.

The study located all sawmills, furniture manufacturers, pallet manufacturers, and so on, in the region and sent a questionnaire regarding their willingness to provide biomass and what quantities were available. Nearly 450,000 tons of oven-dried biomass were found to be located within the 50-mile radius of the plant. The distribution of these biomass resources is shown in Figure 8.7.

The figure clearly shows that essentially all biomass is available within 30 miles of the plant and nearly half of all biomass resources and 70% of the sawdust are available within 20 miles of the plant. This provides an indication that transportation costs may be minimized. While the location and distance

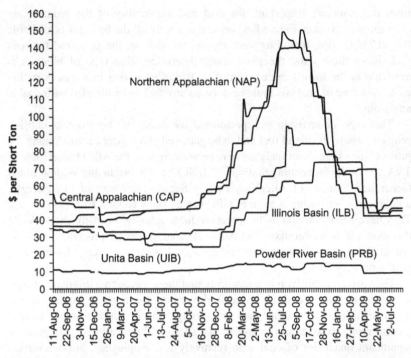

FIGURE 8.6   Spot market coal prices from 2006–2009. Note the dramatic swings in price for Northern and Central Appalachian coals and the very steady price for PRB coals. *Source: [1]*.

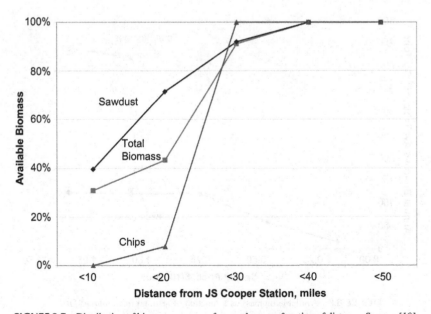

FIGURE 8.7   Distribution of biomass resources from a plant as a function of distance. *Source: [10]*.

from the plant are important, the cost and availability of the biomass are even more important. Figure 8.8 shows that nearly all the biomass is available for $0.75/$10^6$ Btu, including sawdust and woodchips, the preferred biomass fuel. From these data, the plant could determine what type of biomass is available at the lowest price per million Btu after adding in a transportation cost. This type of analysis must be done on any fuel to be blended with coal at any plant.

This type of analysis was performed for many of the biomass cofiring projects, where the blend fuel had to be gathered and concentrated. Among the projects for which these analyses were performed were the Allen Fossil Plant of TVA, the Bailly Generating Station of NIPSCO, the Albright and Willow Island Generating Stations of Allegheny Energy Supply (now part of FirstEnergy), and Seward Generating Station of GPU Genco (now Reliant Energy) [11–13]. They have been common in the forest products industry for many years. While this example is for biomass fuel being blended, a similar procedure must be completed for any waste fuel or any other coal being considered for blending with a primary fuel.

For coal availability, it is noteworthy that there has been a dramatic change in mining technique and mine size. Today mines are larger and more mechanized regardless of whether they are underground or surface mines. Underground mines now feature long-wall and short-wall technologies and produce significant quantities of coal with relatively few employees. Surface mines, particularly those in the Powder River Basin and in the lignite fields, produce

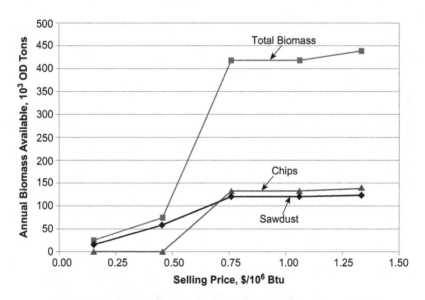

FIGURE 8.8   Available biomass as a function of selling price. *Source: [10].*

massive quantities of coal from a single production unit. As of 2008, the North Antelope Rochelle mine was producing 97.6 million tons of coal annually, and the Black Thunder mine was producing 88.6 million tons/yr [14]. These are massive operations using very large, capital-intensive equipment.

Table 8.1 presents the 20 most productive U.S. mines as of 2008. Note that the 10 largest mines all are in Wyoming and all in the Powder River Basin producing subbituminous coal. The very largest underground mines, producing

TABLE 8.1 The 20 Most Productive Mines in the United States

Rank	Mine	State	Surface/ Underground	Production (10^6 tons)
1	North Antelope Rochelle	WY	S	97.6
2	Black Thunder	WY	S	88.6
3	Jacobs Ranch	WY	S	42.1
4	Cordero Rojo	WY	S	40.0
5	Antelope	WY	S	35.8
6	Caballo	WY	S	31.2
7	Belle Ayr	WY	S	28.7
8	Buckskin	WY	S	26.1
9	Eagle Butte	WY	S	20.4
10	Rawhide	WY	S	18.4
11	Spring Creek	MT	S	17.9
12	Freedom	ND	S	14.6
13	Rosebud	MT	S	13.1
14	Coal Creek	WY	S	11.5
15	Enlow Fork	PA	U	11.1
16	Bailey	PA	U	10.0
17	McElroy	WV	U	9.6
18	Navajo	NM	S	8.9
19	Kayenta	AZ	S	8.0
20	Foidel Creek	CO	U	8.0

Source: [14].

northern Appalachian bituminous coal, are in the 10 to 11 million–ton range, not the 88 to 98 million–ton range.

8.3.2. Fuel Procurement

Procuring blending fuels on a long-term basis requires considerable attention. The three main focal areas that must be addressed to ensure a successful and cost-effective blending application are contracts, the use of third-party suppliers, and payment schedules.

Since the blending fuel normally is purchased for use at one specific power station or industrial facility (e.g., cement kiln, pulp mill power boiler), an initial question to resolve is who is in charge of procuring the blending fuel—the purchasing department or the plant itself. Many criteria go into this decision, such as quantity to be purchased, manpower availability, centralized purchasing of all fuels resulting in lower costs, quality control of the product, delivery schedules and transportation systems, and locations at the plant.

While contracts have many similarities, they have some significant differences as well. If an alternate coal or other fuel is to be blended and/or cofired on a continual basis, there must be a supply available year-round. For example, if sawdust, furniture waste, and other woody biomasses are to be used, these will be more readily available during the winter months. If herbaceous crops are to be used, alternate sources will be needed if the crop is not being grown as an energy crop. Further, attention must be given to the harvest cycle and the ability to manage the crop or crop waste being used as fuel. Corn stover, which is harvested only once a year, is an example of this issue. Storage becomes a key part of the procurement process.

To reduce the number of problems at the power plant or industrial facility, a consistent product material is necessary. This includes moisture content, ash content, heating value, and size. While most fuel suppliers take care to provide a product according to the utility contract specifications, over time some fuel properties may tend to exceed contract limits. In particular, particle size and ash (debris) content tend to be larger than the specification. If the blend fuel is a coal, then attention must be given to the calorific value and its variations as a function of coal moisture. Coal produced and/or shipped in the winter tends to contain more moisture than coal produced in the summer. It may be necessary to include an on-site screen and/or magnetic separator to act as the final guarantee. The reject material must be dealt with, which may include having the supplier take it back, reducing the cost per ton of alternate fuel, or landfilling. This additional handling must be included in the utilities cost of fuel.

One approach to this issue is the installation of on-line coal analysis equipment (see Chapter 3 for a more detailed discussion). The on-line analyzer, properly calibrated and maintained, provides a means for analyzing the extent to which the coal procured meets specification. Many coal mines use this technology to ensure that they are shipping coal within specification.

The utility or industrial user should have the right to sample all fuels at his or her discretion and confirm the seller's analyses. Fuels not in compliance with the specifications should be rejected by the utility or industry or be subject to a reduction in pricing agreeable to both parties.

Many utilities and process industries do not have the personnel or the desire to deal with several alternate fuel suppliers. For this reason, some find it preferable to deal with a single entity rather than multiple suppliers. Suppliers could band together in a cooperative and work through the cooperative to contract with a utility. Also, as an alternative to a cooperative, a private fuel supplier or broker could contract with a utility to supply the fuel and in turn contract with suppliers to provide the necessary fuel. Larger utilities, however, must obtain coal in sufficient quantity that single sourcing is not an option.

At times, it may be necessary to work directly with those who are not familiar with contract language, content, terms, and so on, from large corporations such as utilities. Since these contracts and their associated language are important to both parties, it will be in the best interest of the utility to work with the individual suppliers to be certain that all contract points are understood and agreeable. This may also be the case when working with brokers or cooperatives: The language and the intent of the fuel supply contract must be understood by all parties for a successful program.

To determine the price the utility will pay the supplier for a secondary alternate fuel, three components are primarily considered: avoided cost for the main fuel, economic value cost sharing, and agreed-on percentage of economic value cost sharing going to the supplier. The avoided cost is the value of a given amount of coal, on a 10^6-Btu basis, replaced by the alternate fuel. The avoided cost of a mix of fuels including coal, petroleum coke, and natural gas may be based on a four-year rolling average. If the minimum value of the avoided cost base price drops below an agreed-on level, the utility might consider guaranteeing a minimum price. The base price could be adjusted based on the quality of the alternate fuel. For example, the price could be adjusted upward if the weighted average heating value of the fuel exceeds the contract-guaranteed amount. On the other hand, the base price could be adjusted downward if the ash content of the secondary fuel is above the guaranteed level (refer to Figure 8.8).

Ultimately, the success of a blending or cofiring program is in the generation of power for the utility or industry. As a result, the final confirmation of all fuel specifications should be done at a reputable location (i.e., blending terminal) or at the utility site. The consequences of out-of-specification material could be disastrous.

8.3.3. Fuel Transportation

A key component of the economics of fuel blending is the cost of fuel transportation—either the base coal or the blend fuel. A total of 62% to 68% of all coal travels by rail [15, 16]; the rest is distributed among Great Lakes boats, barges, and trucks. Biomass fuels travel almost exclusively by truck due to the short transportation distances involved (e.g., <50 miles). Further, coal unloading facilities at power plants and industrial facilities must be reserved for the main fuel—typically coal. Plugged chutes in the coal unloading system—particularly if caused by biomass fuel—would cause significant concern at any utility or industrial power plant. (See Figure 8.9.)

With the passage of the Staggers Act in 1980, which effectively deregulated railroads and permitted them to be competitive with the trucking industry, the change in the railroad system over the past 30 years has been felt significantly by coal mines, coal-based electric utilities, and coal-using process industries. Starting in 1972, the Burlington Northern Railroad built a single track into the PRB—now known as the Orin Sub (refer to Figure 8.4). The Chicago and Northwestern Railroad became a participant in the joint line. The action to build into the basin was a significant act of faith; when started, PRB coals were not viewed favorably by the utility community.

With the implementation of the Clean Air Act of 1970, the low-sulfur properties of PRB coals made them a desired commodity for utilities in the Midwest. Demand for PRB coal accelerated to the current level of approximately 400×10^6 tons/yr. The joint line changed dramatically as well. The

FIGURE 8.9 Coal-handling equipment, including a coal load out, at the Black Thunder Mine. This mine produces 88.6 million tons of coal annually. *Source: [7]. (Photograph by Pentrex, Inc. Used with permission.)*

Burlington Northern Railroad merged with the Santa Fe Railroad to become the BNSF, one of the four major Class I railroads in the United States. Ultimately it was purchased by Warren Buffet. The Chicago and Northwestern was acquired by Union Pacific (UP) and with it the rights to the Joint Line. The two railroads—BNSF and UP—have invested hundreds of millions if not billions of dollars in the Joint Line, double-tracking and triple-tracking the entire length of the line to accommodate the demand for PRB coal (Figures 8.10 and 8.11).

At the same time, there have been significant changes in Eastern coal railroads. CSX and Norfolk Southern (NS) effectively divided up the assets of Conrail, a Class I railroad initially created by the federal government from numerous bankrupt railroads (Penn Central, Erie Lackawanna, Reading, Lehigh Valley, and numerous others). Conrail consolidated assets, abandoned many less than productive lines, and became a very efficient carrier. Conrail was created in 1976, and it effectively split up from 1997 to 1999. Before that time, it was the Class I railroad supplying rail services to the Northeast.

Transportation became a key element in the institutional structure for blending coals and blending coal with other fuels such as petroleum coke. By transporting 68% of all coal moved in the United States, the railroad became the backbone of the fuel blending business. Today, the structure of railroading

FIGURE 8.10 Massive equipment used to produce coal at the Black Thunder mine. The truck shown will carry upwards of 300 tons of coal—enough to fill about three railroad cars. *Source: [7]. (Photograph by Pentrex, Inc. Used with permission.)*

FIGURE 8.11   Loaded coal cars leaving Black Thunder mine. Several strings of coal cars are waiting to be loaded. *Source: [7]. (Photograph by Pentrex, Inc. Used with permission.)*

is such that most coal cars are owned by utilities. (Railroads own some, but relatively few, coal cars.) Coal car design has changed as the transportation system has changed. Most modern coal cars are designed to be unloaded in rotary car dumpers; these cars are never uncoupled but are loaded and unloaded in a string. The cars are made of aluminum, so most of the weight is payload [15].

Even though there are rapid discharge bottom dump cars, these are less common than the ubiquitous "Johnstown Coalporter" and competitive designs for rotary unloaded cars. The utilities own the vast majority of cars, and they are tied together in strings of about 120 cars per train. These unit trains of coal cars, which exist in both western and eastern Class I railroads, serve as conveyors from mine to power plant or cement kiln. These technology and institutional changes have been designed to manage delivered coal costs.

At the same time, deregulation under the Staggers Act has caused dramatic increases in the cost of transporting coal by rail, particularly as the railroads have had to invest significant hundreds of millions of dollars annually in track improvements and other infrastructural actions to improve the efficiency of coal transportation. PRB coals carried east of Chicago can have transportation costs equal to 67% to 80% of the total cost of the delivered fuel.

Transportation systems may serve blending coal terminals or utilities that have the capability of blending coals or blending coals with petroleum coke and related fuels. Transportation systems for biomass fuels are fundamentally different from those for coal. Because biomass fuels have densities of 5 to 20 lb/ft^3 (before densification; after densification or pelleting, the densities may be as high as 35–40 lb/ft^3, compared to coal at 55–70 lb/ft^3), the transportation distance is very short. Further, although rail transport is possible, truck transport is most preferred.

8.4. PROCESS MODIFICATIONS

Depending on the coal blends and/or coal–alternate fuel blends, some significant process modifications may be required to efficiently and safely utilize the new blended fuel. This section discusses many of these areas, from the point of view of providing the reader an idea of those areas that must be evaluated before final economic decisions can be made regarding the blended fuel.

Gunderson, Selle, and Harding [17, 18] completed a study for DOE and EPRI in which 12 different units from 9 utilities were surveyed to document the modifications either envisioned or completed as these units began blending or switching from bituminous coals to subbituminous coals. The study incorporated tangentially fired units, wall-fired units, and cyclone-fired units ranging in size from 90 to 1100 MW. All aspects of fuel utilization were evaluated, including coal handling and storage, coal blending, pulverizer performance, boiler, convective pass, and emissions. Collett and Ready [19] provide an excellent overview of many of the issues that must be investigated when evaluating coal–coal blends. Their focus was on many of the handling, storage, and blending issues related to coal blending. Summaries of these issues are presented in the following subsections.

8.4.1. Coal Handling and Storage

In the DOE/EPRI study [17, 18], fuels were transported to each site by rail, barge, or some combination of the two. In general, there were no shipping problems with the subbituminous coals, and in nearly all cases the coal reached its destination without incident. One issue noted was that in shipping the subbituminous coals by barge, care had to be taken to level each barge to prevent spontaneous combustion due to the higher reactivity of the subbituminous coals. Also, the coals were stockpiled and compacted to prevent hot spots.

Additional care and housekeeping were necessary when using the subbituminous coals throughout the plant. They included the stacking and the reclaiming of the coals, as well as transporting the coal to the bunker house. Other improvements completed by some utilities included the addition of or improvement of dust suppression and/or collection systems. In some instances manpower was increased by one person to be able to improve the reliability of the entire handling system.

Other studies by Ethen [20], Weiss [21], Fuji [22], and Eimer [23] noted similar results as Western coals and Eastern coals were blended for use in bituminous coal–designed boilers. Specific examples, such as changing belt angles and hopper angles to prevent spillage and pluggage, were discussed. Case histories from utility installations are provided.

8.4.2. Coal Blending

Proper and accurate blending is without a doubt the most critical concern in the process. It is also the most difficult to accomplish. This has been discussed extensively throughout this book but, to summarize, the better and the more homogeneous the blend, the better the performance in the boiler. To summarize some of the coal blending issues encountered with Eastern/Western coal blends, the most common method of blending coals uses existing reclaim hoppers and feeders to blend the fuels on the conveyor entering the crusher house. This is accomplished by varying belt speeds to achieve the desired blend ratio. This approach may be enhanced by using belt scales, variable-speed drives on the coal belts, and on-line coal analyzers to check the blend.

This approach has been taken by a number of utilities and power plants, as previously described. However, the quality and consistency of the blended product are directly related to the sophistication of the feeder and control systems available. The need for a reliable blended product quality increases as the level of subbituminous coal in the blend rises and the operational margin on critical plant components narrows. It has been the experience of the utility industry that the tighter the blend—the better control of it—the closer to optimum conditions the plant can be run. This is sometimes referred to as running "closer to the edge." There is money to be made with that practice.

8.4.3. Pulverizer Performance

Several types of pulverizers are available, including ball and race mills, roller mills using tires in a track, bowl mills operated at negative pressure, and atritta or deep-set hammermills. With the exception of the atritta mills, pulverizers are designed such that the coal grinds against itself. Experience has shown that coal–biomass blends can be pulverized in atritta mills, but ball mills, bowl mills, and the large MPS (vertical spindle, tire-based) mills are not successful at pulverizing coal–biomass blends [12].

For coal–coal blends, the pulverizer is often the first area where a load limit is encountered. This is particularly true when blending Eastern and Western coals. This is due to two characteristics of Western coal: higher moisture and lower heating value. Because of the higher moisture content, the primary air inlet temperature must be increased in order to dry the coal for proper transport and flame stability. The coal throughput to the mill must also be increased in order to maintain the same megawatt output from the unit. These two factors cause the mills to often be the source of an overall unit derate.

The findings from the DOE/EPRI study [17, 18] indicated a wide variation in the degree and complexity of modifications in pulverizer operations and safety measures. Most of the units lowered the mill outlet temperature from about 150°F to about 135°F. This decrease in mill outlet temperature greatly reduced the incidences of mill puffs and pyrite trap fires. Additional care must be taken when starting up and shutting down mills due to the higher reactivity of the Western coal in the blend. Some utilities have installed inerting systems as an added precaution.

In a study by Woodside [24] at the 157-MW station of Ohio Edison (now part of FirstEnergy) he noted that heavier than normal pulverizer loading experienced while burning the subbituminous coal blends resulted in increased attention by plant personnel. Maintaining pulverizer component clearances to within the design specifications was essential for continual operation with the blended coals.

One of the positive contributions attributable to the addition of more reactive subbituminous coal to the blend is in the area of flame stability and turndown. Blended fuels have no detrimental effects on flame stability or turndown. This enhanced reactivity also provides some explanation as to why a reduction in mill outlet temperature did not appear to be detrimental in terms of flame stability or carbon burn-out.

8.4.4. Furnace Effects

In many instances, the utilization of bituminous/subbituminous coal blends results in a change in the heat distribution in the boiler. The higher moisture content of the blends (relative to bituminous coals) tends to lower the average flame temperature and increase the heat capacity of the flue gas. In addition, the reflective nature of subbituminous ash results in more energy leaving the furnace and entering the convection section. This can result in higher furnace gas exit temperatures that may enhance deposition.

Brobjorg [25] noted a similar experience in the 580-MW cyclone unit at the Northern States Power Company, where the higher furnace gas exit temperature reached while burning a bituminous/subbituminous coal blend rendered the long-term use of the blend as unacceptable due to the increase in convection section fouling.

Levasseur [26] notes that under low-NO_x firing conditions, coal blends of bituminous/ subbituminous coal have a lower probability for adverse changes in lower furnace ash deposits. Utilizing petroleum coke–coal blends with the high vanadium content in petroleum cokes, Bryers [27] states that petroleum cokes containing as much as 10,000 ppm V_2O_5 have been fired free of corrosion and with minimal fouling. There appears to be sufficient absorbing minerals in the coal blend ash to inhibit fouling, slagging, or corrosion when petroleum cokes are blended with coal in an environmentally sound manner.

The more reflective ash layer formed on the furnace walls as a result of blended coals may require additional wallblowers and more frequent wall-blowing. Maintenance of wallblowers is also very important. However, in most cases the deposition on the furnace walls as a result of firing coal blends can be controlled.

8.4.5. Convective Pass

The use of coal blends did not present unmanageable deposition problems in the convection section unless sootblowers were unavailable or if sootblower coverage was inadequate. As mentioned in the previous section, the additional heat entering the convection section may result in additional primary and secondary superheat attemperation. It should also be noted that the furnace backend temperatures generally rise as blends of Eastern and Western coals are fired.

8.4.6. Emissions

With the use of Western subbituminous coals as blending fuels with most Eastern bituminous coals, the fuel sulfur level drops, which may have an impact on particulate emissions if the unit utilizes an electrostatic precipitator (ESP). The higher moisture combined with lower sulfur may alter the flyash resistivity, making it more difficult to collect. However, most units that have precipitators with a specific collection area (SCA) of at least 300 to 350 $ft^2/$ 1000-ACFM perform acceptably under most blend conditions. Low SCA values, as well as the condition and performance of energization, rapping, and control equipment, are of special concern when lower-sulfur coal blends are fired. Other upgrade technologies to counter lower-sulfur coal blends for ESPs include the following [28]:

- Gas flow optimization
- Optimizing existing surface
- SO_3 flue gas conditioning
- Improved rapping technology
- Pulsed energization
- Humidification/gas cooling
- Ionization chemistry
- Precipitator enlargement or replacement
- Switching to fabric filter technology

Any or all of these electrostatic precipitator upgrade technologies should be evaluated and the economics included to arrive at an actual cost of blending bituminous and subbituminous coals. Certain precipitators may lend them-selves to only a select few opportunities for improved performance.

The management of emissions addresses the permitting consideration; even though permitting is usually thought of in terms of airborne emissions, the list of permits required for an operating utility or industrial plant is legion. Permits and permitting activities address airborne emissions, dust management on site, water and wastewater management, worker safety and qualifications (e.g., boiler operator licenses), and numerous other activities.

The most recent regulations requiring attention are the updated Maximum Available Control Technology (MACT) regulations and the Cross State Air Pollution Regulations (CSAPR) that impact power plants in 23 states. These regulations will impact the availability of blend coals, the use of blend coals, and even the availability of coal-fired power plants. Ameren, the major utility based in St. Louis, Missouri, has signed an agreement with Peabody Energy to use increased quantities of PRB coal at several generating stations in that state [29]. Blending to reduce sulfur and NO_x emissions can be taken to the next level using Indonesian coals, Alaskan coals, and other Ring of Fire coals with 0.1% to 0.2% sulfur and low NO_x potential.

At the same time, electric utilities have announced the retiring of more than 23,000 MW_e of coal-fired generating capacity between 2011 and 2020 [30]. American Electric Power alone announced plans to retire approximately 6,000 MW_e of generating capacity at stations such as Tanners Creek, Muskingum River, and Connesville [30]. Duke Energy announced retirement plans for more than 2,000 MW_e, while Progress Energy and the Tennessee Valley Authority each announced plans to retire more than 1,500 MW_e of capacity [30]. These retirements will impact specific coal supplies, creating blending opportunities for the plants that remain in operation. Most of the plants being retired are located in the South, the Midwest, and the Mid-Atlantic.

For the larger utilities and larger plants, particularly those that can accept unit trains of Western coal, blending opportunities are strong. Further, it is our experience that unless a boiler is designed specifically for Western coal, blending is essential. Smaller plants, including those served by trucks rather than rail or barge, will have a harder time using coal blending to manage emissions. Table 8.2 summarizes the issues encountered by many utilities arising from firing blends of bituminous and subbituminous coals.

8.5. FUTURE U.S. AND WORLD COAL PRODUCTION

Since coal is the backbone of solid fuel blending, it is important to consider future production both in the United States and all over the world. World production is particularly significant because China has already eclipsed the United States in coal production despite having fewer total coal reserves. Global coal production has been studied extensively by Hook et al. [31]. Using a logistics model based on the methodology developed by M. King Hubbert, China has analyzed coal resources and reserves, production, and future coal production. Their analysis focuses on coal resources and reserves in China, the United States, Russia, the Ukraine, Kazakhstan, India, Australia, South Africa, and other exporting nations, including Colombia, Indonesia, Poland, Canada, and Vietnam.

TABLE 8.2 Summary of Utility Issues Related to Firing Blends of Bituminous/Subbituminous Coals

Modifications as a Result of Utilizaton of Blends	Unit A	Unit B	Unit C	Unit D	Unit E	Unit F	Unit G	Unit H	Unit I	Unit J	Unit K	Unit L
Derate Accepted in Advance			14%		16%						15%	15%
Coal Handling												
Upgraded Fuel Handling System		Yes		Yes		Yes						
Considering On-Site Blending								Yes			Yes	
Added Feeder Capacity in Coal Handling					Yes							
Added Personnel in Coal Handling		Yes			Yes							
Implemented Water Washdown in Coal Handling					Yes							
Increased Inspections and Improved Housekeeping in Coal Handling			Yes					Yes				Yes
Installed Dust Suppression System		Yes	Yes				Yes		Yes			
Installed Dust Collection System						Yes					Yes	

Installed Fire Detection-Protection System					Yes	Yes	Yes
Modified Bunker Design					Yes		
Installed Monitors for Combustibles in Bunkers				Yes			
Installed Bunker Inerting System							CO_2
Rotated Bunkers to Minimize Retention Time							Yes
Pulverizer							
More Frequent Coal Fineness Testing				Yes		Yes	
Improved Pulverizer Maintenance	Yes				Yes		
Added Pulverizer Inerting System		Steam		Yes	Yes		CO_2
Considering Installation of Inerting System	Yes			Yes			
Revised Pulverizer Start up and Shutdown	Yes						Yes
May Alter Pulverizer Start up and Shutdown							Yes

(Continued)

TABLE 8.2 Summary of Utility Issues Related to Firing Blends of Bituminous/Subbituminous Coals—cont'd

Modifications as a Result of Utilizaton of Blends	Unit A	Unit B	Unit C	Unit D	Unit E	Unit F	Unit G	Unit H	Unit I	Unit J	Unit K	Unit L
Derate Accepted in Advance			14%		16%						15%	15%
Added Induct Air Preheaters			Yes									
Increased Coal Feeder Capacity			Yes									Yes
Lower Pulverizer Exit Temperature Control Point		Yes	Yes				Yes	Yes	Yes		Yes	
Added Pulverizer Air Flow and Temperature Instrumentation								Yes			Yes	
May Install Pulverizer Temperature Monitors		Yes										
Improved Coal-Air System Maintenance				Yes								
Furnace												
Adjusted Air Distribution System				Yes						Yes		Yes
May Raise FD, ID, and Exhuster Fan Capacities							Yes					
Installed High-Temperature Oxygen Probes				Yes						Yes		

Action					
Revised Oxygen-Based Operation Guidelines	Yes			Yes	
Recalibrated Oxygen Measurements versus Furnace Levels to Compensate for Air Leakage		Yes			
Monitor Slag Tap Coninuously					Yes
Improved Wall and Retractable Blower Maintenance	Yes	Yes	Yes		
Increased Sootblower Air Supply Capacity		Yes			
Considering Raising Steam Pressure on Furnace Wall Sootblowers					Yes
Increased Wall Sootblowing Frequency	Yes	Yes		Yes	
Added Furnace Wall Sootblowers					10
Considering Adding Sootblowers			Yes	Yes	
Added Water-Medium Furnace Wall Sootblowers	Yes	Yes			

(Continued)

317

TABLE 8.2 Summary of Utility Issues Related to Firing Blends of Bituminous/Subbituminous Coals—cont'd

Modifications as a Result of Utilizaton of Blends	Unit A	Unit B	Unit C	Unit D	Unit E	Unit F	Unit G	Unit H	Unit I	Unit J	Unit K	Unit L
Derate Accepted in Advance			14%		16%						15%	15%
May Add Water-Medium Furance Wall Sootblowers							Yes		Yes			
Raised Primary Furnace Exit											Yes	
Convective Pass												
Increased Retractable Sootblower Frequency	Yes	Yes	Yes	Yes, Slight		Yes	Yes, Slight	Yes, Slight	Yes, Slight	Yes	Yes	
Revised Retractable Sootblower Sequencing				Yes						Yes		
Added Retractable Sootblowers											4	
Inspect Furnace and Convective Pass More Often	Yes											
Removed PSH Surface											Yes	
Redesigned Economizer							Yes					
Closer Monitoring of Back-End Temperature									Yes			

	Yes
Considering Larger Reheat Attemperation Nozzles	Yes
May Replace Finned-Tube Economizer	Yes
Particulate Collection	
Increased ESP Capacity	Yes
Upgraded ESP	Yes
Will Install Larger ESP	Yes
Installed ESP Inlet Gas Distributor	Yes
Modified ESP Rapper Sequencing	Yes
Increased SO_3 Conditioning Levels	Yes
Added SO_3 Conditioning System	Yes
Improved LOI Sampling Methods	Yes
Considering Ash-Handling System Alterations	Yes

Hook et al. [31] concluded that the Hubbert approach is appropriate for coal and that the world is at or near peak production. Driven largely by coal production in China, Hook et al. predict that worldwide coal production will peak somewhere around 2020 to 2030. They have shown that U.S. coal production will remain essentially constant throughout the twenty-first century. Patzek and Croft [32] have shown similar results.

Hook and Alklett [33] have applied the Hubbert methodology to U.S. coal production, with extensive attention to historical data. They show that coal production may peak around 2030 at about 1.4×10^9 tons/yr and remain relatively steady through the remainder of the century. To achieve this, however, they cite the need for massive increases in production from Montana. While they cite economic disadvantages, including a railroad monopoly and distance from markets, they ignore the technical disadvantage of Montana subbituminous coal—relatively high-sodium concentrations in the inorganic fraction. However, they show a peak production consistent with world coal production.

8.6. CONCLUSIONS

As discussed in the preceding sections, there are many institutional issues related to coal–coal blends, especially when one of the coals in the blend is of two different ranks (see Table 8.2). As with all fuel changes, test burns should be undertaken prior to entering into any long-term contracts with new or different fuel suppliers.

In addition to the technical issues related to coal–coal blends, there are some externalities that may have an important bearing on the ability to successfully utilize coal blends. A principal concern is supporting local industry and the economy. If an alternate fuel is a potential candidate for blending with the primary coal, but perhaps not the first choice, it may be advantageous to utilize this fuel to keep local businesses functioning, since they will undoubtedly be a consumer of the power. In addition, employment can remain in the area, creating goodwill between the utility and the local town. All costs, including potentially lost revenues, must be considered when blending coals and alternate fuels with a primary fuel source.

REFERENCES

[1] USEIA. Annual energy review 2009. Washington, DC: US Department of Energy; 2011.
[2] USEIA. Renewable energy trends in consumption and electricity. Washington, DC: US Department of Energy; 2008.
[3] USEIA. Petroleum coke: consumption for useful thermal output by sector, 1997 through May 2011. Washington, DC: US Department of Energy; 2011.
[4] Dai J, Sokhansanj S, Grace JR, Bi X, Lim CJ. Overview and some issues related to co-firing biomass and coal. Canadian Journal of Chemical Engineering 2008;86:367–86.
[5] Tillman DA, Harding NS. Fuels of opportunity. Amsterdam: Elsevier; 2004.

[6] Box TW. Rehabilitation potential of Western coal lands. Cambridge, MA: National Academy of Sciences/Ballinger Publishing Company; 1974.

[7] Pentrex. Powder River combo DVD: King Coal–Powder River Rails & Mines; and Powder River Basin coal trains. Pasadena: Pentrex, 2003.

[8] Surmen Y, Demirbas A. Cofiring of biomass and lignite blends: resource facilities: technological and environmental issues. Energy Sources 2003;25:175–87.

[9] Johnson J. Changes ahead for old power plants. Chemical and Engineering News 2011; 89(38):22–3, Sep 19.

[10] Harding NS, Van De Graaff C, O'Connor DC. Southeastern US biomass resource assessment. Palo Alto, CA: EPRI Technical Update; 2004.

[11] Tillman DA. Opportunity fuel cofiring at Allegheny Energy. Palo Alto, CA: Final Report, EPRI; 2004. Report #1004811.

[12] Tillman DA. Final Report: EPRI-USDOE Cooperative Agreement, vol I: cofiring. Palo Alto, CA: Electric Power Research Institute; 2001.

[13] Tillman DA. Biomass cofiring: field test results. Palo Alto, CA: Electric Power Research Institute, Report #TR-113903; 1999.

[14] Anon. Map of the month. Trains 2010;70(4):38.

[15] Ekmann J, Le PH. Coal storage and transportation. In: Encyclopedia of energy, vol I. Boston: Elsevier; 2003.

[16] Murray T. Where's that coal train going? Trains 2010;70(4):28–37.

[17] Gunderson JR, Selle SJ, Harding NS. Technology assessment for blending Western and Eastern coals for SO_2 compliance, Final Report, DOE Agreement No. DE-FC21-86MC10637 and EPRI Contract No. RP1891-07; 1995, June.

[18] Gunderson JR, Selle SJ, Harding NS. Utility experience blending Western and Eastern coals: survey results. Engineering Foundation conference on coal blending and switching of low-sulfur Western coals. Snowbird, UT; 1993, Sept 26–Oct 1.

[19] Collett M, Ready L. Coal blending system methods. PRB Coal Users Group Meeting; 2007, April 2.

[20] Ethen JA, Shusterich FL. Blending, Dust control and stockpile management activities associated with large-scale handling of Western coal. Engineering Foundation conference on coal blending and switching of low-sulfur Western coals. Snowbird, UT; 1993, Sept 26–Oct 1.

[21] Weiss HC. Transportation and handling of Western coal. Engineering Foundation conference on coal blending and switching of low-sulfur Western coals. Snowbird, UT; 1993; Sept 26–Oct 1.

[22] Fujii RK, Madan SC, Rupinskas RL. Experience modifying coal-handling systems for switching to low-sulfur Western coals. Engineering Foundation conference on coal blending and switching of low-sulfur Western coals. Snowbird, UT; 1993, Sept 26–Oct 1.

[23] Eimer RW, Hayes RH, Pollmann KB, Diewald DJ. Blending Illinois and Powder River Basin coals for testing on a 585 MW unit. Engineering Foundation conference on coal blending and switching of low-sulfur Western coals. Snowbird, UT; 1993, Sept 26–Oct 1.

[24] Woodside JR. Test burn experiences: Powder River Basin and eastern bituminous coal blends. Engineering Foundation conference on coal blending and switching of low-sulfur Western coals. Snowbird, UT; 1993, Sept 26–Oct 1.

[25] Brobjorg JN, Peterson TC, Mehta AK. Coal quality impacts testing at NSP's King Plant. Engineering Foundation conference on coal blending and switching of low-sulfur Western coals. Snowbird, UT; 1993, Sept 26–Oct 1.

[26] Levasseur AA, Chow OK, Pease BR, Thornock DE. Evaluation of the impacts of fuel switching on utility boiler performance. Engineering Foundation conference on coal blending and switching of low-sulfur Western coals. Snowbird, UT; 1993, Sept 26–Oct 1.

[27] Bryers RW. Utilization of petroleum coke and petroleum coke/coal blends as a means of steam raising. Engineering Foundation conference on coal blending and switching of low-sulfur Western coals. Snowbird, UT; 1993, Sept 26–Oct 1.

[28] Bibbo PP. The impact of coal blending and switching on the performance of air pollution control equipment. Proceedings Engineering Foundation conference on coal blending and switching of low-sulfur Western coals. Snowbird, UT; 1993, Sept 26–Oct 1.

[29] Peabody, Ameren sign long-term PRB deal. Coal Age 2011;116(8):6.

[30] Upcoming, recent coal-fired power unit retirements. Coal Age 2011;116(8):16.

[31] Hook M, Zittel W, Schindler J, Aleklett K. Global coal production outlooks based on a logistic model. Fuel 2010;89:3546–58.

[32] Patzek T, Croft GD. A global coal production forecast with multi-Hubbert cycle analysis. Energy 2010;35:3109–22.

[33] Hook M, Aleklett K. Historical trends in American coal production and a possible future outlook. International Journal of Coal Geology 2009;78:201–18.

A

Åbo Akademi Furnace Model, 287
AccuTrack® system, 62–63, 63f, 64f,
  116–117, 279
Acid rain, 251
Activation energy, 149
  char oxidation, for bituminous coal, 47
  for coal blends, 42–43, 43f, 158–159,
  159t
  of devolatilization, 66
Adaro coal availability, in markets, 36
Agglomeration, 25–26, 35, 231
Agribusiness biomass-processing wastes,
  140–141, 141t. See also Waste
  fuels
Agricultural biomass, 138–139. See also
  Biomass
  proximate analysis of, 140t
  trace metal concentrations in, 147t
  ultimate analysis of, 140t
Air/coal flow loop, 282f
Air Flow Sciences cold flow physical model,
  282f, 283f, 284f
Al₂O₃-CaO-FeO-SiO₂ system, 108
Alabama Power's Gadsden Plant, 187–189
Albright Generating Station, 20f, 262–263
  cofiring at, 178–180, 211t
Alkali metals, 25–26, 35, 55–58, 79–81, 145,
  171
Alkalinity, reactivity of, 171
Allen Fossil Plant, 22
  biomass on coal pile, blending, 176
  NO$_x$ reduction at, 171, 172f
Allen Generating Station (TVA), cofiring
  TDF at (case study), 210–212
Alliant Energy, 35–36
Ameren, 118
American Society for Testing and Materials
  (ASTM) Specification C-618, 226

Antelope coal
  char oxidation of, 102t
  devolatilization kinetics of, 97, 98t, 99f
  reactivity constants for, 101f
Anthracite, 82t, 126
Aromaticity, 50, 51f, 130, 132t, 168–169
Arsenic (As), 10
  concentrations in agricultural material, 221
  in petroleum coke blends, 221
Ash, 129–130
  elemental analysis, 85t, 86t, 87t, 88t, 89t,
  90t, 91t, 92t, 94t, 145
  for biomass fuels, 40t, 132t, 146t
  for coals, 38t, 39t, 131t, 179t, 180t,
  188t, 191t, 193t
  fusion temperature of petroleum coke, 41t
  management, 257, 262
  for power plants, 252–253
  trace metal concentrations in, 148t
  variability for Black Thunder coal, 95t
Atlantic City Electric, 260
Availability of fuels, 300–304

B

Bailly Generating Station (BGS), 222–223,
  224t, 225t
  cofiring at, 183–187
  on-site blending facility at, 3–4, 4f
Baldwin Generating Station, Illinois Power
  cofiring tire-derived fuel at (case study),
  212
Base/acid ratio, 81, 117, 156
Big Brown Plant, 17, 259–260
Binary blends, 37–42, 112
Biofuels, 142–144
  herbaceous, 142
Biomass fuels, 138–146
  ash elemental analysis of, 40t, 132t, 146t
  blending, on coal pile, 176–177

Note: Page numbers followed by f indicate figures; those followed by t indicate tables.

323

Biomass fuels (*Continued*)
    and coal, comparison of, 156
    into coal-fired boilers, separate injection
        of, 177
    cofiring. *See* Cofiring
    pelletizing, 139
    properties of, 126–134
    proximate analysis of, 129t
    reactivity measures of, 146–150
    role in coal-fired plants, 135
    types of, 138–142
    ultimate analysis of, 129t
Biomass–coal blending, 18. *See also*
        Blending
    burning profiles of, 161–169
    systems, 173–175
Biomass–coal cofiring systems, implications
        for, 169–177
Bituminous coal, 82t, 126, 137. *See also*
        Subbituminous coal
    aromaticity values of, 132t
    Central Appalachian. *See* Central
        Appalachian bituminous coal
    characteristics of, 38t, 91t
    Eastern derivative thermogravimetric
        analysis of, 163f, 164f, 165f
    thermogravimetric analysis of,
        162f, 164f, 165f
    firing blends of, 314t
    ignition temperature of, 102
    proximate analysis of, 188t
    reactivity of, 102
    speculative structure of, 134f
    ultimate analysis of, 188t
    Western, 91t
Black Thunder Mine
    coal-handling equipment, 306f
    transportation at, 307f, 308f
Black Thunder Powder River Basin coal, 126,
        137. *See also* Powder River Basin
Black Thunder subbituminous coal, 94t.
        See also Subbituminous coal
    ash elemental analysis of, 131t
    ash fusion temperature variability for,
        95t
    blending scenarios for, 255t
    with Central Appalachian bituminous coal,
        255t
    fuel analyses for, 254t
    with petroleum coke, 255t
    volatile matter evolution patterns for,
        256f

Blending, 1–29, 310
    coal with biomass. *See* Coal with biomass,
        blending
    coal-on-coal, 16–18, 71–123
    coal opportunity fuel, 21–22
    economic advantages of, 31–32
    reactivity of coals, 97–103
    solid fuel. *See* Solid fuel blending
    for steel industry, 13–16
B.L. England Station, 8, 117
Boiler(s)
    capacity, 204
    cyclone, 35–36, 174–175
    efficiency, 170, 176–177, 180, 184, 187,
        189, 194–195, 209–212, 223,
        226–227, 229, 264
    fluidized bed, 34, 175
    pulverized coal, 34, 138, 147–148,
        173–174, 174f, 175f
    stoker, 36
    types of, 26–27
Boiler and Industrial Furnace (BIF)
        regulations, 238
Bottom ash, 23, 34–35, 179–180, 192,
        213–214, 232, 235
    management of, 252
Bubbling fluidized bed (BFB) boilers, 25,
        34–35, 175. *See also* Boiler(s);
        Fluidized bed boilers
Bucket blending, 3–4. *See also* Blending
Buckskin PRB coal, devolatilization kinetics
        of, 101f
Bulk density, 33, 126–127, 139, 144–145,
        144t
Burning profiles, of biomass–coal blends,
        42–43, 43f, 44f, 72–74, 161–169

C

Calorific value
    of agricultural biomass, 140t, 141t
    of coals, 96
    of crumb rubber, 205
    of fecal materials, 92t
    of flexicoke, 219
    of hazardous wastes, 238, 239t
    of orchard and vineyard materials,
        142t
    of petroleum coke, 219, 221–223,
        226–227, 231
    of PRB coal, 259–260
    of tire-derived fuels, 202, 204–205,
        210

CaO-Na$_2$O-SiO$_2$ system, 57f, 78, 105f
    ternary diagram for, 56f
CaO-P$_2$O$_5$ system, 60f
Carbon cycle, 9f, 135
    inventories and fluxes in, 136f
Carbon footprint of plants, reducing,
        134–135
Carbon monoxide, 226, 252
Cellulose, 127–128
    holocellulose, 127–128
    hemicellulose, 127–128
Cement kilns, 239, 244t
Central Appalachian bituminous coals, 38t,
        89t, 126. See also Bituminous
        coal
    ash elemental analysis of, 131t
    chemical fraction analysis of, 154t
    cofiring woody material, impact of,
        158t
    comparison with biomass fuels, 156
CFD. See Computational fluid dynamics
        modeling
Char oxidation, 47, 97,
        102–103, 102t
Chemical composition
    of coals, 96
    of fuels, 127
Chemical fractionation (CHF) analysis, 153
    of Central Appalachian bituminous coal,
        154t
    of lignite, 153t
    of sheep manure, 143t
    of switchgrass fuel, 155t
    of wood fuel, 155t
    of Wyoming subbituminous coal, 154t
Chemical pulping, 125
Chlorine, 7, 10
    alkali chlorides, 173
    biomass, 40t, 126, 128–129, 140, 142, 173
    coal, 38t, 39t, 51–53, 58, 81, 85t, 86t, 87t,
        88t, 89t, 90t, 91t, 92t, 94t, 191t, 257
    corrosion, 26, 55–59, 192
    deposition, 107–108
    occurrence/forms, 152, 193t
    reactivity, 34–35, 104
Circulating fluidized bed (CFB) boilers, 25,
        34–35, 175, 296–297. See also
        Boiler(s); fluidized bed boilers
Clean Air Act (CAA), 10, 17, 48–49, 299,
        306–307
Clean Air Mercury Rule (CAMR), 252
Clear Skies Initiatives (CSI), 252

Coal(s)
    aromaticity values of, 132t
    ash elemental analyses for, 131t, 191t
    and biomass, comparison of, 156–157
    calorific value of, 96
    chemical composition of, 96
    classification of, 82t, 126
    commercial application of, 84–96
    consumption of, 76f
    energy consumption of, 73t
    export of, 74t
    handling of, 309–310
    pile, blending biomass on, 176–177
    pricing of, 6f, 7, 301f
    properties of, 126–134
    proximate analysis of, 128t
    storage of, 309–310
    ultimate analysis of, 128t, 179t, 184t, 188t,
        191t, 193t
Coal-fired boilers
    fouling deposits in, 110f
    separate injection of biomass into, 177
Coal-on-coal blending, 16–18, 71–123
    area of, 113–115
    influence on materials handling issues,
        115–116
    inorganic interaction, quantifying, 111–113
    managing, 113–119
Coal-water slurries (CWS), 22, 264, 266
    cofiring of
        at Marion County Generating Station,
            267
        at Seward Generating Station, 266–267
    pollutant measures, 266t
Coal with biomass, blending, 125–200
Cofiring, 47, 134–135
    biomass with coal, 260–267
    emission aspects, 260–262, 264–266
    case studies of, 262–264, 266–267
        Alabama Power's Gadsden Plant,
            187–189
        Albright Generating Station,
            178–180
        Bailly Generating Station, 183–187
        Naantali-3 generating station, 176–177
        Studstrup Power Station, 190–193
        Virginia City Hybrid Energy Center,
            187
    chemistry of, 157–161
    in Europe, 137–138, 262
    fuel reactivity and, 157–159
    methods and equipment, 176–177

Cofiring (*Continued*)
    reasons for, 135–137
    in the United States, 137–138
    with waste, 264
Cold flow models. *See* Physical model
Combustibles, reactivity of, 146–148
Combustion, 2, 75–77, 97
    coal blending and, 80–81
    computational fluid dynamics modeling
        of, 285–288
    inorganics in, 80f, 81
    and petrography, link between, 74–75
    products of, 288–290
Co-milling, 170, 173, 176
Comminution, 33
Compliance, 103, 203, 233, 250
Computational fluid dynamics (CFD)
        modeling, 274, 279–280, 283–284
    applications of, 290
    basic approach of, 284–290
    of combustion processes, 285–288
    products of, 288–290
Convective pass, 312
Copper smelters, 204
Corn stover, 39t
    ash elemental analysis of, 132t
    bulk density of, 144t
    derivative thermogravimetric analysis of,
        165f, 167f
    proximate analysis of, 129t, 140t
    thermogravimetric analysis of, 165f,
        167f
    ultimate analysis of, 129t, 140t
Corrosion, 170–171, 173
    chlorine, 26, 55–59, 192
    high-temperature testing, 192
    low-temperature, 27, 35, 55–58
    mechanisms, 33, 53–59, 103, 311
    sulfates, 251
Cotton gin trash
    bulk density of, 144t
    proximate analysis of, 141t
    ultimate analysis of, 141t
Creative Power Solutions (CPS), 285–287
Cross State Air Pollution Regulations
        (CSAPR), 299, 313
Crude oil, trace metal concentrations in, 221t
Crushing, 33
Cutter blades, 208
CWS. *See* Coal-water slurries
Cyclone boilers, 23, 24f, 35–36, 174–175.
        See also Boiler(s)

petroleum coke use in, 222–226
Cyclone burners, 24f, 80

D
Dairyland Power, 118
Deposition, 103, 171
    high-temperature, 192
    influenced by fuel blending characteristics,
        53–55
    low-temperature, 27, 35
    mechanics, 33–34, 104, 109
Derivative thermogravimetric analysis
        (DTG), 161–164, 167–169
Devolatilization
    of antelope coal, 97–98, 98t, 99f, 101f
    of blended coal, 98t, 100f, 101f
    of Buckskin coal, 101f
    conditions, 48
    of fuel mass, 50–51
    kinetics, 97–102
    of Long Fork coal, 98t, 99f, 101f
    mechanism of, 78
    rates, 179
    reactivity/ignition temperature associated
        with, 46, 81, 84, 97, 102
    of sawdust, 149f, 150f
    of Texas lignite coal, 101f
    of wood waste, 150f
Discrete element modeling (DEM), 278–279
Dolomite ratio, 81
Dominion Energy, 138
Drop tube reactor (DTR) kinetics, 43–44, 46,
        50, 112, 149–150
DTE Energy, 31, 62, 97, 105–106, 117,
        258–259
    Monroe Power Plant. *See* Monroe Power
        Plant
    River Rouge Power Plant, 11
    Trenton Channel Power Plant, 11
DTG. *See* Derivative thermogravimetric
        analysis

E
Eastern bituminous coal. *See also*
        Bituminous coal
    derivative thermogravimetric analysis of,
        163f, 164f, 165f
    thermogravimetric analysis of, 162f, 164f,
        165f
Economic considerations, of fuels, 300–309
    availability, 300–304

procurement, 304–305
transportation, 305–309
Electrostatic precipitator (ESP), 312
Emission(s), 312–313
  benefits, blending for, 253–260
  case studies, 258–260
  constituents of concern, 250–252
  emission aspects, 254–258
  fossil CO_2, 252
  mercury, 251–252
  nitrogen oxides, 251
  particulates, 250–251
  PRB coal with other solid fuels, 253
  sulfur dioxide, 251, 295
Empirically derived models, 279, 288
Energy consumption, 73t, 125, 178
Energy Policy Act, 252
Energy dispersive spectroscopy (EDS),
    153–155
Energy Institute of Pennsylvania State
    University, 97, 158
Engineering Consultants Group (ECG),
    62–63, 280–281
Environmental aspects, of fuel blending,
    249–269
  ash management, for power plants,
    252–253
  bottom ash, 252
  flyash, 252–253
  case studies, 258–260, 262–264,
    266–267
  cofiring biomass with coal, 260–267
  cofiring with waste, 264
  emission aspects, 260–262, 264–266
  in Europe, 262
  blending for environmental and economic
    reasons, 250
  emission benefits, blending for,
    253–260
  emission constituents of concern, 250–252
  fossil CO_2, 252
  mercury, 251–252
  nitrogen oxides, 251
  particulates, 250–251
  PRB coal with other solid fuels,
    blending of, 253
  sulfur dioxide, 251
  regulatory climate, 249–250
EPA Office of Air Quality Planning and
    Standards, 203
Eulerian method, 287
Europe, cofiring in, 137–138

Eutectic, 55, 81, 105, 108, 183
Exinite, 14t

F
FactSage thermodynamic modeling, 54–55
Fecal materials, 141–142
  proximate analysis of, 143t
  ultimate analysis of, 143t
Firing systems, 22–27, 34–36
  types of, 23–26
FirstEnergy Toronto Plant, cofiring tire-
    derived fuel at (case study),
    214–215
Flame characteristics, of fuel blends, 42–48
Flame temperature, 227–229
  of pulverized coal-fired boilers, 80, 104
Flexicoke, 216–217, 219
Flue gas recycle (FGR), 288–289
Fluid coke, 40t, 216–217
Fluidized bed boilers, 25–26, 34, 175
  bubbling, 25, 34–35, 175
  circulating, 25, 34–35, 175, 230t,
    296–297
Flyash, 21, 49, 93, 96, 146, 178, 184,
    186–187, 212, 213t
  management of, 252–253, 262
  pulverized coal boiler, 34–35
  resistivity of, 63
  unburned carbon in, 179–180, 223–224,
    226–227, 228t, 252, 257, 263
Fossil CO_2, 262
  emission, 252
  PRB coal with, 257–258
Foster Wheeler, 158, 272
Fouling, 26–27, 35, 53–54, 109–111, 110f,
    111f
Free swelling index (FSI), 11, 33, 35
Fresh switchgrass. *See also* Switchgrass
  ash elemental analysis of, 146t
  nitrogen/carbon volatile evolution in, 152f
Fuel, designing, 32–34
Fuel analyses for solid fuels, 254t
Fuel blending, 271–293
  detailed analytical approach to, 291
  modeling for, 290–291
  traditional approach to, 290–291
Fuel chemistry
  biomass, 31–32, 137
  coal, 22–23, 33–36, 54
Fuel consumption, 217, 241
Fuel No$_x$, 40t, 251
Fuel procurement, 304–305

Fuel supply strategy, 36–37
Fusinite, ultimate analysis of, 130t

G

Gadsden Plant, 20f
Gasification, 2, 75–77, 84
Geisler index, 35
Gibbs free energy, 277–278
Global Climate Change regulations, 252
Granular coke, 216
Greenhouse gas, considerations with fuel
    binding, 37
Green power, 31–32
Grinding, 4, 23, 33, 60–61, 116
Gulf Coast lignite, ash elemental analysis
    of, 131t. *See also* Lignite

H

Handling, 33, 61, 139, 205, 304
    blending influence on, 115–116
    coal, 3f, 299
    of petroleum coke, 218–219
    of straw, 191–193
    of tire-derived fuels, 204
Hardgrove grindability index (HGI), 33–36,
    59–62, 226–227
Hardwood sawdust, 127–128. *See also*
    Sawdust
    ash elemental analysis of, 146t
Hazardous air pollutants (HAPs), 7
    fuel blending effects on, 10
Hazardous wastes, 218–219
    fuel characteristics of, 238
    in rotary kilns, 239–241
    waste oil use, 241–243, 244t
Hazelnut shell, ash elemental analysis of,
    146t
Hemicellulose, 127–128. *See also* Cellulose
Herbaceous biofuels, characteristics of, 142.
    See also Biofuels
Heteroatoms, 133
High-density polyethylene (HDPE), 22
Holocellulose, 127. *See also* Cellulose
Hybrid corn seed, bulk density of, 144t
Hydrocarbons, 77–78, 131, 186f, 225f, 252

I

Ignition, 97, 157, 161, 171, 222, 261
    of lignocellulosic materials, 125
    temperature of blended coals and fuels,
        42, 46f, 102

of vegetation materials, 125
    vitrinite, 95–96
Illinois Basin coals, 38t, 89t, 126
    ash elemental analysis of, 131t
    comparison with biomass fuels, 156
Industrial hygiene, 214
Inertinite, 14t, 15–16
    ultimate analysis of, 130t
Inorganic constituents, behavior of, 103–113
Inorganic interaction, quantifying, 111–113
Inorganic matter, 49, 79, 81, 104, 109, 145,
    219, 222
Inorganics, in combustion, 79, 80f, 81
Institutional issues
    with coal blending, 295–322
    economic considerations, 300–309
        fuel procurement, 304–305
        fuel transportation, 305–309
        fuels availability, 300–304
    with fuel blending, 297–300
    future of coal production, 313–320
    process modifications, 309–313
        coal blending, 310
        coal handling and storage, 309–310
        convective pass, 312
        emissions, 312–313
        furnace effects, 311–312
        pulverizer performance, 310–311
Integrated gasification-combined cycle
    (IGCC), 77f, 231

J

Jacobs Ranch coal, 126. *See also*
    Powder River Basin
Japanese Industrial Standards (JIS), 235
Jennison Generating Station, NYSEG,
    cofiring TDF at (case study),
        212–214
Joint Line, 297, 298f
    coal hauling on, 298f

K

K_2O-Na_2O-SiO_2 system, 58f
Kinetics, devolatilization of, 97–102, 98t

L

Lignin, 127–128
    Adler structure of, 133f
    reactivity of, 148–149
Lignite, 82t, 126
    aromaticity values of, 132t

characteristics of, 39t, 85t, 86t
chemical fraction analysis of, 153t
comparison with biomass fuels, 157
Gulf Coast, 128t
Texas, 39t, 52f
western. *See* Western lignite
Lignocellulosic materials, 125, 132–133,
   148–149
Limestone Generating Station, 17
Liptinite, 14t, 15, 130t
Long Fork coal
  blends, 159f, 159t, 160f
  char oxidation of, 102t
  devolatilization kinetics of, 97, 98t, 99f
  nitrogen/carbon volatile evolution in, 151f
  reactivity constants of, 101f, 159f, 159t,
    160f
Low-density polyethylene (LDPE), 22
Luminant, 259–260

M
Maceral index (MI), 93, 95
Macerals, 14–15, 14t, 96
  group, 14–16, 14t, 127
  ultimate analysis of, 130t
  weighted average of, 96
Mathematical modeling, 274
Maximum achievable control technology
    (MACT) standards, 240, 240t, 313
Mechanical pulping, 125
Mercury, 262
  capture of, 96, 104
  concentration
    in biomass, 10, 262
    in coal, 10, 81, 179–180
    in petroleum coke, 219
    in PRB coals, 257
  emission, 118–119, 179–180, 189,
    251–252, 263
  methylmercury, 251
  PRB coals with, 257
Metallurgical coke, 13–14
Methylmercury, 251. *See also* Mercury
Mineral matter, 129–130
Minerals, 53t
Modeling, 271–293
  for blending purposes, 290–291
  limitations of, 291
  methods of, 274–280
    CFD modeling, 279–280
    discrete element method (DEM),
      278–279

numerical zonal models, 275–277
physical models, 275
thermodynamic models, 277–278
purposes of, 272
to reduce physical tests and costs, 273–274
specific applications of, 272–280
Moisture
  in agricultural biomass, 140t, 141t
  in Black Thunder coal, 94t, 128t, 254t, 255t
  in biomass fuels, 129t, 180t, 184t, 193t
  in blended coals, 42t
  in Central Appalachian bituminous coal,
    38t, 89t, 128t, 254t, 255t
  in corn stover, 40t
  in fecal materials, 143t
  in Gulf Coast lignite, 128t
  in fluid coke, 41t
  in flyash, 228t
  in Illinois Basin bituminous coal, 38t, 90t,
    128t, 224t
  in manually harvested switchgrass, 188t
  in mechanically harvested switchgrass,
    188t
  in Montana Subbit coal, 39t
  in Northern Appalachian coals, 88t
  in orchard materials, 142t
  in petroleum coke, 218t, 224t, 228t, 254t
  in Pittsburgh seam coal, 179t
  in Pratt bituminous coal, 188t
  in PRB coals, 86t, 87t, 255t
  in Sandow coal, 85t
  in San Miguel coal, 85t
  in Shoshone coal, 224t
  on So. Beulah coal, 85t
  in sponge coke, 41t
  in switchgrass, 40t
  in Texas lignite, 39t
  in tire-derived fuels, 202t
  in urban wood, 224t
  in Utah coal, 38t
  in vineyard materials, 142t
  in Western bituminous coal, 91t
  in wood waste, 40t
  in Wyoming Subbit coal, 39t
Molybdenum (Mo), 221
Monroe Power Plant, 2–3, 3f, 8, 16, 97,
    105–106, 258, 282–283
  coal-handling and blending facility at,
    31, 32f
  fuel binding, management and control
    of, 62–66, 63f

Monroe Power Plant (*Continued*)
  fuel quality screen, 65f
  zonal modeling at, 277, 278f, 279f
Montana subbituminous coal, 39t. *See also*
    Subbituminous coal
Mulch hay
  proximate analysis of, 140t
  ultimate analysis of, 140t

N

Naantali-3 generating station, cofiring at,
    176–177
National Ambient Air Quality Standards
    (NAAQS), 10, 249
National Environmental Policy Act (NEPA),
    253
Needle coke, 216
New Source Performance Standards (NSPS),
    249
New Source Review (NSR), 249
Nitrogen oxides (NO$_x$), 272
  considerations with fuel binding, 36
  emission, 251
  fuel, 251
  management of, 49–51, 137
  PRB coal with, 256
  prompt, 251
  reduction, 52f, 151f, 160–161, 160f,
    169–171, 172f
  thermal, 251
Northern Appalachian bituminous coal, 88t,
    126
Northern Indiana Public Service Co.
    (NIPSCO), 222
Numerical modeling, 271, 274–275
Numerical zonal models, 275–277

O

On-line analyzers, 62–63
On-line blending, 34. *See also* Blending
On-line coal analysis equipment, 304
Opportunity fuels, 21–22
  types of, 21
Optimal fuel chemistry, 23
Orchard materials, 141
  proximate analysis of, 142t
  ultimate analysis of, 142t

P

PacificCorp, 117–118
Paper
  preparation, waste, 235–237, 236f

production method, 235–237
  refuse-derived, 235
  small-scale model, 237
Particulates, 250–251, 261–262
  considerations with fuel blending,
    36–48
  emissions, 295
  PRB coals with, 257
Peach pits, bulk density of, 144t
Pellet Fuels Institute (PFI), residential/
    commercial fuel standards, 139,
    139t
Pellets, 233, 235
  production of, 235, 236f
  versus unprocessed woody biomass,
    139
  wood, 9, 36
Permitting, 13–14, 299, 312–313
Petcoke, utilization of, 296–297
Petrography/petrographic analysis, 13–16,
    96, 127
  and combustion, link between,
    74–75
  of subbituminous coals, 96
Petroleum coke, 21, 216–232, 257,
    296–297
  ash characteristics of, 219
  ash species mobility, 220–221
  characteristics of, 40t, 218–219
  consumption, by U.S. electric utilities
    (in 2000), 217t
  issues, 219–221
  proximate and ultimate analysis, 184t,
    218–219
  typical ash characteristics of, 220t
  typical fuel characteristics of, 218t
  ultimate analysis of, 184t
  utilization, in boilers, 221–230
    cofiring at Bailly Generating Station,
    222–226, 224t, 225t
    cofiring at Widows Creek Fossil Plant,
    227–229, 228t
    in a cyclone boiler, 222–226
    in pulverized coal (PC) boilers,
    226–229
Physical model, 271, 275, 276f, 280
  applications of, 282–283
  principles of, 280–284
  of pulse-jet fabric filter and spray dryer
    absorber, 274f
  of SCR, 272, 273f
Physical properties, of blended fuels, 62

Pine shavings, chemical fraction analysis
    of, 155t
Plastic(s)
    -based fuels, 22
    waste, composition, preparation, and
        utilization of, 233, 233f, 235–238
Polish coal
    proximate analysis of, 193t–194t
    ultimate analysis of, 193t–194t
Polk County project, 231
Pollutants, formation of, 48–53
Polymer as feedstock for fuel production,
    234t
Porosity, 145
Powder River Basin (PRB) coal, 12, 33–34,
    48–49, 250, 297, 299
    availability in markets, 36
    Black Thunder, 126
    blending of, 16–18
    characteristics of, 86t, 87t
    combustion of, 81
    comparison with biomass fuels,
        156–157
    derivative thermogravimetric analysis of,
        166f, 167f, 169f
    hauling, 114f
    Jacobs Ranch, 126
    mining, 114f
    with other solid fuels, 253
    thermogravimetric analysis of, 166f, 167f,
        168f
    U.S. coal production, 253f
Preblended fuels, 2
Predetermined blends, mixing of fuels in, 34
Process modifications, 309–313
    coal blending, 310
    coal handling and storage, 309–310
    convective pass, 312
    emissions, 312–313
    furnace effects, 311–312
    pulverizer performance, 310–311
Prompt Gamma Neutron Activation Analysis
    (PGNAA), 117–118
Prompt NO_x, 251
Proton magnetic resonance thermal analysis
    (PMRTA), 54
Proximate analysis, 145
    of agricultural biomass, 140t, 141t
    of biomass fuels, 40t, 129t
    of coals, 38t, 39t, 85t, 86t, 87t, 88t, 89t, 90t,
        91t, 92t, 94t, 128t, 184t, 193–194
    of fecal materials, 143t

    of orchard materials, 142t
    of petroleum coke, 41t
    of vineyard materials, 142t
Pulp(ing), 11, 71, 80, 125
    chemical, 125
    fruit, 138
    kraft, 125
    mechanical, 125
    mill powder boiler, 304
    spent liquor, 295–296
Pulverization, 4, 33, 219
    performance, 310–311
    petroleum coke, 226–229
Pulverized coal (PC) boilers, 2, 23, 34–35,
    138, 147–148, 173–174, 174f,
    175f. *See also* Boiler(s)
    cofiring in, 157
    petroleum coke use in, 226–229
Pyrolysis
    biomass, 50, 132–133, 146–149, 158–159,
        162–163, 168–169
    coal, 17, 78, 81, 97–98, 162–163, 169, 277
    conditions, 127
    kinetics, 42–44, 84, 97
    products, 105–106

R

Reactivity, 146–150, 152–155
    of alkalinity, 171
    of blended coals, 102
    and cofiring, 157–159
    of combustibles, 146–148
    determination of, 149–150
    factor, 75
    structure and, 148–149
Reflectance, 15, 72–75, 93, 95–96
Refuse-derived paper and plastics (RPF)
    densified fuel, 235
Reliant Energy, 117–118
Renewable energy, 13, 178, 262, 295–296,
    297f
Renewable Portfolio Standards (RPS), 47,
    135
Replaceable cutter tire shredder, 207,
    207f
Resource Conservation and Recovery Act
    (RCRA), 238
Reynolds number, 281
Rice hulls
    bulk density of, 144t
    proximate analysis of, 141t
    ultimate analysis of, 141t

River Rouge Power Plant, 11
Rotary kiln, 11, 58–59, 80, 210
Russian coal
    proximate analysis of, 193t–194t
    ultimate analysis of, 193t–194t

S
Sawdust
    ash elemental analysis of, 132t, 146t,
        193t–194t
    ash fusion temperature, 193t–194t
    blends, 159f, 159t, 160f
    bulk density of, 144t
    hardwood, 146t
    nitrogen/carbon volatile evolution in, 151f,
        160–161, 160f
    proximate analysis of, 129t, 193t–194t
    reactivity constants of, 159f, 159t,
        160f
    reactivity of, 149f, 150f
    ultimate analysis of, 129t, 180t, 188t,
        193t–194t
Scanning electron microscopy (SEM),
    153–155
Selective catalytic reduction (SCR) systems,
    230
Selective noncatalytic reduction (SNCR)
    systems, 26
Seward Generating Station, 263–264
Sheep manure, chemical fractionation
    analysis for, 143t
Shoeshone, 137
Shot coke, 40t, 216
Shredder, 192, 206–208, 206f, 207f, 209t
Silica-alumina-calcium oxide system,
    108
Slagging, 26–27, 53–54, 104, 107f, 108f
SmartBurn, 288
Slurries, coal-water, 22, 264, 266
    cofiring of, 266–267
    pollutant measures, 266t
SO$_x$ (sulfur oxide), 251
    emissions, 127
Softwoods, 127–128, 133, 133f
Solid cutter shredder, 206–207, 206f
Solid fuel blending, 1–5, 31–70. See also
    Blending
    area, 2–5
    bucket, 3–4
    considerations, 37
    and corrosion, 55–59
    deposition, 53–55

developing, 36–48
    economic advantages, 31–32
    economic considerations, 34–35
    environmental considerations, 35–36
    firing method considerations, 34–36
    flame characteristics of, 42–48
    in furnaces/boilers, 34–35
    historical and technical considerations,
        48–53
    ignition of, 42–48
    impact on physical characteristics, 59–62
    management and control of, 62–66
    market considerations, 36
    objectives for, 34
    reactivity of, 42–48
    system considerations, 1–2
Southern Company, 118
Sponge coke, 40t, 216
Spring Creek coal, 35–36
Staggers Act in 1980, 306
Steel industry, blending for, 13–16
Stochastic models, 279
Stoker boilers, 36. See also Boiler(s)
Stoker firing, 23–25
Storage
    of coals, 309–310
    process modifications, 309–310
Stream velocity, 45
Studstrup Power Station, 137, 190–193
Subbituminous coal, 12–13, 82t, 126.
        See also Bituminous coal
    aromaticity values of, 132t
    Black Thunder. See Black Thunder
        subbituminous coal
    characteristics of, 39t, 46f
    firing blends of, 314t
    ignition temperature of, 102
    Montana subbituminous coal,
        characteristics of, 39t
    petrology of, 96
    reactivity of, 102
    Wyoming subbituminous coal
        characteristics of, 39t
        chemical fractionation analysis of, 154t
Sulfur coal
    proximate analysis of, 184t
    ultimate analysis of, 184t
Sulfur dioxide (SO$_2$), 7–8, 48, 251
    blending with PRB coal, 256
    emission, 7, 118, 180, 224, 260
    management of, 62, 136–137
    reductions, 170, 260, 295

Switchgrass, 18, 19t
  ash elemental analysis of, 132t, 146t
  bulk density of, 144t
  characteristics of, 39t, 142
  chemical fraction analysis of, 155t
  cofiring of, 173
  derivative thermogravimetric analysis of,
    169f
  fresh. *See* Fresh swichgrass
  proximate analysis of, 129t, 188t
  thermogravimetric analysis of, 164f, 168f
  ultimate analysis of, 129t, 188t
"Sweet" (low-sulfur) crude oils, 219
Synthesis gas, 77, 77f, 84, 231

T

Tanoma Coal Company, 2
Tennessee Valley Authority (TVA), Widows
    Creek Fossil Plant, 2
Terminal velocity, 45
Ternary blends, combustion characteristics
    of, 37–42
Ternary diagrams, 55, 105–106, 108
Texas Genco, Limestone Generating Station,
    17
Texas lignite coal, 39t
  devolatilization kinetics of, 101f
  nitrogen evolution in, 52f
Thermal analytical techniques, 161
Thermodynamic models, 277–278
ThermoElectron Coal Quality Manager
    (CQM), 117–118
Thermal NO_x, 251
Thermodynamic modeling, 277–278
  FactSage, 54–55
  Gibbs Free Energy-based, 271–272
Thermogravimetric analysis (TGA), 46–47
  biomass, 161–168, 291
  burning profile, 72
  coal, 161–168, 291
Tipping fees, 208–209
Tire crumb (crumb rubber), 21–22, 204–205
Tire-derived fuel (TDF), 21–22, 201–216
  ash constituents for, 203t
  case studies, 210–215
    at Allen Generating Station, TVA,
      210–212
    at Baldwin Generating Station, Illinois
      Power, 212
    at FirstEnergy Toronto Plant, 214–215
    at Jennison Generating Station,
      NYSEG, 212–214

  combustion considerations, 210
  crumb rubber, 205
  physical characteristics, 204–205
  preparation costs, 208–210, 209t
  preparation of, 206–208
  with steel, 205
  without steel, 205
  trace element emissions from, 204t
  typical analyses of, 202t
  typical composition, 202–204
Total suspended particulates (TSP), 49
Trace elements, 146, 147t
  emissions, 10
Transportation, 305–309, 307f, 308f
Trenton Channel Power Plant, 11
Trifiring, 16, 186–187, 210–212, 226

U

Ultimate analysis
  of agricultural biomass, 140t, 141t
  of biomass fuels, 40t, 129t
  of coal macerals, 129t
  of coals, 38t, 39t, 85t, 86t, 87t, 88t, 89t, 90t,
    91t, 92t, 94t, 128t, 179t, 184t, 188t,
    191t
  of fecal materials, 143t
  of orchard materials, 142t
  of petroleum coke, 41t
  of vineyard materials, 142t
Unburned carbon (UBC), 223–224, 263
United States, coal in
  cofiring, 137–138
  future coal production, 313–320
  NO_x emissions in, 296f
  productive mines in, as of 2008, 303t
  SO_2 emissions in, 296f
Urban wood waste. *See also* Wood waste
  ash elemental analysis of, 132t
  bulk density of, 144t
  devolatilization reactivity of, 150f
  proximate analysis of, 129t
  ultimate analysis of, 129t
Usibelli coal, availability in markets, 36
Utah coal, characteristics of, 38t

V

Vanadium, 221
Variability of coal, 11–13, 37, 66, 93, 94t,
    116, 132t
  ash fusion temperature, 95t
  inherent, 62, 84–93, 117–118
  in-seam, 2

Vineyard materials, 141
  bulk density of, 144t
  characteristics of, 144
  proximate analysis of, 142t
  ultimate analysis of, 142t
Virginia City Hybrid Energy Center
    (VCHEC) plant, cofiring at, 187
Visibility
  degradation, 251
  impaired, 250–251
Vitrinite, 14t, 15
  ultimate analysis of, 130t
Volatile matter
  biomass, 15, 127, 156, 168–170, 219, 222,
    231
  coal, 15, 21, 50, 78–79, 127, 169–170,
    256f, 257f, 259–260
  fixed carbon ratio, 147–148, 156, 163
  plastics, 15
  tire-derived fuel, 201
Volatile organic compounds (VOCs), 250

W
Walnut shells, bulk density of, 144t
Waste-based fuels, 22
Waste fuel–coal blending, 201–248
  hazardous wastes, 218–219
    combustion of, in rotary kilns,
      239–241
    fuel characteristics of, 238
    waste oil use, 241–243
  petroleum coke, 216–232
    ash characteristics, 219
    ash species mobility, 220–221
    cofiring at Bailly Generating Station,
      222–226, 224t, 225t
    cofiring at Widows Creek Fossil Plant,
      227–229, 228t
    in cyclone boiler, 222–226
    handling and pulverization, 219
    proximate and ultimate analysis,
      218–219
    in pulverized coal boilers,
      226–229
    typical ash characteristics of, 220t
    typical fuel characteristics of, 218t
  tire-derived fuel, 201–216
    ash constituents of, 203t
    case studies, 210–215
    combustion considerations, 210
    crumb rubber, 205
    physical characteristics, 204–205

preparation costs, 208–210
    preparation of, 206–208
    with steel, 205
    without steel, 205
    trace element emissions from, 204t
    typical analyses of, 202t
    typical composition, 202–203
  waste plastic and paper preparation,
    235–237
    large-scale model, 235–237, 236f
    production method, 235–237
    small-scale model, 237
  waste plastics composition, 233, 233f
  waste plastic utilization, 237–238
Waste fuels, agribusiness biomass-
    processing, 140–141
Waste oils
  emission factors for combustion of,
    243t
  representative properties of, 242t
  trace metal concentrations in, 242t
Waste plastics
  composition of, 233, 233f
  large-scale model, 235–237, 236f
  preparation of, 235–237
    production method, 235–237
    small-scale model, 237
  utilization of, 237–238
Weathering, 163, 168–169,
  251–252
Wepco decision, 299
Western bituminous coal, characteristics of,
  91t. See also Bituminous coal
Western lignite, trace metal concentrations
  in, 148t. See also Lignite
Wheat straw, ash elemental analysis of,
  146t
Widows Creek Fossil Plant (WCF), 226–227,
  228t
Willow Island Generating Station, 22
Windbox modeling, 283, 285f
Wood waste, 18, 138–139. See also Biomass
  characteristics of, 39t, 142, 144
  cofiring, impact on Central Appalachian
    bituminous coal, 158t
  derivative thermogravimetric analysis of,
    163f, 164f, 166f
  proximate analysis of, 184t
  thermogravimetric analysis of, 162f,
    166f
  trace metal concentrations in, 147t
  ultimate analysis of, 184t

urban. *See* Urban wood waste
    unprocessed, 139
Wood fuel. *See* Pine shavings
Wyoming subbituminous coal. *See also*
    Subbituminous coal
    characteristics of, 39t
    chemical fractionation analysis of, 154t

X

X-ray diffraction, 16
X-ray fluorescence (XRF) on-line analyzer,
    62–63

Z

Zonal models, 275

Printed in the United States
By Bookmasters